Small Gasoline Engines
Training Manual

by

Ted Pipe

Howard W. Sams & Co., Inc.
4300 WEST 62ND ST. INDIANAPOLIS, INDIANA 46268 USA

International Standard Book Number: 0-672-20994-2
Library of Congress Catalog Card Number: 73-83366

Printed in the United States of America.

Preface

This book was originally written to fill a particular need for complete, not-too-technical service instructions covering the types of small horsepower engines used on lawn mowers, out-board motors, and many other similar applications. It was specifically written for a national service organization to serve as the basic text for a training program offered to their servicemen and trainees. Earlier editions were used in such training for many years, and this present edition is also serving the same purpose.

I mention this because its beginnings have greatly influenced the contents, format, and characteristics of this edition. In preparation for each of the four earlier (private) editions, the author worked closely with the training program instructors and students to review their problems and to alter the book as necessary to provide a thorough, basic understanding of the subject. Illustrations were added wherever obscurities in the text seemed to require them—so that now there are over 440 (an exceptional number for a book of this type). This text was expanded until we felt that it contained all available, practical information needed by anyone interested in servicing or maintaining these engines.

It is for this reason that you will find here a very complete coverage of the operating principles and construction features of all types of small gasoline engines—including the popular 4-stroke-cycle types, the (now) less used 2-stroke-cycle types, the foreign types presently being imported for many uses, and the new rotary type. Included is a chapter covering the various types of starters and power transmissions that are employed in different applications with these engines; another chapter is devoted to trouble-shooting and quick-service techniques.

The reader will not find exact descriptions and/or illustrations of particular engines. This book is not a service manual; each manufacturer provides such specific information on his own engine. Rather, it is intended to give the reader the knowledge he needs to understand manufacturers' manuals and specifications, and to intelligently approach any minor (tune-up or adjustment) or major (overhaul) service problem he may encounter.

In conclusion, I would like to give thanks to the many friends and associates who contributed so much to the making of this book. Also, I sincerely hope that you—the reader—will find it helpful.

TED PIPE

Contents

INTRODUCTION 7
  History of Gas Engines—Why Understanding Is Needed

SCOPE OF OUR STUDIES 12

CHAPTER 1

DEFINITION OF AN ENGINE 13
  An Engine Transforms Potential Energy Into Mechanical Power—Energy and Power —Mechanical Power

CHAPTER 2

EXPLANATION OF COMBUSTION 19
  Two Types of Combustion Engines—The External-Combustion Engine—The Internal-Combustion Engine—How Useful Kinetic Energy Is Created—Where the Heat to Create Kinetic Energy Comes From—Combustion Is Started in Several Ways—Internal Combustion Is Actually a Controlled Explosion

CHAPTER 3

EXPLANATION OF A RECIPROCATING-PISTON ENGINE 28
  The Types of Internal-Combustion Engines—The Basic Design of a Piston-Type Engine—The Principles Involved in Intake—An Explanation of Compression—An Explanation of Ignition—The Principles Involved in Exhaust—How a Piston-Type Engine Accomplishes the Five Events—Single-Cylinder *vs* Multicylinder Engines—A Flywheel Bridges the Gap Between Power Strokes—Other Engine Terms—Accessory Systems Required by a Piston-Type Engine

CHAPTER 4

TIMING THE FIVE EVENTS 40
  The Valves of Four-Stroke-Cycle Engines—The Valves of Two-Stroke-Cycle Engines —Twin-Piston Engine Design and Operation—Ignition Timing—Firing Order

CHAPTER 5

OTHER BASIC ENGINE REFINEMENTS 52
The Power-Producing Components—The Piston and Connecting-Rod Assembly—The Crankshaft and Counterbalances—Bearings and Bushings—Gaskets—Gears and Chains—Principal Engine Structural Assemblies—Tolerances and Fits—A Word About Metals

CHAPTER 6

THE ROTARY COMBUSTION ENGINE 67
The Engine We Will Discuss—Advantages of This Engine—Mechanical Design of the Wankel Engine—How the Five Events Are Accomplished—Performance Factors—Sealing—Timing—Lubrication—Cooling—Multirotor Engines

CHAPTER 7

THE FUEL SYSTEM 97
A Word About Gasoline—How the Gasoline-Air Mixture Affects Combustion—The Complete Job and Parts of a Fuel System—The Principle on Which Any Fuel System Operates—The Application of These Principles to the Fuel Supply—*Systems Components*—Fuel Tanks and Feed Systems—Air Cleaners—The Intake Manifold—Throttle Valves and Governors—Rich and Lean Mixtures—A Carburetor Controls the Gasoline-Air Ratio—One More Principle—The Airfoil—How Engine Speed Affects Carburetion—Choke Systems—*Float-Type Carburetors*—Float Systems—Idling Systems—High-Speed Systems—Economizer Systems—A Metering Rod System—A Power (Economizer) System—Accelerating Systems—Idling and High-Speed Adjustments—*Diaphragm-Type Carburetors*—Comparison With Bowl Types—The Two Types—The Fuel Pump (When Used)—Carburetor Operation—Additional Features—Adjustments—*Suction-Type Carburetors*—Comparison With Other Types—Carburetor Operation—Adjustments—*Foreign-Make Carburetors*—Slide-Throttle Direct-Proportioning Carburetor—Variations—Adjustments—Carburetor Maintenance

CHAPTER 8

THE IGNITION SYSTEM 147
Types of Ignition Systems—Battery Systems—Components of a Generator-Battery Ignition System—DC Generators—An Alternator-Battery System—The Cutout—A Voltage Regulator—A Lead-Acid Storage Battery—An Ammeter—An Ignition Switch—An Ignition Coil—A Distributor With Condenser—Spark Plugs—Wiring Harness and Wires—*Magneto Systems*—A Flywheel Magneto—Rotor (and Magnematic) Magnetos—Rotating-Coil Magnetos—Magneto-Dynamos (AC Magnetos)—An Energy-Transfer Magneto System—Magneto Adjustments and Maintenance—An Impulse Coupling—*Solid-State Ignition Units*—Terminology—Units and Operation of a Capacitor-Discharge System—A Flywheel Alternator System—Electronic Ignition for Alternator- or Generator-Battery Systems—Electronic Equipment Adjustments and Maintenance—*Practical Checking Guide*

CHAPTER 9

THE EXHAUST, LUBRICATING, AND COOLING SYSTEMS 182
Exhaust Systems—Exhaust Manifolds—Mufflers—Exhaust System Maintenance—*Lubrication-Systems—In General*—A Word About Lubrication—What Motor Oils and Greases Are—The Functions of a Motor Oil—Should an Engine Consume Oil?—How Long Should Oil be Good For?—*Lubrication Systems—Two-Stroke-Cycle Engines—*

General Types—Automatic-Mix Types—*Lubrication Systems—Four-Stroke-Cycle Engines*—General Requirements and Types—A Splash System—A Constant-Level Splash System—An Ejection-Pump System—A Dry-Sump System—A Force-Feed System—The Component Parts of an Oil System—Oil Pumps and Linkages—Oil Strainers—Oil Filters—Oil Breathers—Oil Seals—*Heat—Its Relation to an Engine*—Measuring Heat—Heat Is an Expander—Heat Is an Evaporator—Heat Contributes to Chemical Changes in Matter—Heat Builds up in an Engine—The Undesirable Effects of Heat in an Engine—Heat Travels in Three Ways—The Problem of Engine Cooling—*Air-Cooled Engines*—An Open-Draft and Exposure Type—A Forced-Draft Type—System Maintenance—*Water-Cooled Engines*

CHAPTER 10

STARTING AND POWER TRANSMISSION 210

Starting and Starters—The Need for a Starter—Starting Setups—Requirements of a Starter—Types of Starters—Manual Rope-Type Starters—Manual Crank and Kick Types—Mechanical Windup Starters—Electrical Friction-Clutch Starters—Electric-Centrifugal and Bendix-Type Starters—Electric Starter Motors—*Power Transmissions*—Terms Used to Define Engine Output—The Factors Involved in Power Transmission—Friction Can Be Put to Use—A Few Often-Employed Transmission Machines—Transmission Machine Maintenance

CHAPTER 11

TROUBLESHOOTING AND QUICK-SERVICE TECHNIQUES 238

Classifying Engine Troubles—Shortcuts to Get a Balky Engine Going—Diagnosing Troubles—A Few Quick-Service Hints—Tune-Up Tools and Equipment

INDEX 251

Introduction

The term *combustion engines* originally designated all engines which utilized the combustion principle to convert the energy stored in fuel into mechanical power. Similarly, fuel meant anything which could be exploded, and in the beginning some rather weird (as they may seem to us today) engines were designed. One of the earliest failures of which we have record was an engine, built during the 1680's in France, which attempted to explode gunpowder in a manner to obtain continuous motion. Soon after the trial run, it was decided that gunpowder would *never* make a satisfactory fuel!

HISTORY OF GAS ENGINES

It wasn't until a century later (1794) in England that the first real gas engine was proposed, to be driven by a flame-ignited explosive mixture of vaporized spirits of turpentine and air. Then, in 1820, an actual gas engine was built (operating on a hydrogen-air mixture) and ran "quite regularly" at all of 60 revolutions per minute.

These early gas engines were monstrous. The combustible gas-air mixture was exploded below the piston, in its cylinder, to thrust the piston freely upward as high as it would go. At the top of the piston's *free-riding* stroke a rack and pinion gear became engaged; afterward, the piston, falling back down (partly from gravity and partly because there was now a partial vacuum in the cylinder under it) rotated a shaft to create power. Engines like this were put into commercial service in the 1830's and were even used for driving boats. They were called *gas-vacuum* engines.

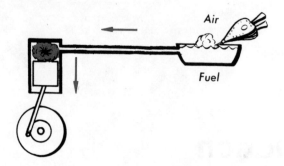

The first internal-combustion engine.

By 1860, a workable *internal combustion engine* was introduced. An engine was developed in France in which the *charge* (a gas-air mixture) was exploded at the top of an enclosed cylinder to force a piston, linked to a crankshaft, downward, and to thus *directly* produce usable power. Unfortunately, this engine was based upon a sequence of occurrences known as the Lenoir cycle, which omitted the occurrence of compression. Combustion of the *uncompressed* charge wasted considerable fuel and failed to produce a satisfactory ratio of power.

Compression, as one of the occurrences required for practical efficiency of engine operation, was first introduced by another Frenchman (Alphonse, Beau de Rochas) who, in 1862, described *five necessary events*. These five "events" were incorporated by a German engineer, Dr. Nicholas August Otto, in an engine he invented in 1876 and successfully exhibited two years later. This world-famous *Otto silent gas engine* was the forerunner of all modern four-stroke piston engines and made development of the automobile possible. The same *de Rochas' four-stroke* or, more commonly called, *Otto-cycle* principle has been used in a majority of all gasoline engines built to date, including many of the small engines which are the subject of this course.

Other memorable pioneers hastened popular acceptance of the Otto-cycle engine (now preferably called the *four-stroke-cycle* engine to designate the piston action). In 1883, Gottlieb Daimler, credited with being the "father" of the automobile, developed a *small* engine that ran on a light oil fuel similar to the gasoline of today. Previously, engines had used coal gases or heavy oil, weighed as much as 1100 pounds per horsepower, and ran at not over 200 rpm. Daimler's engine weighed only 88 pounds per horsepower and ran at 800 rpm. Following Daimler, Charles Benz of Mannheim and de Dion Bouton also developed gasoline-type engines which further increased the power-to-weight ratio.

Development of *two-stroke-cycle* (piston) engines, also using the "five-events" principle, closely paralleled that of the four-stroke engines. In 1876 Sir D. Clark produced an engine that used an auxiliary piston and cylinder (like a small hand pump hung on the side of the working cylinder) to feed the charge to the working piston (thus eliminating need for two of the strokes). However, this "Clark-cycle" engine was not efficient. The first basic two-stroke engine as we know it today was not invented until 1891, by an Englishman named Day. Like many of the small engines we will study here, the Day engine utilized crankcase pressure instead of an auxiliary piston—and there were both two- and three-port types.

Even following the introduction of Otto's engine, many inventors favored one-directional rotary motion, as opposed to the reciprocal motion of his pistons, for the development of power with what appeared to be less waste of energy. (It was not until later that the flywheel principle and conversion of energy became fully appreciated.) Historically, rotary devices (such as windmills) and positive-displacement rotary machines (such as gear pumps) predate the earliest piston devices. No wonder that there is a multitude of rotary power machine designs—both steam and internal-combustion types—which provide a greater variety of basic concepts than those recorded for reciprocating-piston machines. In retrospect it seems unfortunate that most of this endeavor was concentrated on empirical attempts to adapt existing gear, rotary-vane and turbine mechanisms, with no adequate research in the fields of pressure sealing, compression ratios, intake-combustion-and-exhaust timing, cooling, etc.—all of which were being scientifically developed by the proponents of reciprocating-piston engines.

The first *RC* (*rotary combustion*) engine which promised feasible competition with the (by then) economically entrenched reciprocating-piston types was not tested until 1957. Based upon a rotary compressor invented in 1954 by Felix Wankel, this DKM (rotary-piston engine) was developed by Wankel and a staff of engineers at the NSU automobile plant in West Germany. Adequate volumetric efficiency and compression ratio were obtained by predetermined eccentric movement between two rotors within a stationary housing, the essential factor being the two rotor shapes—one, a triangle, contained within the other in a chamber having an epitro-

choidal (modified figure 8) perimeter. A subsequent "kinematic inversion" (mechanical reversal) patented by Dr. Walter Froede (chief NSU engineer) greatly simplified the DKM model by converting the outer rotor into a stationary housing, and thus eliminating the original housing. The resulting *KKM (circuitous-piston engine)* model became the prototype for industry development.

Despite the more than 80-year lead of the reciprocating-piston types—and due, in large measure, to the improved research, materials and production methods achieved during this period —the "infant" KKM has several apparent advantages. Future KKM-type engines, now being developed for numerous applications using both light and heavy fuels, may very well have performance characteristics better than those of equivalent reciprocating-piston engines. The trend seems to be toward a future in which this, or a counterpart rotary, may sooner or later replace the conventional piston engine as we know it today.

WHY UNDERSTANDING IS NEEDED

Because reciprocating-piston engines are—and for some time will be—used for a number of small engine applications, we must defer to these "seniors" in what we have to say here. From the foregoing you can see that all the basic principles of the Otto-cycle were discovered and put to use *prior to the 20th century.* By 1900 there were workable four-stroke-cycle and two-stroke-cycle engines.

The chief accomplishments of the past century have been scientifically engineered refinements, refinements that have taken millions of man-hours of study and labor to develop. Better metals and other materials, greater appreciation and usage of ever-increasing scientific knowledge, superior production know-how and techniques— these have been the major contributions of our technical era. Now we measure tolerances to the 10-thousandth of an inch where once "an easy fit" would do. Today, the engine of 1900 (even of 30 years ago) seems simple as a toy—truthfully, almost any handyman with a few tools could quickly make one of these old-timers run good as new. Not so today! What simple tool (or knowledge) could, for instance, check out the dwell angle of distributor breaker contacts or a compression loss.

Let's put it this way: In a simple old-time engine you could *sense* everything that needed fixing. For instance, you could tighten a bearing until it felt tight, or fit a piston in its cylinder so it appeared snug. In today's engines you can no longer depend upon your horse sense—*you have to know what you are doing and have the proper tools and instruments to help you.* If you should attempt to fix a modern engine in the happy-go-lucky manner of a 1900 handyman, it won't run much better (if at all) than one of the "old-timers." A good part of over 80 years of refinements might as well be discarded!

Scope of Our Studies

We shall study the principles of operation of *four-stroke-cycle* and *two-stroke-cycle reciprocating-piston* engines, and of the Wankel/NSU engine which, it now appears, will be the basic model for future *four-phase-cycle rotary-combustion* engines.

Our studies of the "stroke" engines will be limited to those commonly used to propel lawn mowers, compressors, garden-type tractors, generator sets, lightweight motorcycles and scooters, snowmobiles, outboard-motor boats, etc. These are generally referred to as *small* engines, meaning that they are lightweight and portable and that their individual outputs (rated in horsepower) are small in comparison with the horsepowers of automotive engines. Small engines of the stroke type generally have just one cylinder and most are air- (rather than water-) cooled. Their fuel supply, ignition, lubrication, exhaust, cooling (if any) and power-transmission accessories are generally of the simplest and most compact designs. Because these engines and accessories are all now in use, and necessary data is thus available, we shall also discuss tune-up, troubleshooting and accessory maintenance and repairs.

Our studies of the Wankel/NSU engine shall be limited to a discussion of the basic principles involved and the various ways in which these principles are being developed to adapt this engine for the above small-engine applications. A *small* engine of the Wankel type may have two rotors with limited total displacement, or may contain a single small rotor. Either design is the equivalent of a four-stroke engine in the sense that it operates on the same four-cycle (five "events") principle, and its accessories parallel those used with the four-stroke engines. For this reason, much of our tune-up and accessory maintenance and repair information will be applicable also to these rotary engines.

To be exact, *our* stroke-type engines are classified as *single- (or possibly more) cylinder, reciprocating-piston-type, internal-combustion engines*. The strokes are sometimes loosely referred to as "cycles" (i.e., two-cycle or four-cycle) but now that we must differentiate between these and the rotary types it is better to use the terms "two-stroke" and "four-stroke" (since the word "stroke" is necessarily associated with piston movement while the word "cycle" can be associated with either piston or rotor movement). In order to explain this classification the first three chapters will be devoted to separate discussions of: (1) engine, (2) internal-combustion, (3) reciprocating-piston-type with single cylinder (or more).

The rotary engine has been broadly designated *rotary combustion* (*RC*), and the particular rotary that is our model has been identified by its originators as *Kreiskolbenmotor* (*circuitous piston engine*), abbreviated *KKM*. To be specific we shall refer to it as the *Wankel KKM-Model RC Engine* and, for our purposes, we can classify it as *single- (or more) eccentric-rotor, stationary-epitrochoidal-chamber, internal-combustion engine*. Explanations not already contained in the aforementioned chapters will be found in Chapter 6.

CHAPTER 1

Definition of an Engine

AN ENGINE TRANSFORMS POTENTIAL ENERGY INTO MECHANICAL POWER

The word *engine* is defined as: *A machine designed to transform potential energy into mechanical power.* You know already, of course, that the engines that we are concerned with "burn" a fuel to turn a crankshaft and do work. But since our definition uses the terms "potential energy" and "mechanical power," you will have a better understanding of an engine if you know the meanings of these terms.

ENERGY AND POWER

All the universe consists of matter and energy.

Matter is the word physicists use to describe everything in our universe that has weight and occupies space. A bar of steel, speck of dust, the flesh of your hand, water, wood, cloth, the air you breathe—these are all examples of matter. Matter exists in such a way that it generally can

be seen, felt, tasted, and/or smelled. Yet all matter is "dead" in the sense that it has no warmth, no light, and no motion of its own.

Energy is the indefinable something that puts "life" into matter—that gives it warmth, light, and motion. We cannot see energy; we cannot weigh it or measure it in terms of how much space it occupies. Yet it does exist. It is as real as the warmth and light of a bonfire, the electrical spark that jumps the gap of a spark plug, or the turning of a wheel.

Since energy has no measurable weight nor dimensions of its own, we think of it as occupying and acting on matter. When it acts on matter it changes the substance and condition of matter. The actions of energy have shaped our earth into what it is, have formed some matter into dirt, some into water, crude oil, air, iron, wood, etc. These actions we call the *natural* actions of energy (the works of nature); and by performing these natural actions, energy demonstrates itself to us.

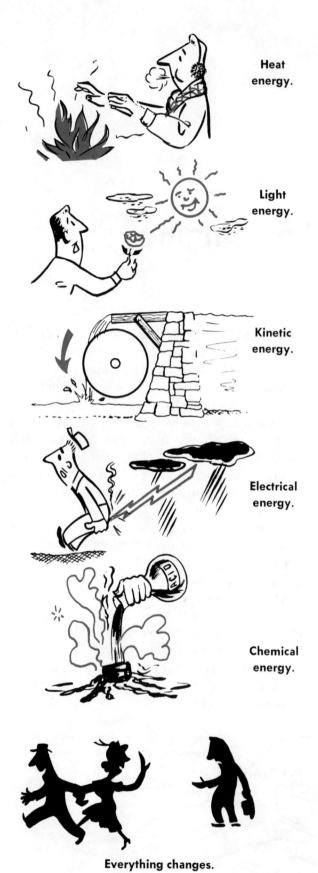

Heat energy.

Light energy.

Kinetic energy.

Electrical energy.

Chemical energy.

Everything changes.

The Active Forms of Energy

Because we can see, feel, taste, or hear the effects of energy in different ways, we say that there are five commonly known, *active forms* of energy:

> *Heat Energy*—Warmth.
> *Light Energy*—Light and color.
> *Kinetic Energy*—Motion; the movement of a body of matter.
> *Electrical Energy*—Shock, and the magnetic effect on certain types of matter.
> *Chemical Energy*—The change brought about in one substance by another.

All the preceding active forms of energy are interrelated. Any one form of energy is generally accompanied by one or more of the other forms. Also, one form usually changes, partly or entirely, into one or more other forms while it is acting on (and changing the form of) matter. This makes our world a constantly changing one. Winds blow, lightning flares, waters flow, rock disintegrates into pebbles or sand, iron rusts—and millions of other changes are ever occurring.

Nature Conserves Matter and Energy

In spite of all these changes, both matter and energy are indestructible. Nothing is lost; only the nature of matter and the forms of energy change. For instance, a piece of wood disappears in flames and we think of it as completely gone. But the matter that was wood has combined with air and formed an equivalent quantity of ash, water, and gases; the energy that accomplished it has shown itself as flame (light energy) and heat (heat energy). All will continue

Nothing disappears forever.

A release of latent energy.

in these new forms until another change takes place.

Energy Is Limitless

Nature has, indeed, made abundant use of energy to fashion our world of today. But, fortunately for us, only a fractional part of all the energy that exists on earth has ever been released by nature in the forms of energy that we know. By far the greatest portion of existing energy is still stored away within matter in much the same manner that electrical energy is stored inside a battery. In fact, atomic experiments lead us to the conclusion that if all this stored energy were to be released at once, the earth and all of the matter which it represents would disintegrate into space. But because we do not know how to release the energy from most forms of matter, we refer to it as hidden, and call it *latent energy.*

Potential Energy Is That Part of the Supply We Know About

Whenever we do isolate a form of matter from which we can release part or all of the energy, we say that it contains *potential energy.* Crude oil and/or the gasoline derived from it are examples of matter which contains potential energy, for we have learned how to release a small part of the energy stored in these substances.

Gasoline has potential energy.

It is the potential energy of gasoline (or other crude-oil derivative) that our engines are designed to release and to convert into mechanical power.

MECHANICAL POWER

We have previously mentioned kinetic energy as being one of the forms of energy and have called it "the motion of a body of matter." Motion is something we can see, measure, and easily understand. In fact, kinetic energy is doubtless the first form of energy that man recognized and put to use. Our basic science, physics, was originally founded on our knowledge of matter and kinetic energy.

Only a useful push counts.

NOTE: Physicists speak of a moving body of matter as having mass, volume, dimensions and velocity. We shall use these terms here.

Mass is the weight of a body at sea level. Everything has many different weights since the higher a body is above sea level, the farther it is from earth's center of gravity, and the lighter it becomes. To compare the quantity of matter in one body with the quantity in another it is necessary to know what each would weigh at sea level; that is, to know their respective masses.

Volume designates the total amount of space a body occupies (regardless of its shape). The *dimensions* of a body are its length, width, height, or the area of a side, etc. Its *velocity* is the speed at which it is moving at a particular time.

We shall also use the word *object* to mean any single body of matter—whether it is a piston, a speck of dust, or a cloud of gases, etc.

Mass.

We know from observation that when an object is in motion and strikes a second object it has the capacity: (1) to move the second object (if it was stationary); (2) to turn or stop it (if it was moving); or (3) if the first object cannot visibly move, turn, or stop the second, it has (at least) the tendency to do so. A billiard ball striking another one will visibly move, turn, or stop it; but if it were to strike the side of an automobile, we could not see any effect other than the dent it would leave. This dent would prove that the ball tended to move the whole automobile, but could only move the metal in the dent.

Volume.

Velocity.

Because the kinetic energy of a moving object gives it such a capacity, we say that the moving object creates *force*. Wind creates force; it can turn a windmill. So does freezing water (it can crack an engine block as it expands into ice), a thrown baseball, a moving car, even a feather dropped from a shelf. All such moving objects, whether large or small, create force. Yet the amount of force obviously differs. We need to distinguish between the force of a speeding car and that of a falling feather.

Experiments have shown that the quantity of force (total push) at any instant is equal to

Objects.

the mass of the object times its velocity. We call this the *momentum* of the object. That is: *total push = momentum = mass × velocity*. Momentum is not the size, shape, or weight of an object alone. Velocity must also be considered. A running 280-lb man can be stopped instantly by a 1-lb cannon ball because the ball has a superior momentum due to its vastly greater velocity.

Mass is measured in pounds; velocity in feet per second. We could measure momentum in pounds times feet per second, but there is no convenient device for doing so. Instead, we measure the effect of an object in motion by measuring the amount of force (in pounds) required to stop it. For example, a 1-lb ball hanging motionless on a cord attached to a spring scale will weigh 1 lb, but if it is raised a few feet and dropped to the end of its cord it will snap the scale pointer down to register somewhat more than 1 lb. Hence, we can say the ball has exerted a force of so many pounds applied to

Momentum equals mass times velocity.

the scale. This force equaled its momentum, registered in pounds.

If an object's momentum is applied to move another object in a useful manner, we speak of the visible result as *mechanical energy.* Mechanical energy is, for instance, the energy made evident by a waterfall turning a paddle wheel.

Force equals momentum in pounds.

A Machine Uses Mechanical Energy

Not every moving object, however—not even some of those which have considerable momen-

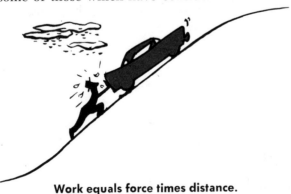

Work equals force times distance.

tum—can be classed as providing us with mechanical energy. The uncontrolled forces of nature seldom accomplish any useful purpose. On the contrary, they often do us harm. Therefore, the only moving objects which can rightly be said to provide us with mechanical energy are those which we have learned to control and learned to harness. And the means by which we do harness a moving object is called a *machine.* A machine is any contrivance that harnesses (uses) mechanical energy to accomplish a purpose. We call this purpose *work.*

Machines Accomplish Work

A simple crowbar on a pivot is a machine; mechanical energy applied to one end will result in controlled movement of an object at the opposite end. So is a sewing machine; it uses mechanical energy applied to its needle and bobbin to sew. An automobile, a bicycle, a rowboat, and an engine are machines. Machines accomplish work. Each moves a designated object (the sewing thread, the automobile passengers, etc.) a measured distance to accomplish a purpose. In its mechanical sense, *work is the useful movement of some object.*

To measure work we must know: (1) how far the object has been moved; and (2) the force required to move it. In short, *work = distance (feet) × force (pounds)* and is measured in foot-pounds (*ft-lb*).

Continuous-Operation Machines Produce Mechanical Power

Some machines, like a lever or chain hoist, have to be operated by an external source of

Lever.

Inclined plane.

Screw.

Pulley.

Wheel and axle.

Gears.

Six basic types of machines.

A compound machine.

One horsepower.

energy. Others, like a steam engine or a gasoline engine, operate continuously (once started) until stopped. Machines in the second class have the capacity to move an object continuously, not just one certain distance. Such a machine is said to produce *mechanical power*.

Mechanical power is rated in terms of the foot-pounds of work that the machine (running at a constant rate) will accomplish each second.

We also (and more commonly) use the term *horsepower* (*hp*) instead of foot-pounds per second. By definition, 1 hp = 550 ft-lb per second or 33,000 ft-lb per minute (the actual power of an English dray horse as measured by James Watt, inventor of the steam engine, when he worked the horse for a reasonable length of time).

Mechanical power.

CHAPTER 2

Explanation of Combustion

TWO TYPES OF COMBUSTION ENGINES

A combustion engine is defined as one that uses the heat energy of fuel. There are said to be two types: *external* combustion engines and *internal* combustion engines.

THE EXTERNAL-COMBUSTION ENGINE

The external type uses the heat energy released from a fuel to heat some form of matter and thus obtain mechanical energy from this *other* form of matter. Our best example of this type is the steam engine with boiler. The boiler burns a fuel (wood, coal, oil, etc.) to heat water (the other form of matter) and generate steam. The steam, which is "alive and moving," then has the mechanical energy to run the working parts of the engine.

THE INTERNAL-COMBUSTION ENGINE

The internal type of combustion engine burns its fuel (gas, gasoline, oil, etc.) *inside* of the engine, in such a manner that the fuel, itself, becomes "alive and moving" and has the mechanical energy to run the working parts of the engine. Small two- and four-stroke-cycle engines are of this type. They burn gasoline, which further distinguishes them from internal-combustion engines designed to burn some other fuel.

Actually, when the potential energy stored in gasoline is released (by its burning), this energy shows itself in three of the familiar forms previously mentioned—light energy, heat energy, and kinetic energy. The light energy is negligible and useless to an engine. The heat energy is considerable (up to 4000° F is possible inside a running engine) and it serves some useful purpose (though much of it is excess). It is the kinetic energy alone that becomes the mechanical energy used by the engine.

In fact, an internal-combustion engine is one that uses the *kinetic energy of fuel*, and the heat energy released in the process is mostly a byproduct. This point is important. The heat energy is not only mostly useless; it is a hindrance. One of the major problems of an internal combustion engine is the dissipation of the excess heat.

To make this fully understandable we shall have to explain how the combustion of the fuel (gasoline) inside an internal combustion engine actually takes place.

HOW USEFUL KINETIC ENERGY IS CREATED

The Three States of Matter

Because we see and feel objects either as solids, liquids, or gases, we say that there are three states of matter—*solid, liquid,* and *gas*. All of the objects we know appear to us in one of these three states, yet we know from experience that one and the same object can change from one state to another.

The best example is water. Everyone has seen it change into ice or into a vaporous mist. Gasoline evaporates, tar melts in the sun, liquid foods

External combustion.

Internal combustion.

turning even solid steel into gas.) Any gas can be changed to a liquid and/or a solid. Some gases have been solidified in laboratories, but the key to such change is heat or cold, and it takes a great deal of heat or cold (absence of heat) to change most forms of matter. Consequently, in the world of nature, most matter retains one familiar state.

All Matter Consists of Basic Substances Called Elements and Compounds

All matter, whether solid, liquid or gas, can be broken down into smaller parts. A cup of oil can be separated into drops of oil; an engine

Mechanical analysis.

can be frozen solid, metals can be melted into liquids, air can be chilled into a liquid, etc. There are numerous examples of how nature and man change the state of matter.

Matter can be made to change states under the proper conditions. Many solids can be made liquid, then vaporized into gas. Other solids can be directly vaporized into gas. (Atomic explosions, for instance, vaporize all adjacent matter,

block can be ground up into a pile of metal filings; etc. These are typical divisions of matter into parts which anyone can make with proper equipment. By other means (called *chemical analysis*) it is possible to break matter down into much smaller parts. And by so doing we learn the basic substances of which it is composed and how these substances are combined and held together to create all the millions of different forms of matter in the world.

By analysis we have learned that all matter belongs in one or another of three distinct groups—mixtures, compounds, and elements.

A *mixture* is any form of matter that can be separated into two or more separate substances.

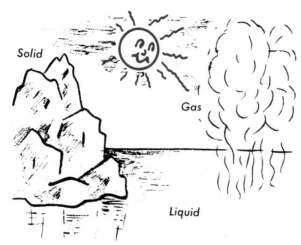

Solid

Gas

Liquid

Three states of matter.

The parts

The mixture

CEMENT

Concrete

A mixture.

Explanation of Combustion

For instance, concrete is a mixture because it is made from, and can be again separated into, the separate substances known as sand, gravel, cement, and water. Paper, cloth, hash, paint, and steel are other examples of mixtures.

To separate a mixture into its parts we use what we call a *mechanical process*. This is any process (like pulverizing and sifting or settling in water, melting, etc.) which uses energy but does *not* release energy during the process. It creates *no heat, no electricity*, etc., and, as a rule, no gas is given off. In short, no chemical

Mechanical separation.

change takes place; the separate substances that compose the mixture (like the sand, gravel, cement, and water of concrete) are the same after being separated as they were when combined.

NOTE: Mixtures are held together by *adhesion*. Hash, for instance is held together by the adhesive (gummy) quality of the juices in it. Adhesion is a form of energy. It is a form of the same energy that we call *cohesion*, which will be explained later.

Using mechanical separation we can break down any mixture into its parts; and if any of these parts are also mixtures, then we can further break these down into their parts, etc. In breaking matter down, however, we reach a point at which the parts can no further be

A basic substance

A basic substance.

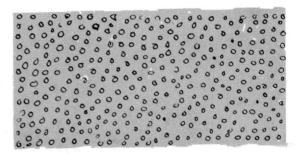

Not even a 100th part of a single grain of sand.

broken down by mechanical separation. And when we find parts like these we say that they are *basic substances*.

All matter is composed of such basic substances. Sometimes we find these basic substances combined to form mixtures; sometimes we find them alone. And we have learned that all these basic substances belong to one or the other of the two groups called *compounds* and *elements*.

The State of a Compound or Element Depends on the Cohesion of Its Molecules

Through chemical analysis we've learned that all compounds and elements (basic substances) are composed of particles that we call *molecules*. Molecules cannot be seen, even under our most powerful microscopes. It is estimated, for instance, that if we could expand one drop of water to the size of our earth, the molecules in it would be the size of baseballs. Yet, small as they are, we know a great deal about them.

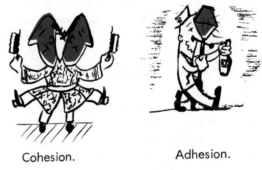

Cohesion.   Adhesion.

Cohesion and adhesion.

We know that molecules are separated from each other by spaces called voids (meaning that, so far as we know, there is nothing there); also,

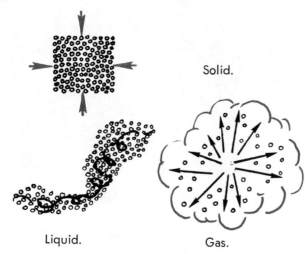

Solid.

Liquid.   Gas.

Three states of matter.

will bounce as far apart as permitted by the molecules of surrounding substances (which, of course, will bounce most of them back again).

NOTE: Two gases mix very quickly because their molecules are in such rapid motion. Liquids, even some solids, evaporate because their moving molecules bounce out into the surrounding air (which is why we can smell them). And for the same reason, a gas and liquid, a gas and a solid, two liquids, a liquid and a solid, or even two solids will (when placed in contact) mix, by themselves, in varying degrees. The degree depends on their respective molecular activity, and on the adhesion and cohesion of their molecules.

Molecules

Evaporation—solid to gas.

that they are active (*in motion*). We know, too, that these molecules are held together by a form of energy that we call *cohesion*.

NOTE: Cohesion is defined as the attraction between *like* molecules. All the molecules of any one basic substance are alike, while those of different basic substances are not alike. Consequently, only the molecules of a basic substance are held together by cohesion; the molecules of any one element or compound are held to those of another element or compound (to form a mixture) by the form of energy called adhesion (as previously mentioned).

Cohesion is similar to (perhaps the same as) gravity. Like gravity, its force is weakened by distance and by any opposing movement (such as the opposing movement you create when you throw a baseball into the air). Consequently, the farther apart the molecules of a basic substance are, the weaker is their cohesion. Also, their cohesion is further weakened if the molecules become more active (that is, if they move around more so that they bump together and bounce apart).

Elements or compounds that we call solids are solid because their molecules are sufficiently inactive (and therefore close enough together) to be held in a rigid group by their cohesion. The molecules of a liquid are inactive (and close) enough to be held together, but they are more loosely held in what we call a sliding contact instead of in a rigid group. Those of a gas are extremely active—so active, in fact, that they

How Heat Changes the State of a Basic Substance

Heat energy opposes the form of energy we call cohesion. While the latter tends to keep molecules stationary and together, heat energy tends to make them more active so that they bounce farther apart. Hence, as heat is increased, cohesion is decreased (by the greater distance between the molecules). For each basic substance there is a degree of heat (*melting point*) at which the decrease of cohesion converts the substance from a solid to a liquid. There is also a degree of heat (*boiling point*) at which the substance becomes a gas. On the

Molecules

Liquid to gas

The boiling point.

other hand, when heat is being taken away instead of being added, we call these same points the *condensation point* (gas to liquid) and the *freezing point* (liquid to solid).

How Heat Creates Kinetic Energy

Whether or not a substance is heated sufficiently to change state, heating it does result in its expansion (as the molecules bounce farther and farther apart). If and when it reaches a gaseous state, its molecules have so much activity that, as previously said, the gas will expand to fill any container it is in. However, every gas *is* in a container of some kind. Even the "free" gases of air are kept from expanding entirely away from earth by the force of gravity, which thus serves to keep them contained.

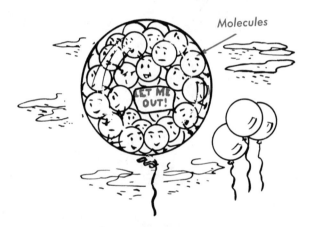

Molecules

Pressure.

Consequently, continued heating of a gas must result either in a pushing outward of its container (as a balloon is expanded by heat), or in a tendency of the gas to push it outward (if the container is strong enough to resist). We refer to this tendency as the *pressure* of the gas, and say that its pressure increases as a gas is heated. If contained, solids and liquids will similarly exert pressure on the container.

NOTE: Not every substance expands regularly as it is heated. One obvious exception is water. Due to various causes, water reverses this process and contracts when being heated from 32°F to 39.2°F. Conversely, it expands when being cooled between 39.2°F and 32°F. Hence we say that water expands just as it freezes; but at all other temperatures it acts just as described previously.

Since the expansion and/or pressure effects caused by heat are the results of molecular movement, they are simple types of kinetic energy. When present in sufficient amount, they are usable types which may be transformed into mechanical energy.

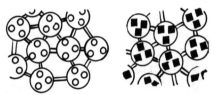

Heat energy to kinetic energy (by way of expansion pressure) then to mechanical energy.

WHERE THE HEAT TO CREATE KINETIC ENERGY COMES FROM

The Structure of Elements and Compounds

Small though they are, the molecules that make up our various elements and compounds have been broken down into their parts. We call these parts *atoms*. Nature has provided 92 different atoms, and these, for the chemist, are the building blocks with which all matter is constructed.

Elements.

Compounds.

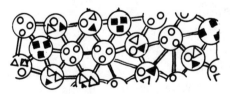

A mixture.

The atoms that make molecules that make . . . .

The molecules of some substances are composed of just one atom or of two or more identical atoms. Such substances are called *elements*. Since there are 92 different natural atoms, there are 92 different natural elements. Typical elements are copper, aluminum, tin, zinc, iron, lead, gold, oxygen, hydrogen, sulphur, carbon, and nitrogen.

All other basic substances have molecules composed of two or more unlike atoms, and these substances are the *compounds*. Since almost any of the 92 different atoms may be combined to form a compound molecule, the number of compounds is practically limitless. Nature has created many thousands, and chemists have learned to regroup atoms and create many others. Typical compounds are water, salt, alcohol, sugar, ether, turpentine, sulfuric acid, and marble.

NOTE: With cyclotrons and atomic fission we have even split atoms and learned that these are composed of parts which we call *electrons* and *protons*. We have found that all electrons and all protons are apparently alike, that, in the final analysis, all matter is composed of just these two tiniest subdivisions. Atoms differ only because each kind has a different quantity and arrangement of electrons and protons, which gives each kind of atom distinctive characteristics of its own. The elements differ because each has the characteristics of the atoms which compose it. The compounds are still different because, in combining to form a compound molecule, the characteristics of the atoms blend, forming new characteristics. Finally, elements and compounds further blend their characteristics when forming mixtures.

By altering atoms we have created twelve new elements for a total of 104 different elements.

How Atoms Regroup to Form Different Substances

The atoms of a compound molecule are held together as molecules are held together—by cohesion. However, each molecule has a definite pattern. If the pattern is altered, for instance by the loss or addition of a single atom, the whole molecule changes its characteristics and

A chemical change.

becomes some different substance from the compound that it was. We call such a change a *chemical change.* Chemical changes are brought about by *chemical processes,* such as burning, oxidation, electrolysis, or dissolving in acid or water. A *chemical change is always accompanied by the release of energy (heat, light, or electrical energy).* It usually results in the formation of gas. The presence of heat hastens most chemical changes.

Regrouping of atoms.

All atoms combine readily with others of the same kind; some combine quite readily with certain other kinds; and some do not combine readily (or at all) with certain other kinds. Consequently, there are many atom groups (molecules) which are altered easily, while there are many others which are not altered easily. When two or more molecules of different substances are brought together, if their atoms will combine more readily into groups that are different from the existing ones, they may do so.

As a result, one, two, or more new substances will be created in place of the original ones.

Oxidation Is a Regrouping of Atoms

Our most plentiful element is oxygen. It is present in air (about 20%), in water (33%), and in many other forms of matter. Oxygen atoms combine very readily with a great many other kinds of atoms; but it so happens that few of these are present in air. In fact, air is a mixture of many elements and compounds, most of which do not combine too readily with oxygen. Consequently, the supply of oxygen in the air is mostly unattached in the sense that it is free to combine with any atoms that are attracted to it.

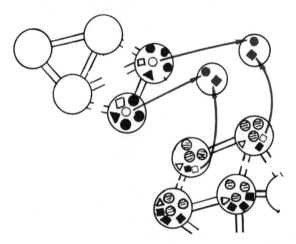

Oxygen combines readily with the atoms of some substances.

As a result, the oxygen in air is constantly combining with substances on earth to form various compounds. It combines with iron to form rust; with copper or lead to form tarnish; with a dead animal or vegetable matter to decompose it; etc. We call all such processes whereby the oxygen of air *slowly* combines with other substances *oxidation*. Oxidation occurs at normal earth temperature and produces a small degree of heat energy.

NOTE: The supply of oxygen in air is also being constantly replenished by the release of oxygen from substances that are being altered by other types of chemical change. Most of it comes from vegetation, which "breathes" out oxygen much the same as we breathe it in.

Combustion Is Rapid Oxidation

Some substances which either oxidize very, very slowly, or not at all under normal temperatures, will oxidize quite rapidly when they become hot enough. In fact, they will oxidize rapidly enough for a *considerable* amount of heat and more-or-less light energy to be released. Moreover, the heat that is generated will, of course, increase the temperature of the substance. This increases the rapidity of the oxidation, until the substance is oxidizing (or, as we say, *burning*) quite rapidly.

The flash point.

The accelerated oxidation, or burning, is also called *combustion*. Substances which will unite with oxygen in this manner are said to be *combustible or flammable*, and are referred to generally as *fuels*. The temperature at which such a substance will start to burn is called its *kindling point* or its *flash point*.

COMBUSTION IS STARTED IN SEVERAL WAYS

One method of raising the temperature of a fuel to its flash point is to *ignite* it by bringing it into contact with something that is already burning and therefore releasing considerable heat.

A second method is to keep it confined so that the small amounts of heat released by its slow oxidation (if it oxidizes at normal temperature) will accumulate and build up to the required

Ignition.

temperature. When this happens we call it *spontaneous combustion.*

A third method is to raise the fuel's temperature by *compression.* Compression is the reduction of space in which a substance is confined. Gases are easily compressed; liquids and solids, for all practicable purposes, are considered impossible to compress. Whatever the state of a substance, if it is at a certain temperature its

Spontaneous combustion.

modecules are moving with a certain amount of activity. If compression occurs and the temperature has not been simultaneously reduced, this molecular activity is immediately restricted by the smaller space so that the molecules bump oftener and more violently. If violent enough, the bumping of molecules will knock atoms from them. This is a chemical change which releases heat energy. Therefore, if the compression is sufficient, it causes the temperature of the substance to rise, and, if the additional heat does

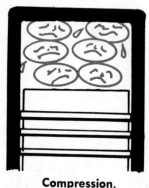

Compression.

not escape fast enough, the substance will reach its flash point.

NOTE: What we call *friction* also involves a bumping or rubbing together of the molecules of the two objects. Friction creates heat for the same reason that compression does. This is why you can strike a match or produce a spark with flint and steel.

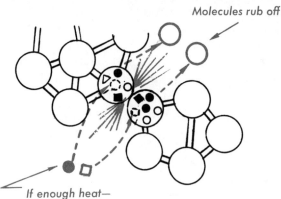

Molecules rub off

If enough heat—
then chemical changes

Friction.

INTERNAL COMBUSTION IS ACTUALLY A CONTROLLED EXPLOSION

You will fully understand the behavior of fuels if you will keep these facts in mind:

A fuel must be a substance containing atoms that will unite readily with oxygen under proper conditions. And the more of these it contains, the better it is as a fuel.

For rapid oxidation (burning) to start, the fuel must first be heated until it reaches its flash point.

If a fuel is to burn at all, there must be a ready supply of oxygen to combine with its atoms (support the combustion). The faster oxygen can be combined with all the available atoms in the fuel, the faster it will burn.

As has been said, combustion releases (creates) considerable heat. Also, as previously said, heat activates molecules, changes the state of matter, causes substances (especially if they are changed into gas) to expand considerably and exert pressure.

Now take a situation in which a good fuel is thoroughly mixed with oxygen (or air containing oxygen) so that most of its available atoms are in contact with oxygen atoms. Bring this

An explosion.

mixture suddenly to its flash point so that it burns all at once. The terrific heat created helps to make the combustion complete and instantaneous (by raising any slower-burning atoms present to their flash point). It simultaneously activates all the molecules present: those which comprise any unburned portions of the original mixture together with the molecules newly formed by the oxidation process.

To all the preceding add the fact that internal combustion takes place in a confined space. The heat is confined so that it builds up to a very high temperature—enough not only to change most of the fuel substances present to their gaseous states, but even to superheat these gases so that their tendency to expand (their pressure) suddenly becomes more than enough to break out of the confined area with a violence that we call an *explosion*.

In an engine the breaking out is accomplished by the pushing of the piston outward in the cylinder.

CHAPTER 3

Explanation of a
Reciprocating-Piston Engine

THE TYPES OF INTERNAL-COMBUSTION ENGINES

There are three basic types of internal-combustion engines: the *turbine,* the *jet,* and the *piston* (both reciprocating and rotary types can, in basic classification, be referred to as having pistons).

In a turbine type, the kinetic energy of the fuel is transformed into mechanical energy by directing the moving (exploded) mass of gases into a multiblade wheel (similar to a windmill) to revolve it. The motion is continuous in one direction. Some aircraft use such engines.

The jet type operates on the principle that two objects, each free to move, will, if bumped together, rebound away from one another in opposite directions. In a jet, the engine serves as one object, the mass of exploded gases serves as the second object, and the practically continuous explosions of the fuel serve as the bumps which bounce these two objects apart (to propel the engine forward while the mass of gases is propelled backward, out the open rear of the engine). Similar engines, carrying their own oxygen and called *rockets,* propel craft even through outer space where there is no atmosphere for a propeller to use for traction.

A piston-type engine transforms the kinetic energy of exploded fuel into mechanical energy by confining the expanding gases in a cylinder (or chamber) so that they will thrust a piston (or rotor) in a desired direction. In a reciprocating-piston engine, each explosion of fuel results in a straight-line outward movement of the piston which we call a *power stroke.* This power stroke is a single, forceful thrust that is limited

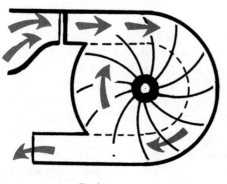

Turbine.

Piston.

Jet.

Types of internal-combustion engines.

in duration and also limited to the distance that it is practicable to allow the piston to travel in its straight-line movement. In order to develop mechanical power (the capacity for continuous work) it is necessary for such an engine to accomplish two things:

1. It must transform the straight-line piston movement into a movement which can be harnessed to various types of work.
2. It must provide a series of power strokes that will continue as long as desired.

THE BASIC DESIGN OF A PISTON-TYPE ENGINE

We harness the straight-line piston movement by transforming it into a rotary movement (of a shaft) that becomes continuous in one direction of rotation. The piston is connected to one end of a simple lever that has a fixed fulcrum point at the other end. During its power stroke the piston revolves this lever through a half circle (180°), at which point the lever stops the piston movement. With the power stroke thus ended, the rotary momentum of the lever carries it on around through the remaining 180° of a full circle, and the lever returns the piston to its starting point.

The Engine Parts That Provide Harnessed Movement

In a piston-type engine, the *cylinder* is a cylindrical bore inside a *cylinder block*. The *piston* is a metal cylinder fitted into this cavity and encircled by *piston rings* which seal it tightly (so that the expanding gases cannot escape around it), while also serving as bearings for the piston to slide on when it moves in the cylinder. One end of the cylinder is sealed closed by a *cylinder head* which contains a mated cavity called the *combustion chamber* (in which fuel is exploded). This cylinder head may be a separate part bolted to the cylinder block with a *cylinder-head gasket* in between (to seal it tightly), or it may be an integral part of the cylinder block. The opposite end of the cylinder opens into an area enclosed by a pan, called the *crankcase*, that is fastened to this end of the cylinder block.

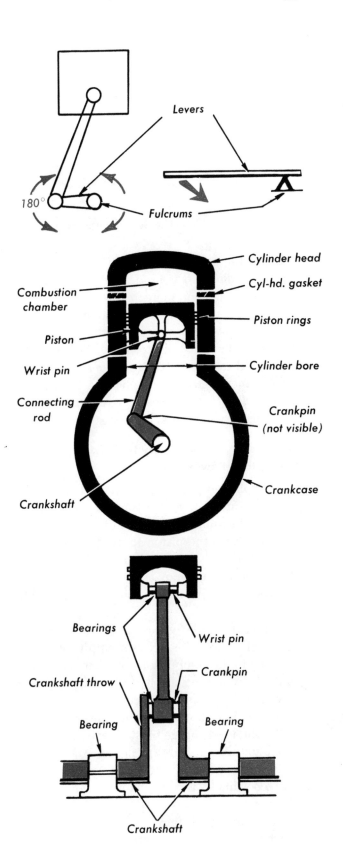

Basic parts of a piston engine.

| Intake. | Compression. | Ignition. | Power stroke. | Exhaust. |

The five "events" necessary to provide "continuing" power strokes.

Instead of a simple lever with a fixed fulcrum point, a *crankshaft* is used, held in a stationary support by one or more *main bearings* which allow it to revolve. The lever is an offset (*throw*) of the crankshaft (which "juts out" at a 90° angle to the shaft centerline). The piston is connected to the shaft (*crankpin*) of this offset by a *connecting rod* that is held to the crankpin by a *connecting-rod bearing*. A *wrist pin* with bearings serves to attach the piston to this rod.

The Five Events Necessary to Provide Continuous Power Strokes

As explained previously, the mechanical parts of an engine function to transform each power stroke into rotary movement of the crankshaft, which returns the piston to its starting point at the end of each full revolution (360°). To produce continuous movement, then, it is only necessary to repeat this performance time after time.

For each single power stroke to occur there must, of course, be a fresh supply of fuel (*a charge*) in the combustion chamber. This charge must be compressed, to obtain its maximum power. When it is ready, combustion must take place, to be followed by the power stroke. Finally, the used charge must be expelled to make room for the next fresh charge. And these requirements must be satisfied in a manner that will coordinate the mechanical movements of the piston and crankshaft.

It is to fulfill these requirements that de Rochas and Otto (as mentioned in the Introduction) introduced the *five events* that must occur in any piston-type internal-combustion engine.

These five events are the same today, for all such engines, as they were when first described by de Rochas and used by Dr. Otto. They constitute what we call a cycle (or Otto cycle), which means one complete sequence of events from one power stroke to the next. They are as follows:

1. *Intake*—The filling of the cylinder to capacity with a fresh charge (of fuel).
2. *Compression*—The compression of this charge into as small a space as practicable.
3. *Ignition*—The firing of the compressed charge.
4. *Power Stroke*—The resulting outward thrust of the piston.
5. *Exhaust*—The discharge of the burned gases from the cylinder.

Completion of these five events in this sequence brings the engine full circle from one starting point to the next. Continuous repetition of this cycle of five events keeps the engine running.

THE PRINCIPLES INVOLVED IN INTAKE

The Hydraulic Principle

As was said earlier, whenever a substance is compressed—or is kept from expanding normally when heated (which is the same thing)—it develops an internal force called pressure, which is due to the "pushing outward" of its activated molecules. In liquids and gases the molecules are free to flow past and around each other in any direction, as evidenced by the fact that liquid will take the shape of any container it is poured into, while gas will even expand to fill

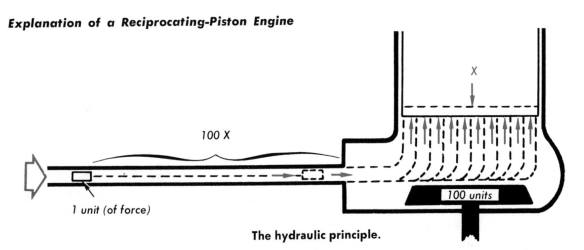

The hydraulic principle.

every crevice of its container. Hence, if a liquid or gas is *pressurized* by heat or compression its molecules will push outward with equal force (pressure) in all directions, and will (as we say) exert an identical pressure against every spot on the container that is keeping the liquid or gas compressed. This is known as the *hydraulic principle.*

We measure pressure (like any force) in pounds. To make it easier to measure, our instruments are usually calibrated to register the pounds of pressure acting against one square inch of a container's surface. That is, the pressure of a liquid or gas is so many pounds per square inch. Then we arrive at the total pressure against an area, such as one wall of a container, by multiplying the pounds per square inch by the total number of square inches. Hence we can state the hydraulic principle simply by saying: *If a force equal to X pounds per square inch is applied to any surface of a contained liquid or gas, the liquid or gas will push outward against every other surface of the container with a pressure equal to X pounds per square inch.*

NOTE: The hydraulic auto hoist used in service stations works on this principle. A container under the hoist is filled with oil, and the hoist column fits into the top of this container like a large piston. The flat bottom of this column, which rests on the oil, has (let us say) an area of 100 sq. in. A small piston, with (for instance) a bottom area of 1 sq. in., is set up so that it can pump additional oil into the container. The hoist may be operated by hand, by compressed air, or by a motor.

Now, every time the small piston pumps oil into the container with (for example) a force of 150 lb act-

ing on the 1 sq. in. of oil surface it contacts, the oil will transmit a pressure of 150 lb against each of the 100 sq. in. of the column bottom to push the column up with a total force of 15,000 lb.

Of course, for each inch the column moves up it will take 100 times as much oil to fill the space it has vacated as will be added to the container by each inch that the small piston moves down. Therefore, the small piston will have to move down 100 inches (or pump in this much new oil) for each 1 inch that the column rises. In short, the operation is the same as if we were to use a lever that is 100 inches long at the pumping side of the fulcrum and only 1 inch long at the work side of its fulcrum.

The Meaning of Density

We know that the molecules of different substances differ from one another in size and mass (weight). They differ because they contain different quantities of atoms, and because the various kinds of atoms differ from one another in size and mass. Also, the atoms and molecules of one substance may be spaced farther apart at a given temperature than those of another substance. In short, even though two different sub-

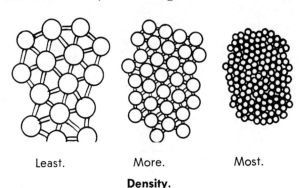

Least.   More.   Most.

Density.

stances are at the same identical temperature, one may have more and/or heavier molecules per cubic inch than the other. In other words, one is *denser* than the other (meaning it has more mass per cubic inch).

In order to compare various liquids and solids, we use the density of water at 4°C and call this "1.0." A cubic foot of water at 4°C weighs 62.4 lb. By weighing a cubic foot of any other substance at 4°C we can learn how much more or less it weighs than water. The ratio of the weight of a substance to the weight of water (each having the same volume) is called the *specific density*. For instance, if a solid should weigh 93.6 lb per cu. ft (one and one-half times that of water), we would say that its specific density is 1.5. When comparing the densities of gases to one another we use air (or sometimes hydrogen) as the standard, instead of water.

Now, when a substance is heated it expands, so that there are fewer molecules per cubic foot. On the other hand, when it is compressed, there will be more molecules per cubic foot. Therefore, heating or compressing a substance changes its density. For instance, water at 20°C has a specific density of only 0.99825.

Denser liquids or gases (since they weigh more per cubic foot) have a tendency to settle toward the bottom, below less dense liquids or gases. And solids which are less dense will float in a liquid or gas (as cork floats on water, or as a balloon full of hydrogen floats in air).

Atmospheric Pressure and Density

The air (or atmosphere) surrounding our earth is many miles high. Gravity pulls each and every molecule of the gases which compose this atmosphere earthward. In fact, as previously noted, if it weren't for gravity these active gas molecules would expand clear away from earth, leaving us no air to breathe! Because of this, each layer of atmosphere—all the way to the top—is pushing downward. The sum total of *this downward push (or force)* approximately amounts to 15 lb per sq. in. (*psi*) on the surface of the earth at sea level. In short, this is the weight of a column of air, one inch square at bottom, resting on the ocean.

Our atmosphere is a gas in a container, a container bounded by the earth at bottom and by the downward pushing molecules at its top. Therefore, it is subject to the hydraulic principle, and its molecules push against every other object with equal force in all directions. Since the downward force at sea level is 15 lb per sq. in., this is the pressure with which the atmosphere pushes against everything located at sea level. For instance, it pushes against every square inch of a sailor's body with a force of 15 lb and would crush him except that the human body is constructed so that it can withstand such a pressure.

Also, because the air toward the bottom of the atmosphere is weighted down by the air above

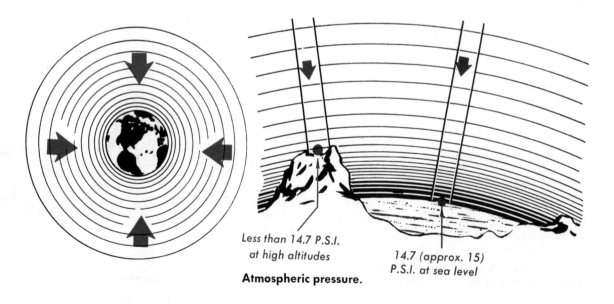

Less than 14.7 P.S.I. at high altitudes

14.7 (approx. 15) P.S.I. at sea level

Atmospheric pressure.

it, it is slightly compressed. It is therefore denser than the air above.

> NOTE: As you climb a mountain the column of air above you becomes shorter (less heavy) so that the air becomes less dense and its pressure less. Water boils more quickly (at lower temperature) because its heat activated molecules can escape (vaporize into steam) with less effort; there are fewer and "less forceful" air molecules for them to pass among when making their escape.
>
> For the same reason, as you go below sea level (in a deep mine, for instance), the air becomes denser, the pressure greater. And when a diver goes deep into the ocean, he finds the pressure greatly increased. In addition to the air there is now a considerable column of water above.

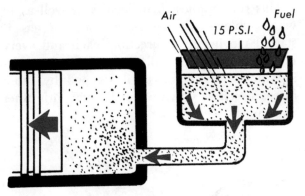

Intake.

The Vacuum Principle of Intake

If a piston were designed to completely fill its cylinder when in it, and it were pulled partially out of the cylinder without any air leaking past it, the space that it had occupied would be completely empty (what we call a 100% *vacuum*). Actually, the piston of an engine does not fill its cylinder (it never goes up into the compression chamber), and there is always a little leakage around it, even when the piston rings are in good condition. Therefore, when a piston is moved outward, only a partial vacuum is created.

This partial vacuum, however, contains many fewer molecules of gas than would normally occupy the space. In short, there is room for many more molecules. If an opening is provided to connect this relatively empty space with the outside atmosphere (which we must remember

Compression.

is under 15 lb per sq. in. pressure at sea level), as much air as the 15 lb per sq. in. pressure can force in will flow into this space. This is what we call the *suction* of a vacuum. The amount of air (number of molecules) that will be sucked in per second depends on the outside pressure (which, as already noted, will be less at higher altitudes) and on the density of the outside air (which is also less at higher altitudes).

In an engine, intake is effected in this manner. The piston is pulled outward by revolution of the crankshaft, to create a vacuum. Instead of pure air, however, a mixture of air and gasoline (under atmospheric pressure) is forced in through an opening into the "emptied" cylinder. At sea level, more mixture will be forced in during a given time than at higher altitudes; less than at lower altitudes.

AN EXPLANATION OF COMPRESSION

To repeat in a slightly different manner what was previously said, compression increases the force of combustion.

It increases the force, first, because compressing the charge very quickly superheats it *throughout* so that every molecule of the charge is approaching its flash point at the instant combustion starts. Hence, when combustion does occur, it is practically instantaneous and complete for all of the charge.

It increases the force, second, because the tightly packed, highly activated molecules (all striving to move apart with considerable pressure, even before combustion occurs) serve as millions of tiny springboards to thrust the piston back outward.

An engine will run on uncompressed charges. The earliest engines were so operated. However, the power output was exceedingly low and

Electrical ignition.

wasteful of fuel. We have found that even though some energy is used to build up compression, the power advantages gained far exceed the power spent. Modern engines are being designed for higher and higher compression.

AN EXPLANATION OF IGNITION

As already mentioned, most piston-type engines do not use compression as the means of firing the charge. They do not, chiefly because the very high degree of compression required to create combustion would make it necessary to build overly bulky and heavy engine parts to withstand the strain of the high pressure involved. There are other reasons, also, concerned with the type of fuel, its handling, performance factors, etc.

> NOTE: Piston-type engines of the *Diesel class* (which burn so-called heavy distillates like kerosene and fuel oil instead of gasoline) do fire the charge by compression alone.

In gasoline engines we use an electrical spark, *created across the electrodes of a spark plug,* to ignite the charge made ready by compression. Electrical energy (as will be explained later) is made to jump the gap between the two electrodes. In jumping the gap it oxidizes the molecules of the charge that are in its path with sufficient rapidity to burn them. Their burning then spreads throughout the charge.

Actually, then, combustion starts at a single point. This point (the location of the spark plug) is carefully plotted for maximum efficiency—that is: (1) so that the combustion will spread according to a plan that will ensure continued,

even pressure all over the piston top throughout the power stroke, and (2) so that the full charge will be ignited and expanded during the incredibly short interval of time allowed.

THE PRINCIPLES INVOLVED IN EXHAUST

The Law of Inertia

We have found that any body that is at rest tends to remain at rest, while any body that is in motion tends to continue moving along the same straight line; and it requires an equal and opposing force to change either condition. This is called the *law of inertia.* It means simply that if you start an object moving (assuming there is no friction or gravity to interfere), it will take the same force to stop it that was used to start it.

This rule applies to molecules as well as to whole objects. Consequently, once a gas is started moving in a direction, each and every molecule of the gas will tend to continue moving in this direction, until stopped by other forces (such as gravity or collisions with other molecules).

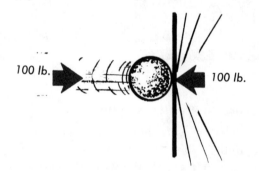

The "law of inertia."

Inertia Helps Produce a Better Exhaust

The inertia principle is used to discharge burned gases from an engine cylinder. As was said, an engine piston does not fill its cylinder, even when all the way in. If it did fill the cylinder, it would be easy to squeeze out all the burned gases. However, it never enters the combustion chamber; yet it is desirable to completely empty the cylinder and combustion chamber of burned gases, to make room for a fresh charge.

As the crankshaft returns a piston to its starting position, the fast-moving piston causes the burned gases in the cylinder to move at a high velocity. Each molecule of the burned gas is given a very forceful push straight toward the combustion chamber. We provide an opening in the cylinder head, and the rushing molecules flow out of this opening. If the opening is properly positioned, nearly all of the molecules in both the cylinder and the combustion chamber will flow out—even though the piston does not actually "squeeze" all of them out.

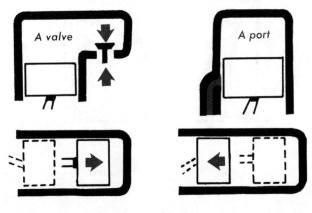

Strokes.

HOW A PISTON-TYPE ENGINE ACCOMPLISHES THE FIVE EVENTS

NOTE: As already indicated, a piston-type engine accomplishes intake and exhaust through passages into the cylinder that must be opened and closed at precise intervals. In a four-stroke-cycle engine these passages are through the cylinder head into the combustion chamber; *valves* are used to open and close them. In a two-stroke-cycle engine, the passages may be through the cylinder head or through the cylinder, and they may have valves, or may be simply covered and uncovered by piston movement. In the latter case we call them *ports*.

Valves and ports vary considerably in design and construction. We will discuss this in detail in *Chapter 4.*

Definition of a Stroke

A *stroke* is the movement of a piston from all the way in to all the way out of its cylinder (or vice versa). By "all the way in" and "all the way out" we mean as far as it is designed to move. In most engines, "all the way in" means the inside end of the piston is flush with this end of the cylinder bore, leaving the compression chamber unoccupied, while "all the way out" means the outer end of the piston is almost at the outer end of the bore. Hence a stroke is the full travel of a piston in one direction and represents a 180° revolution of the crankshaft.

The Four-Stroke-Cycle Engine

A *four-stroke-cycle* engine is so called because it requires four full strokes (*two* complete crankshaft revolutions) to accomplish the five events.

The first is an out stroke. While the piston is making this stroke an *intake valve* is open to accomplish intake, in the manner already described. Following the intake, the intake valve closes, and the piston travels back in (an *in stroke*) to accomplish compression. With compression readied, ignition occurs, and the piston is thrust back out to produce power. Finally, the piston makes its fourth stroke, back in, while an *exhaust valve* is open to accomplish exhaust, as previously described. This second valve closes when exhaust is completed, and the engine is

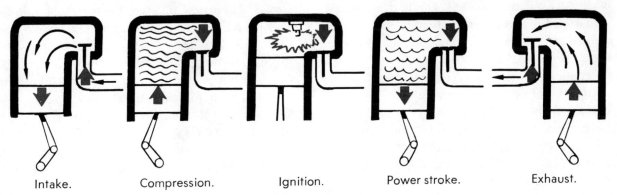

Intake.   Compression.   Ignition.   Power stroke.   Exhaust.

The 4-stroke-cycle engine.

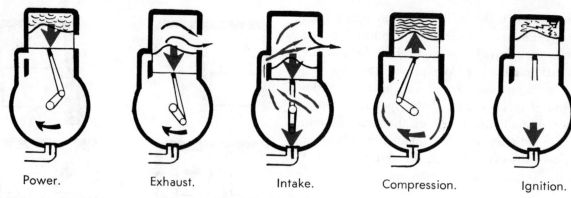

Power.   Exhaust.   Intake.   Compression.   Ignition.

The 2-stroke-cycle engine.

ready to repeat the whole performance (start a new cycle).

Summing up, a four-stroke-cycle engine has these distinct events: (1) an intake stroke; (2) a compression stroke; (3) ignition; (4) the power stroke; (5) an exhaust stroke.

The Two-Stroke-Cycle Engine

A *two-stroke-cycle engine* is called so because it requires only two full strokes (*one* complete crankshaft revolution) to accomplish the five events.

The first one is an *out stroke;* but this stroke is a power stroke (started by ignition) and it also accomplishes exhaust and part of intake. A second (*in*) stroke completes the cycle by finishing intake and accomplishing compression. When compression is readied, ignition occurs and the following out stroke starts the next cycle. Neither intake nor exhaust require separate strokes, as they are accomplished during the power and compression strokes.

Two principles that we have previously discussed make it possible to eliminate intake and exhaust as separate strokes. The first is the law of inertia; the second is the tendency of a compressed gas to expand through any opening provided.

Remember, molecules (or any mass) set in motion tend to continue moving in the same direction. Remember, also, that combustion sets the burned gas molecules in motion. Since the only direction they can move is outward, pushing the piston ahead of them, the exploded gas molecules all tend to continue right on moving outward. In a four-stroke-cycle engine the piston

has to overcome this tendency, reverse the molecules' direction of motion, and exhaust them through a valve in the cylinder head during an *in*-stroke. In a two-stroke-cycle engine the *exhaust* port is located near the outer end of the cylinder and is opened (uncovered) by the piston toward the end of the power stroke. The moving gas molecules simply continue on and out through the opening provided.

As already said, the area at the outer end of any cylinder block is enclosed by a crankcase. In a two-stroke-cycle engine this crankcase is tightly sealed. Hence, the enclosure is similar to a second combustion chamber, or extension of the cylinder, located at the outer end. The piston, when moving in for the compression stroke, creates a vacuum within this enclosure. An *intake valve* (into the crankcase) is open while this is happening, and a fresh charge is sucked into the crankcase enclosure.

During the power (out) stroke, this intake valve is closed so that the fresh charge in the crankcase enclosure is partially compressed by the moving piston. Just at the end of its power stroke the piston opens (uncovers) an *intake port* from the crankcase enclosure into the cylinder. As it does, the partially compressed charge expands through this port into the cylinder (not all of it; but enough for the purpose). This port is closed again just after the start of the compression stroke (at the same time that the intake valve is opened).

The exhaust port is opened a fraction ahead of, and closed a fraction behind, the intake port. Consequently, the fresh charge expanding into the cylinder from the crankcase enclosure helps to evacuate the last of the exhaust gases from the cylinder. And the inner end of the piston is usually shaped so that the fresh charge will strike it and be deflected along a path that will best accomplish this purpose.

After closing first the intake and then the exhaust port, the piston continues traveling in to complete the compression of the fresh charge now in the cylinder. Ignition occurs and the cycle is repeated.

SINGLE-CYLINDER VS MULTICYLINDER ENGINES

How More Than One Cylinder Is Used

Until now we have been talking about one piston and one cylinder. Most small engines do have just one of each and are called *single-cylinder engines*. Other small engines have two pistons with two cylinders, and each piston goes through a complete cycle of operation that is independent of, but timed with, the cycling of the other piston. Such an engine is a *two-cylinder engine*. A few in the small engine class even

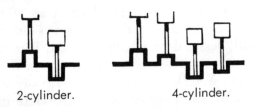

2-cylinder.   4-cylinder.

Multicylinder engines.

have four such pistons and cylinders. Large engines (for autos, trucks, etc.) may have any even number from four to sixteen (or more) such pistons and cylinders.

> NOTE: There is a type of single-cylinder two-stroke-cycle engine called twin-piston (or twin-cylinder) because it uses two pistons operating together to accomplish a single cycle, not separately to accomplish two independent cycles. Don't confuse this engine with a two-cylinder engine. It will be discussed later.

When two or more pistons and cylinders are used in a small engine they are connected to the same crankshaft. (In large engines two crankshafts joined by gears may be used.) The positions of the several piston crankpins around

the crankshaft are arranged so that the power strokes delivered by the independent pistons will be evenly spaced throughout one revolution of the crankshaft. In a two-cylinder engine the two crankpins are at opposite sides of the crankshaft, 180° apart. In a four-cylinder engine they are 90° apart.

For a two-stroke-cycle engine having two cylinders, this means that one piston is accomplishing its power stroke while the other is accomplishing its compression stroke. Power is being produced during the whole two-stroke cycle and is practically continuous. If the engine is a four-stroke-cycle one, however, having two cylinders will produce power only during half of each cycle; it requires four cylinders for such an engine to be producing power constantly.

> NOTE: For obvious reasons, more than two cylinders are not used with two-stroke-cycle engines, nor more than four with small four-stroke-cycle engines. In large engines, six cylinders have been found to provide smoothness consistent with economy, while eight appear to provide the practicable maximum of smoothness.

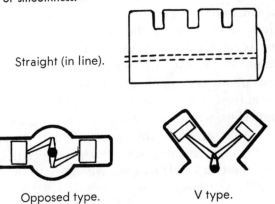

Straight (in line).

Opposed type.   V type.

Cylinder position.

Where Extra Cylinders Are Positioned

In general, three different designs are used when building two or more cylinders into an engine. The first is a *straight in-line* design which places the cylinders all in a row, either in the same or separate cylinder blocks. The second is a V design, in which two cylinders (or each pair if there are four or more) are positioned in a "V" with the pistons connected to a common crankshaft at the center. The third, *opposed cylinder* type is a flattened V. Other types used are: X, double V, single- or double-row radial, etc.

A FLYWHEEL BRIDGES THE GAP BETWEEN POWER STROKES

The compression, intake, and exhaust strokes all use energy; only a power stroke produces energy. Therefore, when an engine (such as any single-cylinder engine or a two-cylinder, four-stroke-cycle engine) does not produce constant power, some other source of energy must be provided to keep the crankshaft turning during the intervals between power strokes. In fact, energy is required in any engine just to carry the crankshaft past the two points (called *dead centers*) at which piston travel must stop to reverse direction between strokes. Then, too, any work that an engine is doing will probably require continuous power, delivered as smoothly as possible.

The Function of a Flywheel

To produce this interval of energy and "smooth out" engine operation, we use a *flywheel*—a more-or-less heavy (massive) wheel that is rotated by and usually mounted on the crankshaft. When revolved at any certain rpm (revolutions per minute) it tends—thanks to the law of inertia—to continue revolving at this same rpm. Thus, when accelerated to a running speed by the power-stroke thrusts, it produces a constant energy (momentum) which is sufficient to keep the crankshaft rotating without noticeable variations of speed.

Flywheel Design

For an object moving along a straight line, momentum equals velocity times mass. In a wheel, however, the whole mass does not move at the same speed. A spot farther out from the center (axle) makes a larger circle for each revolution than a spot nearer the axle. Therefore, any portion of the mass that is at the rim of a large wheel will move at higher velocity than portions nearer the center, or higher than a portion on the rim of a smaller wheel on the same axle. Consequently, in a flywheel, the

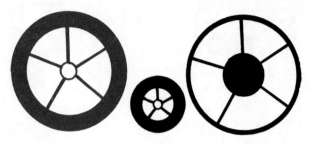

More momentum.   Less momentum.

Flywheel types.

momentum depends not only on the weight and rpm of the wheel, but also on the diameter of the wheel and its shape (that is, where the bulk of the mass is located, whether at the rim or nearer the center). To get the most momentum out of a flywheel, it usually is built with the bulk of its mass at the rim. To keep it from jerking (as a lopsided pinwheel jerks), this mass is carefully distributed evenly around the wheel (the wheel is *balanced*).

The longer the intervals between power strokes, the bigger the job is that a flywheel must do. A flywheel is important to any engine —its size, shape, weight and balance are all critical. With single-cylinder engines (and to a slightly lesser degree with two- and four-cylinder engines) these features are *extremely important*. Yet the overall weight of any small engine is also a critical factor. Therefore many small engines do not have a separate and distinct flywheel; the flywheel function is built into other necessary parts. Common practice is to design a magneto (to be discussed later) to also serve as a flywheel.

OTHER ENGINE TERMS

Early engines were designed with the cylinder bore(s) running horizontally. Also (as mentioned in the Introduction), there were types in which the bore(s) were vertical, with the combustion chamber(s) at bottom. When the first auto engines were developed with vertical cylinder bores having the combustion chambers at top, they were called *inverted-vertical* engines to distinguish them from the others. And since our shop talk has grown up around this engine design, most of the terms we use today are re-

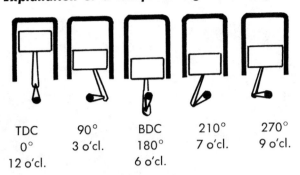

TDC	90°	BDC	210°	270°
0°	3 o'cl.	180°	7 o'cl.	9 o'cl.
12 o'cl.		6 o'cl.		

Engine positions.

lated to the positioning of parts in them. This, despite the fact that a great many makes of small engines, in particular, are actually horizontal types or other variations from the inverted-vertical auto type.

That is why we call the cylinder head a *head* and refer to the *top or bottom* of the cylinder block, cylinder bore, and piston. An *in* stroke is an *upstroke;* an *out* stroke is a *downstroke;* and we call a piston's uppermost position *top dead center (TDC)*, and its lowest position is *bottom dead center (BDC)*. The rotation of a crankshaft and connecting rod, as viewed from one end of the shaft, is expressed in degrees, with 0° representing TDC, 180° representing BDC, and the other degrees on around to 360° (which is back at 0 again) laid out according to the direction of rotation. Or, the movement of a clock hand may be used (with 12 o'clock at 0°, 3 o'clock at 90°, etc.).

Henceforth we shall use these terms and shall refer to the positions of all other engine parts also as if we were talking about an inverted-vertical type, even though the engine under discussion may be a horizontal one.

ACCESSORY SYSTEMS REQUIRED BY A PISTON-TYPE ENGINE

We know that any engine of the types described must have a ready and continuous supply of fresh fuel properly mixed with air for combustion. The electric spark needed for ignition must be provided when needed. Burned fuel must be disposed of. The friction of moving parts must be minimized to prevent their overheating and wearing. And, finally, the excess heat of combustion must be carried off lest it accumulate and burn up the engine parts or cause precombustion.

To provide these various additional requirements, every engine has to have five accessory systems. We shall discuss each of these in later chapters, but we list them here:

1. The fuel system.
2. The ignition system.
3. The exhaust system.
4. The lubrication system.
5. The cooling system.

CHAPTER 4

Timing the Five Events

Up to now we have discussed only basic engine designs and operating principles. The engines we have described could have been built 75 or more years ago. Our engines today differ from them not in principles of operation, but in the thousands of small details (refinements) developed meanwhile to make better use of these principles.

It is not our purpose here to trace the history of these developments. It would require volumes to do that. Rather, we shall present only the major refinements with which you will be concerned and shall discuss these in their present state of development.

THE VALVES OF FOUR-STROKE-CYCLE ENGINES

Because each of the two passages (intake and exhaust) of a four-stroke-cycle engine is opened and closed again during only one-fourth of the cycle, the operation cannot be performed by the piston, so it must be performed by separate mechanical valves timed to function with each

fourth piston stroke (every second revolution of the crankshaft). Moreover, each valve must open so as to pass all the fresh charge (or exhaust gases) desirable during the incredibly short time allowed. (In an engine rotating at 3000 rpm approximately 1/50 second is allowed for each valve to open and close again, and each is operated 1500 times a minute.)

Valve Types

Several types of valves have been used in various makes of engines. Two of these types, rotary valves and sleeve valves, are not used in small engines, so we will discuss them but briefly. A rotary type consists of a revolving shaft or plate (*rotor*) having a through opening which will become aligned with openings at each side when the rotor is in proper position. The sleeve type consists of a sliding plate (*sleeve*) flat or tubular in shape, having an opening which will become aligned with openings at each side when the sleeve is moved into proper position. Both types have the disadvantage of being fully open

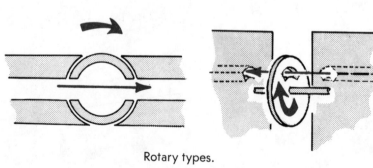

Rotary types.

Sleeve type.

Valve principles.

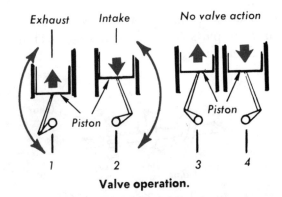

Valve operation.

only during a fractional part of the time between start and finish of an operation.

Small engines use *poppet* (sometimes called *mushroom*) *valves*. The valve seats (fits tightly) in the rim of a circular opening, much like a pot lid. It is held closed by a coil spring (*valve spring*) and is opened by being thrust straight off the *seat*. Its circular opening is large enough (in diameter) so that the valve needs to be unseated only a fraction to provide all of the passageway required for passing a fresh charge (or exhaust gases). Therefore it can reach maximum opening very quickly; it can hold this opening during most of its interval of operation, then close tightly, just as quickly.

The basic operating mechanism of any poppet valve consists of the coil *valve spring*—to close it and hold it closed—and a rotating *cam* to open it. Cams vary in design, but each is simply a wheel (on a shaft) that has one or more projections (*lobes*) to provide an eccentric motion when the cam is revolved. Most engine cams

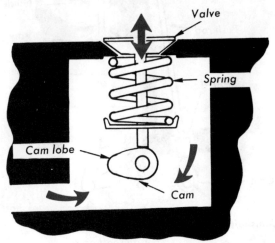

Poppet-valve principle.

have a single lobe and a single plunger or cam follower (called a *tappet*) that rides up and down as the cam is rotated. This tappet actuates the valve.

Placement of Poppet Valves

In general, there are three different popularly used valve placements, and these have required

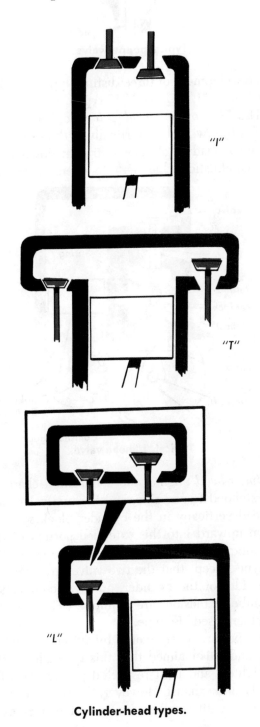

Cylinder-head types.

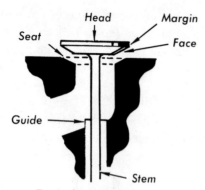

Typical poppet valve.

affords a more direct valve drive mechanism with fewer parts. The T type has not been used in small engines.

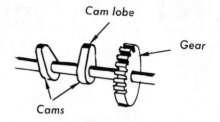

Typical one-cylinder camshaft.

the development of three distinct types of cylinder heads: "I," "L," and "T" types.

The *"I" (or overhead valve) cylinder head* places the two valves vertically in the top of the cylinder head; each moves straight down (into the combustion chamber) to open. In an "L"

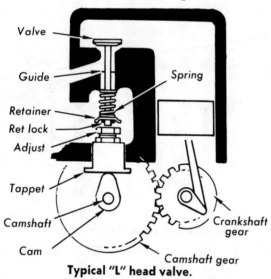

Typical "L" head valve.

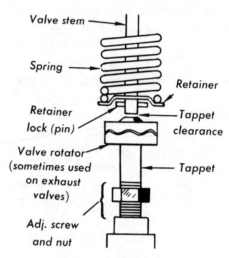

Detail of typical adjustment.

cylinder-head engine, the combustion chamber is extended to one side, and the two valves are placed vertically in the cylinder block so as to open upward into this extended portion of the chamber. A *"T" cylinder-head engine* is like the L type, except that the two valves are at opposite sides of the cylinder, and two combustion chamber extensions are required. (There is also a seldom-used "F" type.)

I types are most favored by auto manufacturers, and it is claimed that this type affords the straightest and least restricted passageways for intake and exhaust. However, the L type is most popular with small-engine manufacturers as it

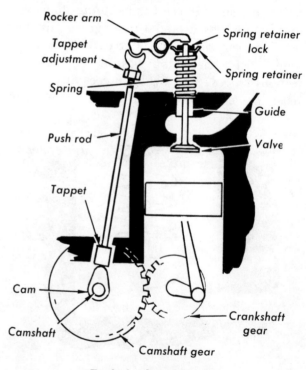

Typical valve-in-head.

Poppet Valve Operations

Accompanying illustrations show the principal parts of a valve and the operating mechanisms used with I, L, and T-type cylinder-head engines.

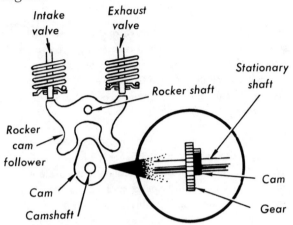

Typical "L" head variation.

Tolerances

When you consider that a valve must operate at speeds already mentioned, and under temperatures hot enough to melt many metals, it is no wonder that a valve must be in good condition to function properly. Then, too, if a valve does not open and close perfectly and on schedule, the engine operation is sabotaged by its failure.

All the parts of the valve itself—the *head, margin, face, stem,* and *seat*—are critical. A warped head will raise the face at one side so that it cannot seat properly. If the margin becomes too thin (usually, less than half its original thickness after the face is perfected), the valve will not be able to withstand the terrific heat (up to 2500°F is usual) and will "burn out." This is particularly true of an exhaust valve, which is exposed only to the superheated exhaust gases (whereas an intake valve is partially cooled by passage of the fresh charge).

A valve face and seat must be *perfectly* matched. To ensure this, both are finely ground to exact dimensions, then *lapped* by using an extremely fine abrasive called "grinding compound" between the two parts, and gently rubbing them together until they shine. A valve must seat well enough to withstand pressures

up to 500 psi without leaking. Manufacturers always specify the width and location of the "shine" required to ensure proper seating.

The valve stem must be straight; if bent, the valve may stick in its guide or fail to close tightly. The *guide,* which usually has a sleeve-type *insert* to serve as a bearing for holding the stem, must be true to prevent wobble. Then the overall length of parts (valve stem, tappet—and rocker arm and push rod, if used) which move to open the valve when the cam raises the tappet, must be very accurately adjusted. Proper adjustment always calls for a *tappet clearance* between the tappet and stem (or push rod)—a different clearance for intake and for exhaust—and is made by resetting the adjusting screw and/or nuts provided (or by grinding off the valve stem end if no other adjustment is provided).

NOTE: Tappet clearances are extremely critical and when parts wear must be reset to manufacturer's specifications. Too much clearance would make the valve open late and close early; too little clearance would have the opposite effect. The clearance is provided to allow for the expansion of parts when engine is at running temperature; a cold engine will actually have a little too much clearance.

Timing

Obviously each valve must operate in accord with a planned sequence coordinated with piston operation. We call this the *valve timing.*

The one, two, or more cams on a camshaft usually are either integral parts of the shaft or of the camshaft gear. A cam must be properly machined to lift the valve tappets correctly when this shaft (or the gear alone) is rotated in proper time with the crankshaft. Hence, timing is simply a matter of meshing the camshaft gear with the crankshaft gear so that the two gears

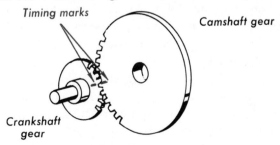

Timing gears.

are correctly related. Marks on the two gears usually are provided to identify the gear teeth which must be meshed.

NOTE: Sometimes the two gears are connected by a timing chain, similarly marked.

A camshaft gear is twice the diameter of and has twice as many teeth as the crankshaft gear. This causes the camshaft to be rotated at half the rpm of the crankshaft. As a result, each valve will be operated once for each four piston strokes.

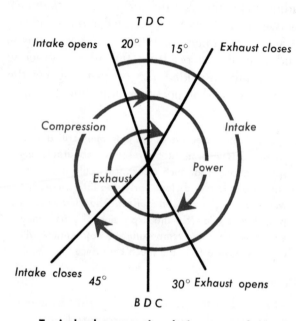

Typical valve operation during one cycle.

In theory, an intake valve should open at TDC and close at BDC of the intake stroke; the exhaust valve should open at BDC and close at TDC of the exhaust stroke. In practice, it is impossible to effect full, instantaneous valve opening. Also, because of inertia, a fresh charge does not start flowing instantaneously. It takes a fraction of a second to start it moving. Also, when moving as fast as gas does in and out of a cylinder (up to 250 miles per hour), its molecules will continue to crowd into the cylinder even after the compression stroke is started. The same applies to exhaust gases which, as previously mentioned, will continue to flow out even after the piston has stopped pushing them.

Consequently, cams are designed to open each valve a trifle ahead of dead center, and to close

it a trifle behind dead center. This means that the intake opens during the end of the exhaust stroke and doesn't close until after compression has started; exhaust opens during the end of the power stroke and doesn't close until after the intake stroke has started.

NOTE: The fact that the intake and exhaust valves open and close as explained is referred to as *valve overlap*. Incorrect tappet adjustment affects overlap severely. As little as a .001-inch error in tappet clearance can alter the opening and closing of a valve by up to 4°.

The faster an engine is running, the greater the overlap must be to compensate for the inertia of the gases. Consequently, overlap cannot be absolutely correct for both idling speed and normal running speed of an engine. For this reason, engines will often run inefficiently and even jerkily at idle—since engineers naturally design overlap for maximum efficiency at normal engine speed. With proper tappet clearance, however, this disadvantage is negligible.

Poppet Valve Maintenance

A valve is subject to considerable wear. Under normal conditions, the valve face and seat will, in time, become pitted and untrue enough to leak. The first few times (and if not too bad) the face and seat can be reconditioned; but if more than this goes wrong, the valve must be replaced with a new one. Valves are always replaceable. In some engines the valve seats (especially the exhaust seat) are also replaceable inserts; in others they are not replaceable. Valve guides also are usually replaceable (by inserts), though in some cases they cannot be replaced. When seats and guides are not replaceable, and if they are too badly damaged to be repaired by other means, the castings of which they are a part must be replaced.

NOTE: Inserts are sometimes available as repair parts, even when not used on the original engine. Valves with oversize stems are also sometimes available.

All other valve parts illustrated are replaceable. New ones generally are used instead of attempting repairs.

A valve face and seat must be relapped and all necessary adjustments must be made whenever a new part is substituted or an old one is repaired. Due to normal wear, the tappet clearances should be reset occasionally, even when no other work is required. Valve springs, too, will often become fatigued (lose their spring) and need replacement, even before they are visibly damaged.

THE VALVES OF TWO-STROKE-CYCLE ENGINES

A two-stroke-cycle engine requires two valves per cylinder, the same as a four-stroke cycle engine, plus one additional valve to admit fresh charges into the crankcase. Hence, there are three valves for a one-cylinder engine, and five for two cylinders.

Valve Types

The first two valves (corresponding to the intake and exhaust valves of a four-stroke-cycle engine) are always openings through the cylinder which are covered and uncovered by the piston. These are called the *transfer* and *exhaust ports*. For the third valve, a *third port*, a *reed valve*, a *rotary valve*, or a *poppet valve* may be

used. Hence, we generally speak of a two-stroke-cycle engine as being a three-port, reed-valve, rotary-valve, or poppet-valve type.

NOTE: Twin-cylinder engines (to be discussed later) are exceptions to the above rule.

The Transfer and Exhaust Ports

These ports are usually located at opposite sides of the cylinder. The exhaust port opens to the "outside," while the transfer port opens into a passageway to the crankcase enclosure (where the slightly compressed fresh charge is stored). Piston length is such that the piston will still fully cover both ports when at TDC, but will fully uncover both ports when at BDC. The exhaust port usually is just a trifle higher in the cylinder than the transfer port, so that, on a downstroke, exhaust begins a fraction ahead of intake. This also results, on an upstroke, in closing of the exhaust port a fraction later than closing of the transfer port, which permits the fresh charge in the cylinder to push out practically all the remaining exhaust gases. Either one or more

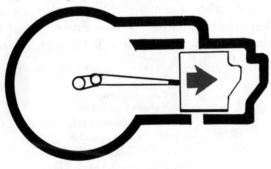

Both ports closed.

Exhaust port open, intake port partially open.

Exhaust and intake ports.

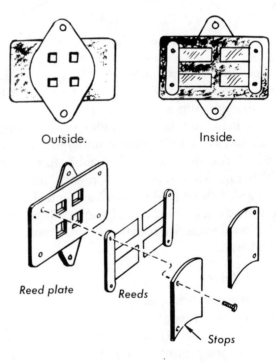

Outside.   Inside.

Reed plate   Reeds

Stops

Exploded view showing stops.

**A typical valve with two reeds—
with and without stops.**

openings may be provided to serve the purpose of each port.

Reed Valves

A reed valve is simply a flap, fastened along one edge, that will flop open under pressure to expose an opening. The flap is called a *reed*. It is mounted on a *reed plate* provided with a suitable opening. This plate is fastened to the crankcase (at any convenient position) with the reed inside. When the pressure inside is highest the reed is pressed against the plate to close the opening, but when the pressure inside becomes less than atmospheric pressure outside the valve opens.

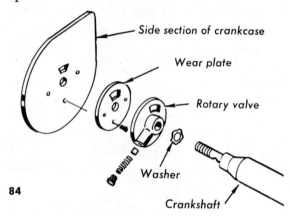

84

Principal parts of a rotary valve.

One, two, or more reeds of thin, springy material may be used. The holes in the reed plate are smoothly surfaced so that the reed(s) will close them tightly. On high-speed engines, which are likely to develop considerable suction inside (and, therefore, considerable force with which to open a reed), each reed is provided with a stiff metal *stop* that limits its opening (to prevent damage to the reed and resulting malfunction). Stops are usually omitted from slow-speed engines.

Rotary Valves

A rotary valve usually is the integral type, meaning that it is attached directly to one end of the crankshaft. However, some are run on a separate shaft geared to the crankshaft. The valve proper is called the *rotary valve*, and it rotates against a *wear plate* which, in turn, is stationarily secured to the crankcase side. When

the valve and plate holes are aligned, the valve is open; and when they are not, the valve makes a tight seal against the wear plate.

The valve may be secured to the crankshaft in any manner. One manner is illustrated. This method employs a key and keyway so that the valve will turn with the shaft, yet be free to move slightly forward or backward on the shaft. A spring washer on the shaft between the valve and a shoulder of the shaft serves to continually thrust the valve firmly against the wear plate on the crankcase.

Poppet Valves

If a poppet type is used for the third valve, its design and operation is essentially the same as already described for the poppet valves of four-stroke-cycle engines. A cam or lever, actuated by crankshaft rotation, is required to unseat (open) the valve during each piston upstroke. It is closed by a spring.

Operation With a Reed, Rotary, or Poppet Valve

As already mentioned (Chapter 3), the third (intake) valve opens to admit a fresh charge into the crankcase while the piston is moving up (to create a suction in the crankcase), and closes during the piston downstroke so that the charge in the crankcase will be slightly compressed.

1 = Power stroke
2 = Exhaust
3 = Compression in cylinder
4 = Intake into cylinder
5 = Intake into crankcase
6 = Compression in crankcase

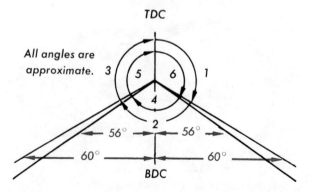

Typical reed, rotary, or poppet valve 2-stroke cycle.

Operation With a Third Port

When a third port is used for intake (in place of either a reed or rotary valve), it is positioned to open the crankcase to the supply line, so that a fresh charge will be sucked into the crankcase while the piston is completing its upstroke. Since this port is opened and closed by piston movement, it will be open for the same duration of the piston downstroke as for the piston upstroke. But it must be closed to allow the fresh charge in the crankcase to be partially compressed by the downstroke.

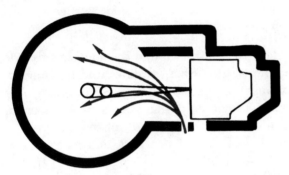

Third port open.

Consequently, this intake port is located low in the cylinder, as illustrated. To compensate for the short time during which it is opened, it usually is wider than the other two (or two openings may be used); and it usually is located 90° around the cylinder from the other ports (instead of alongside the exhaust port, as shown in the illustration for convenience).

There are several variations of the three-port design embodying openings through the piston itself, which allows the charge to be transferred through the hollow piston from one cylinder port to another. This transferring operation accomplishes essentially the same result as the type of third port illustrated. Such a design may also place the transfer port slightly higher in the cylinder than the exhaust port (contrary to the design illustrated), to compensate for the longest passageway from crankcase to cylinder.

In a three-port, two-stroke-cycle operation, the first major portion of the upstroke simply creates a vacuum in the crankcase. The fresh charge is admitted into the crankcase only during the remaining portion of the upstroke (instead of during the entire upstroke, as is done when a reed or rotary valve is used). Also, compression of the charge in the crankcase does not start at the beginning of the downstroke. It is delayed until the third port is again closed.

Timing and Maintenance

There are no timing problems with ports or a reed valve, and the only timing required for a rotary valve is to position it correctly on its shaft. If a poppet is used, timing requires the same adjustments as for a four-stroke-cycle engine.

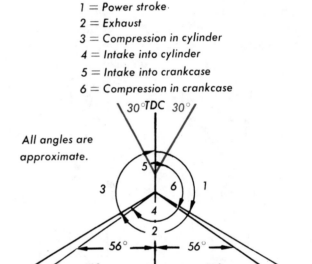

1 = Power stroke
2 = Exhaust
3 = Compression in cylinder
4 = Intake into cylinder
5 = Intake into crankcase
6 = Compression in crankcase

All angles are approximate.

Typical 3-port, 2-stroke cycle.

With ports, the only maintenance required is to keep them clean. There are no replaceable parts (other than the piston and cylinder block).

A reed valve does have replaceable parts, and these are such that it is better to replace damaged ones than to attempt repairs. The openings through the plate and reed must be kept clean, and the contact surface of the plate must be smooth to effect a tight seal. A reed must have the flexibility intended. It will not function if bent or fatigued (proper spring gone). Also, reed stops, if used, must not be bent from the original shape.

All parts of a rotary valve are also replaceable, and it usually is better to replace rather than repair them. Again, the openings must be clean and the contact surfaces smooth to ensure a tight seal. The wear plate and rotary valve must not be bent, pitted, etc. If there is a spring device to hold them in contact, it must not be fatigued.

Maintenance for a poppet valve is the same as previously described.

TWIN-PISTON ENGINE DESIGN AND OPERATION

The twin-piston engine (also called *phased-piston or split single-cylinder*) is a three-port, two-stroke-cycle engine that uses a second (auxiliary) piston to open and close the transfer port. This piston is called the *follow piston;* the other one is called the *lead piston.* In the illustration, the lead piston is shown as larger, but in practice the two usually are of equal size.

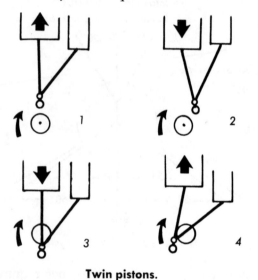

Twin pistons.

Both the intake port and the exhaust port are controlled by the lead piston. The two pistons share the same combustion chamber and cycle simultaneously, though each has its own separate cylinder bore. That is, the two pistons go up and down together; they both operate to compress the fresh charge in their combustion chamber, and both contribute to the power stroke.

The lead piston is connected to the crankshaft in the usual manner, but the follow piston is connected to the connecting rod of the lead piston, slightly above the crankpin. Hence the lead piston does actually lead the follow piston

slightly, as shown by the accompanying illustration. This permits the transfer port to be opened on the downstroke of the follow piston an appreciable amount after the exhaust port is opened by the simultaneous downstroke of the lead piston. This also allows it to be closed an appreciable amount later than the exhaust port on the upstrokes of the two pistons (even though these two ports are at practically the same heights in their two cylinders).

NOTE: Many engines of this type have dual exhaust ports (at opposite sides of the lead piston). In actual operation, the lead piston will uncover 80% of these exhaust ports before the follow piston begins to uncover the transfer port.

This twin-piston design accomplishes two desirable results. First, it allows the transfer and exhaust ports to be placed lower in the cylinder than with other designs. As a result, more of each downstroke is used for power prior to opening of the exhaust port, and more of each upstroke (following closing of the transfer port) is used for compression. Second, and most important, the exhaust is scavenged and the cylinders are recharged by a direct *uniflow* movement of the incoming gases. These incoming gases of each new charge enter by way of the transfer port, flow upward in the follow-piston cylinder and over into the lead-piston cylinder, then down toward the exhaust port. They push all of the exhaust gases ahead of them while moving continuously along this one direct path. This utilizes the inertia of the gases to the fullest possible extent.

Due to these power and compression-stroke advantages and to the effectiveness of the uniflow scavenging, engines of this design have exceptionally high efficiency ratings (approximately twice that of other design two-stroke-cycle engines).

IGNITION TIMING

Combustion Lag

For convenience we have until now been comparing internal combustion to an explosion, thus giving the impression that the gases in an engine combustion chamber fire all over, all at

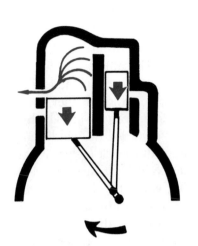

Exhaust starts.

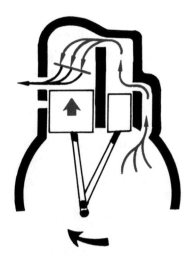

Transfer aids completion of exhaust.

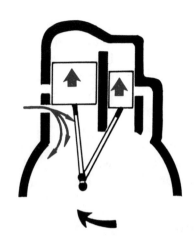

Intake starts.

Twin-piston operation.

once. While this might be true if we were to compress the air-gasoline mixture to its flash point, it is not true of electrical ignition. Though ignited internal combustion may appear to occur in one single flash, the burning actually starts at the electric spark and fans out through the balance of the mixture.

Combustion lag.

For this reason, there is a slight lag between the instant in which the spark occurs and the instant when an explosion reaches its peak force (the *peak of combustion*).

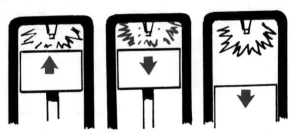

Too soon.   Just right.   Too late.

The peak of combustion.

NOTE: This is true even in a compression-fired diesel engine. The air alone is compressed, and the fuel is injected (sprayed) into the cylinder only at the final instant so that it will start burning at the exact split-second of time desired. It cannot be completely burned until the full charge has been injected into the cylinder.

Combustion Lag and Engine Speed

This *combustion lag* affects an engine in several ways which must be compensated for. The compensation is called *advancing the spark* (that is, timing the spark to occur ahead of TDC, while the piston is still completing its compression stroke). When the spark is moved back to occur at or behind TDC, we therefore call this *a retarded spark*.

A retarded spark is desirable when starting an engine or when it is idling. During starting, if the peak of combustion (or even the initial force of combustion) were to occur before TDC, the comparatively small force applied to start the engine would be overcome. The piston would then be driven back down before it could reach TDC, in what we call a *kickback*. Even at idling speed, there is insufficient momentum to carry a piston through to TDC against the full peak of combustion. An idling piston is still moving slowly enough for the combustion lag to be negligible in comparison.

As an engine is "revved" (speeded up), however, the speed with which a piston travels soon reaches a point at which the piston will move an appreciable distance during the split-second interval of combustion lag. At 5000 rpm, a piston can travel up to 10% of its stroke during the interval of combustion lag. Then, too, at such a speed the piston's momentum is quite enough to overcome the relative slight force exerted by the gases burned prior to the peak of combustion. Therefore, it is desirable to advance the spark to that point at which the peak of combustion will occur when it will do the most good, that is, just after TDC.

Other Factors

Engine speed alone is not, however, the only factor to be considered. Previously we said that at 5000 rpm a piston's momentum is enough to complete the compression stroke, even though the charge has started to burn and expand. But this may not be true of every engine, nor even of the same engine under all conditions. If the amount of compression is high, or if the momentum imparted by the flywheel (especially in a one-cylinder engine) is insufficient, obviously the engine already has about all it can do to complete the compression stroke (without the additional burden created by burning some of the charge before this stroke is completed). The same is true if a heavy load is connected to the engine. Also, the quality of the charge (proportion of air to gasoline, burning speed of the gasoline, etc.) and the temperature of the engine help to determine just how much of the engine's momentum will be used up by advancing the spark.

Consequently, the determination of just when and how far the spark should be advanced for top engine performance is a very critical problem. A *too-retarded* spark will waste much of the power in the fuel, because the peak of combustion will occur after the piston is already well on its way down, and the expanding gases will barely have started to exert their maximum push before the exhaust valve is opened to let them finish expanding outside. But *a too-advanced* spark will rob the engine of an excessive amount of its momentum and can even kill (stop) the engine. It will also cause overheating of the engine.

Provision for Combustion Lag

An engine manufacturer has to take all the factors into consideration. He must consider the idling and normal running speeds of the engine, its compression ratio, the kind of fuel to be burned, the average temperature at which it will operate, and the type of load it will be subjected to. The timing of the spark (when it is to occur) is then planned and built in in the same way that valve timing is built in by the design of the crankshaft and cams. An *initial spark timing adjustment* is provided so that a mechanic can set the timing mechanism in proper coordination with the crankshaft rotation.

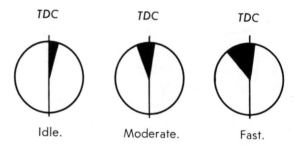

From spark to peak of combustion.

In addition to the initial spark timing adjustment, some engines also have provision for retarding or advancing the spark as required for optimum performance at various operating speeds. Automotive engines are generally equipped with fully automatic devices which retard the spark for starting and idling, advance it in relation to the engine speed, and also reduce the amount of advance in relation to the load. Other large engines have a similar *spark*

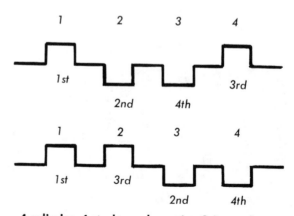

4-cylinder, 4-stroke-cycle engine firing orders.

lever with which the operator can retard or advance the spark (by sound) as seems best to him. Still others are equipped simply with devices without the load-compensating feature.

Small single-speed engines as a rule have fixed timing. That is, no provision is made for advancing or retarding the spark. The manufacturer simply designates a single setting which will provide the desired timing at the operating speed for which the engine is designed. Multi-speed small engines may have provisions for two spark settings or more (up to a full range of settings to correspond with various engine running speeds). Settings may be obtained manually or by an automatic device. No load-compensating features are used, however, as these are unnecessary as a rule in engines designed for specific, known loads.

The methods by which the initial spark timing adjustment is made and (if this feature is included) by which the spark is retarded or advanced will be discussed in Chapter 7.

FIRING ORDER

If the pistons in an engine with four or more cylinders were to deliver their power thrusts to the crankshaft starting at one end and progressing straight through to the other, this would tend to roll the crankshaft (just as you can roll a loop from end to end of a rope by snapping it). To offset this tendency, the *firing order* (sequence) of the cylinders is arranged so that the pistons do not deliver their thrust in an end-to-end sequence. Four-cylinder firing orders (numbering the cylinders from one end to the other) are usually 1-3-4-2 or 1-2-4-3.

CHAPTER 5

Other Basic Engine Refinements

THE POWER-PRODUCING COMPONENTS

Several factors enter into the design of combustion chambers, cylinders, and pistons. These are their relative shapes and sizes, the length of stroke of a piston, etc. Two of these factors, cooling and the materials used, will be discussed later. Now, we shall discuss cylinder intake and exhaust scavenging, piston displacement, and compression ratio.

Cylinder Intake and Exhaust Scavenging in a Four-Stroke-Cycle Engine

As we have previously noted, the valves of a four-stroke-cycle engine open directly into the combustion chamber. In a valve-in-head type (I-cylinder head) the flow of gases in and out is as direct as possible. With L- and T-head types, intake and exhaust gases flow around, through the combustion chamber extension. With these types of cylinder heads in particular, much care is generally given to the exact contours of the

chamber and its extension(s) to provide as smooth and continuous a flow of gases as possible. And in all types, the practice is to avoid corners and pockets in which gases can become trapped.

Pistons are made flat or very slightly convex (bulged) on top, to ensure uniform distribution of the force of each combustion.

Several typical designs are illustrated. Cooling and cleaning (carbon removal) also influence design. Then, too, intake openings and valves are sometimes made larger than exhaust openings, to ensure their being large enough to admit a full charge during intake.

Cylinder Intake and Exhaust Scavenging in a Two-Stroke-Cycle Engine

Port-opening locations in a two-stroke-cycle engine are governed by the requirements not only of intake and exhaust, but also of compression and the power stroke, as any raising of the

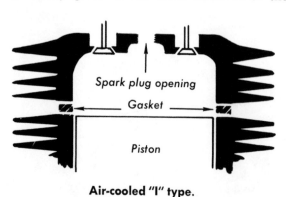

Air-cooled "I" type.

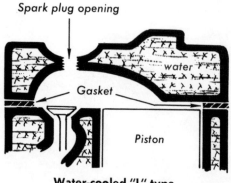

Water-cooled "L" type.

port locations in the cylinder to gain more cylinder intake and exhaust time will necessarily take an equal amount of the time of each stroke away from compression and power. Consequently, to gain width of opening the transfer and exhaust ports are often made elliptical, or "twin" ports are used.

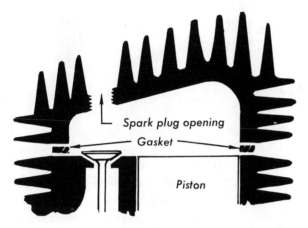

Air-cooled "L" type.

Since cylinder intake usually overlaps exhaust at start and finish, a means must be provided to keep the intake gases "chasing" the exhaust gases out of the combustion chamber and cylinder, rather than blowing themselves out the open exhaust port. For this reason, the piston top is usually shaped approximately as illustrated, and the combustion chamber is often shaped the same as the piston top. Again, cooling and cleaning also enter into the design.

Piston Displacement

In moving down from TDC to BDC a piston vacates (leaves behind) a certain volume of space in the cylinder bore. This is called the *piston displacement* and is measured in cubic

Typical air-cooled 2-cycle piston and combustion chamber.

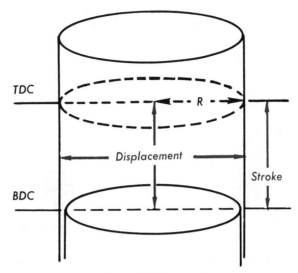

Piston displacement.

inches. Note that the area occupied by the combustion chamber is *not* counted.

To find the displacement of the piston in any engine, you would mark the position in the cylinder of the piston top when at TDC, move the piston down to BDC, then measure the distance it has traveled (its length of stroke). This length of stroke times the area of the cylinder bore is the displacement. Hence the formula is: *Stroke* $\times R^2 \times \pi =$ *Displacement* (where R^2 is the square of $\frac{1}{2}$ of the cylinder bore, and π is the mathematical constant 3.1416). Example: If the stroke is 2 inches and the bore is 2 inches then $2 \times 1^2 \times 3.1416 = 6.2832$ cu. in. of displacement.

Displacement is important because it determines the maximum amount of fresh charge that can be sucked into the cylinder under normal conditions of atmospheric pressure. This, in turn, naturally determines the maximum amount of power the engine will develop when run wide open. Manufacturers always state displacement as an indication of the engine's size.

Compression Ratio

In moving up from BDC to TDC a piston will reduce the total amount of space above it—in the cylinder and the combustion chamber—to considerably less than it was at the BDC piston position. If the cylinder and combustion chamber are full (with a fresh charge), the charge is compressed to a fraction of its original volume.

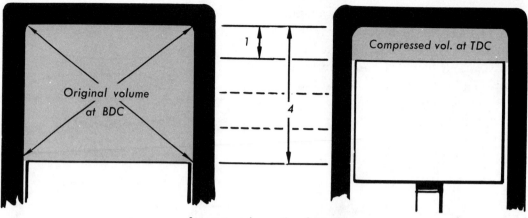

A compression ratio of 4-to-1.

The amount by which it is compressed is called the *compression ratio.*

Compression ratio is defined as *the maximum volume inside the cylinder at the beginning of its compression stroke divided by the minimum volume remaining at the end of the compression stroke.* If the volume at BDC is 8 cu. in. and the volume at TDC is 2 cu. in., the compression ratio is 8-to-2 or, more simply, 4- to-1. This can be written 4:1.

As previously noted, increasing the amount of compression of a charge prior to combustion increases the percentage of the fuel's total energy that can be obtained during combustion. With modern fuels improved for this purpose a higher compression ratio does result in increased engine efficiency. Small engines with ratios of 6-to-1 (even larger) are not uncommon. Automotive engines have still higher ratios. Manufacturers usually state compression ratio as an indication of an engine's efficiency.

The compression ratio does not indicate an engine's total power output. Total power output depends on displacement (the amount of fuel burned with each combustion) and on the engine's rpm, which, as explained in Chapter 9, has considerable to do with the power developed by a running engine. Arbitrary increase of compression ratio without taking valve and ignition timing into consideration may *decrease* engine power. An engine is designed with a certain displacement and for operation at a certain rpm for maximum power.

As the compression ratio is increased, the strain on engine parts becomes increasingly severe. The fitting together of parts and such adjustments as valve and ignition timing become ever more critical. Consequently, each manufacturer very carefully determines the correct compression ratio for his engine.

Never alter an engine's compression ratio without having full understanding of its design features and knowledge of the total end results which will be obtained.

THE PISTON AND CONNECTING-ROD ASSEMBLY

The manufacturing of pistons, rings, and connecting rod parts are often separate businesses in themselves, and volumes have been written about the refinements of these parts. We can only tell you a small part of the story here. Be assured, however, that when a manufacturer has selected a particular piston or types of rings for his engine, everything about them (size, shape, weight, expansion and wearing qualities of the metal, etc.) is critical. Substitutions or alteration can play havoc with engine performance.

The parts of a typical piston with rings and connecting rod are shown on the following page. A piston's function can best be summed up as "the transmitting of the fuel energy to the crankshaft, through the connecting rod." To do this efficiently it must withstand the heat of combustion without expanding to the point of freezing (sticking) in the cylinder bore. It also must travel up and down in the bore smoothly at terrific speed, without letting any damaging amount of the expanding or compressed gases leak past it. In short, it must be made of durable, selected material, must be as lightweight as possible in order to reduce the amount of its momentum that must be overcome each time it is stopped and reversed in direction, and must fit snugly in the bore.

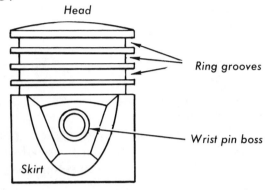

Typical "solid" 4-stroke-cycle engine piston.

Pistons

In general, pistons are made of cast iron or an aluminum alloy. Iron is less desirable because of its weight, but aluminum has the disadvantage of expanding considerably with heat so that a close fit at all engine temperatures is impossible. To overcome this disadvantage the industry has developed a *constant-clearance* (*split-skirt*) aluminum piston that closes up its own slots as it expands so that the skirt fits snugly at all temperatures. Such a piston may be made by casting or other manufacturing process.

Rings

Rings are classed in two types: compression rings and oil rings. Compression rings occupy the top groove or grooves and are designed to stop gas leakage past the piston; oil rings occupy

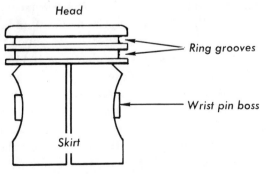

Typical constant-clearance 2-stroke-cycle engine piston.

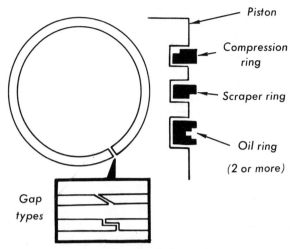

Typical conventional piston rings.

the bottom groove or grooves and are designed to allow crankcase oil (in four-stroke-cycle engines) to circulate around the piston and lubricate its stroke in the bore. There are many different designs of each kind. Conventional rings are one-piece with a break or gap in the circle to allow room for expansion of the ring when heated. But rings are also made in one solid piece, in two, three, and even four pieces, and sometimes with special expanders to be placed under them in the piston groove.

NOTE: Two-stroke-cycle engines do not require oil rings, as will be explained later. The second-from-top (or third) ring is sometimes a scraper, or carbon control ring, different from the top (compression) ring.

Two or more rings may be used. They are positioned so that the gaps are staggered around the piston (to minimize leakage when rings are cold and the gaps are open). The gaps may be straight vertical cuts, but more often are slanted, staggered, or of some other design. Oil rings may have holes through them for oil flow control.

As a result of the metals used, a cylinder bore will often become highly glazed (finely polished) by the action of rings scraping it through countless piston strokes. The rings themselves *wear in* (become perfectly seated) in positions and manners peculiar to the particular rings. For this reason, it is important that rings and the piston, too, always be reinstalled, after dis-

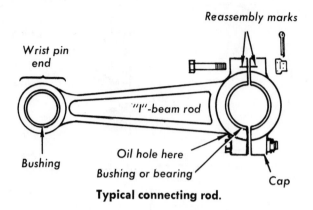

Typical connecting rod.

assembly, in the exact position as before. Also, if rings are so badly worn as to require replacement, new rings must usually be allowed some time to wear in.

Rods and Attaching Parts

A connecting rod usually is forged in the form of an I-beam, and the design depends on the method of mounting it at each end. There is a clamp type having a split boss at the upper end to receive the wrist pin. A bolt or cap screw tightens this boss around the wrist pin, which may have a groove for the bolt end, to secure the rod rigidly to the pin. If this is done, then the wrist pin floats in sleeve-type bearings in the piston bosses.

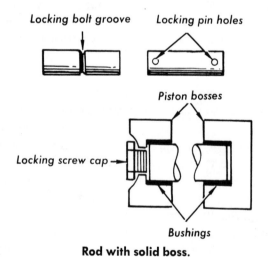

Rod with solid boss.

Another type of rod has a solid (not split) boss at the top end that holds a sleeve-type bearing in which the wrist pin floats. In such case the wrist pin usually is fastened into the piston in a

manner to prevent it from rotating or moving endways. However, a floating-type mount may be used. The chief consideration is that either the pin must float in the piston bosses, or it must float in the end of the rod. End play (end-to-end movement) is undesirable as it would allow the rod to become cocked on the piston instead of being straight.

Different devices are used to secure a wrist pin in the piston. Any of these will be obvious during disassembly of a particular engine. On the whole, wrist-pin design calls for the largest possible diameter of pin (to provide maximum bearing surface) and as lightweight a pin as possible. Tubular pins are much used for this reason. To the same end, connecting rods are designed to be as strong and lightweight as possible.

Standard practice for the lower end of a connecting rod is to provide a cap that is bolted to the rod to secure it around the crankshaft crankpin. This cap and the rod end may be fitted with the two halves of a split sleeve-type bearing, or with a bearing of any convenient type that can be assembled around the crankpin between the rod and cap. The type of bearing used is governed by the kind of application for which the engine is designed. Connecting rod caps are always matched to their rod ends; if there are more than one, they are usually numbered and mated to their rod ends by location marks. It is important that each be replaced on the correct rod, in the correct position.

The Assembly as a Whole

It must be remembered that the sum total weight of a piston, rings, rod, cap, wrist pin, bearings, and fastening parts is the total weight (or mass) that determines the momentum of this assembly during an upstroke or downstroke. Engine performance is calculated on the basis of this known mass. Changing any part—even substituting a heavier cap bolt or nut—can alter engine performance by altering the inertia of the assembly. In engines with two or more cylinders, not only this overall mass, but also the balances between the different piston assemblies are vital. A piston assembly that is lighter or heavier than the rest can "whip" a crankshaft out of shape.

In a small engine in particular, none of these assembly parts are worth repairing, and they are replaceable. Pistons, rings and wrist pins are generally available in oversizes. If badly worn the cylinder bore can be rereamed and honed (perfectly rounded and trued) to a slightly larger size—in which case, either oversize rings or an oversize piston and rings must be used. Manufacturers always specify wear limits beyond which any part so worn should be replaced. On the other hand, as all these parts wear together and become better fitted to each other in time, it is poor practice to install new parts indiscriminately. Most often, if one of two matched parts is replaced, so must the other (as a wrist pin and connecting rod; or a piston and its rings).

THE CRANKSHAFT AND COUNTERBALANCES

A crankshaft is the backbone of an engine. It must receive the piston power thrusts and rotate with them, and, in addition, it must convey the engine power to the load, rotate the flywheel, camshaft, if there is one, and other parts of the engine which are timed to operate in conjunction with the pistons (such as spark-generating and timing parts, cooling fan, etc.). It has also to fit into the space allowed, and run smoothly.

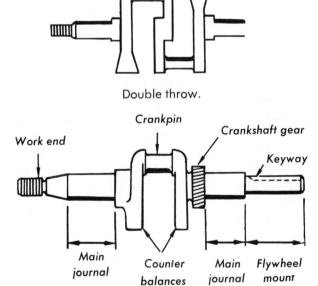

Double throw.

Single throw.

Crankshafts.

Shaft Design

Considering the varied requirements put on crankshafts by the many different engine types, it is no wonder that there are hundreds of different crankshaft designs; each is very precisely engineered and fitted to its particular make and model of engine.

The principal parts of a typical crankshaft are illustrated. In general, at least two journals are provided (even for a single-throw shaft), and shafts are made with journals between each pair of throws as well as with fewer journals. A journal is the bearing surface of the shaft that rides in the main bearings which support the shaft.

NOTE: The word "throw" is used to name the offset that holds the crankpin. Technically, however, it is the distance from the shaft center to the crankpin center.

Whether or not such parts as the crankshaft gear (to drive a camshaft), flywheel mounting flange (if used), etc., are integral parts of a particular shaft (instead of being separate, replaceable parts) depends on the manufacturer. In small engines such parts are usually integral.

The main bearings in which the journals ride may consist of split sleeve-type bearings each held by a cap mounted on the supporting frame similar to the connecting-rod crankpin bearings. Quite often in small engines, however, roller-type bearings are used.

Counterbalance Function

Under power, a piston delivers a one-sided whip to the crankshaft, like whacking one edge of a spinning hoop. This whip sets up a strain in the shaft. To offset this, counterbalances, (heavy, weighted members), are built into a crankshaft opposite each throw. By equalizing

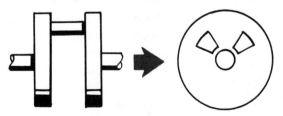

Counterbalance designed also as a flywheel.

the overall weight of the piston assembly, a counterbalance develops an inertia at the opposite shaft side which relieves the strain on the shaft. This keeps the shaft running smoothly.

When they are heavy enough, counterbalances also aid the flywheel function. In fact, in some small engines they are so designed that they will perform both the counterbalance and the flywheel functions together.

BEARINGS AND BUSHINGS

All rotating members of an engine must be supported in at least one, preferably two, or more bearings. If bearings were not used, one or both of the contacting surfaces (depending on their relative durability) would soon wear out-of-round and be useless. A bearing, which is cheaper and easier to replace than the shafts, takes the wear, instead. Also, due to its construction, it will increase engine performance and durability. There are two main classes of bearings—*the sleeve type* and the *roller type*.

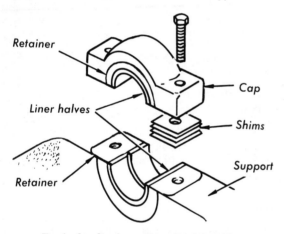

Typical split-sleeve (or plain) bearing.

Sleeve-Type (Plain) Bearings

Usually called a *plain bearing,* this type may consist of a one-piece *retainer* (hard-metal backing piece) with a *linear* (soft-metal bearing surface) mounted in a *support* (some stationary part of the engine structure) so that the shaft can be slipped into place in it. More often, however, plain bearings are of the *split bearing type.* That is, the retainer (and liner) is made in two mating halves to be clamped around the shaft journal. In this case, only the bottom half is

mounted in the support; the top half is a bearing cap that is bolted to the bottom half.

When split plain bearings are used, waferlike metal spacers (*shims*) are often provided for the purpose of obtaining an exact fit of the two halves around the shaft. In such case the manufacturer specifies the desired tightness with which the assembled bearing is to grip the shaft journal. The shims are added or removed during assembly until this tightness is obtained when the cap bolts are tight.

Liners are made of soft metal that will wear faster than the shaft. Babbit or lead-bronze alloys are mostly used. The liner is often sweated into the retainer and is not separately replaceable. Retainer-liner assemblies are replaceable. The retainer may be press-fitted or shrink-fitted (explained later) in its support. In some cases, however, liners are made without retainers, as separate insert pieces to be clamped in place around the shaft by installation of the cap.

Lubrication is required. Oil holes, slots, or grooves of some type are generally provided to ensure proper circulation of oil between the liner and shaft journal.

Damaged bearings are not repairable and must be replaced. In the case of split bearings, both halves must be replaced together. Retainers should be removed and installed in the manner specified by the manufacturer.

Roller-Type Bearings

A roller-type bearing is constructed in three parts: an *outer race,* the *rollers,* and an *inner race.* The two races retain the rollers between them, and the inner race remains stationary on the shaft while the outer race remains stationary in the support. All movement takes place between the rollers and the races, all of which are made of very hard metal. Rollers may be spheres (and the bearing is then called a *ball bearing*), cylinders (*roller bearings*), thin cylinders (*needle bearings*), or tapered cylinders (*tapered roller bearings*).

Bearings of these types cost more to produce than the sleeve types, but they are considered the finest from the standpoint of smooth, friction-free operation and durability. Each bearing assembly is usually preassembled and replace-

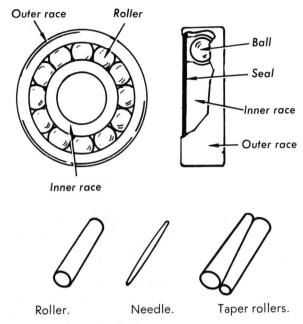

Typical roller-type bearing parts.

able only as a unit, though sometimes bearing parts are assembled around their shaft in separate pieces.

As a rule, both the inner and the outer race are secured in position by being press-fitted or shrink-fitted in place. When so fitted they should be removed or reinstalled in the same manner.

All types of roller bearings require lubrication. Some are packed with grease and sealed and do not require oiling. If needed, however, oil holes or channels are provided to ensure that oil will reach the bearing rollers. Damaged bearings are always replaced with new ones.

Bushings

A *bushing* is a removable metal liner that does not require a separate retainer but which is otherwise similar to a sleeve-type bearing. Although usually in one piece, it may be made in

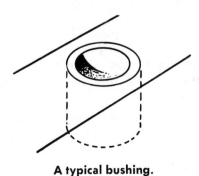

A typical bushing.

halves. Moreover, bushings generally are installed in a manner that permits easy replacement (usually pressed in or held by some mechanical means). They may be used to hold sliding shafts (for instance, a valve-stem guide is a bushing).

A bushing may be of any metal softer than the shaft. Cast iron and bronze are much used (even wood—though not in an engine). Some types require lubrication, and holes or cups may be provided; but there are types of oilless bushings or bearings which are impregnated with graphite and do not require oiling.

Damaged bushings usually are replaced with new ones, but sometimes can be repaired by reaming out smooth to a slight oversize provided the correct oversized mating part is available. They should be removed or installed in the same manner originally used by the engine manufacturer.

GASKETS

An engine uses fuel, air, oil, and water (if water-cooled). These substances must be retained within or excluded from certain areas of the engine, with *no* leakage. To accomplish this, the mating surfaces of parts which form an enclosure are machined (cut or ground to specified dimensions and smoothness) so they will mate with required tightness. Then a gasket is placed between them during assembly to ensure a leak-proof seal.

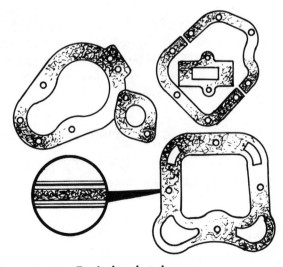

Typical gasket shapes.

The job that any particular gasket has to do depends on two things: (1) the force with which leakage past it is being "attempted," and (2) the perfection with which the two mating parts have been machined and mated. Consequently, gaskets are made of different materials and in different types. Also, materials must be selected to withstand other conditions such as heat, the corrosive action of acids, etc. However, all gaskets have one thing in common. The gasket must be capable of molding itself to the contours of the two mating parts—to fill every scratch, dent, or other imperfection—when squeezed between them by tightening of the bolts that hold them.

Cylinder-head gaskets usually consist of two sheets of copper with a layer of asbestos between. Exhaust opening gaskets may be of a special material composed of copper wires, asbestos, and graphite. Sometimes, sheet-metal grids (screens) are used, with a softer material filling the spaces. Cork, rubber, synthetic rubber, paper, plastic, and spun-fiber materials are also used. Each type fulfills some special requirements of the engine manufacturer. Substitution is a poor policy unless you understand all the requirements.

Once squeezed into place then subjected to the running temperature of an engine, a gasket not only molds itself to the mating surfaces, it usually becomes *set* to this shape. Therefore, if a gasket is removed, it is very important that it be reinstalled in exactly the same position. It is also important, for obvious reasons, that a gasket and the mating surfaces be as clean as possible.

Torn, lifeless (set and dried out), or otherwise damaged gaskets must be replaced by new ones. In fact, the best practice is always to use new ones, if cost factors make this practicable.

GEARS AND CHAINS

It is a basic law of nature that the amount of work accomplished by a machine always equals the amount of work (force or energy) put into the machine, less any loss due to friction, inefficiency, etc. If this were not so—if you could get more work out than you put in—then you would have true perpetual motion; but there is *no such*

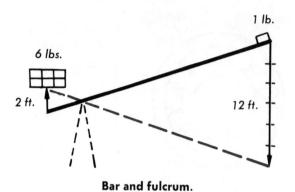

Bar and fulcrum.

thing. Going back to the principles discussed in Chapter 1, no energy (or force resulting from energy) is ever lost; but neither is energy (or force) ever created out of nothing.

At the end of Chapter 1 we illustrated six basic machines. First (and the most basic of all) is a lever. In simplest form this is simply a bar with a *fulcrum* (pivot point). Moving one end of the bar down rocks the opposite end up. If you apply force to one end (do work at this end), then the opposite end will move and do an equal amount of work (less friction, etc.). Using the formula "work = distance × force" we can then say: work (distance × force) at one end = work (distance × force) at the other end. In short, *distance times force at one end equals distance times force at the other end of a lever.*

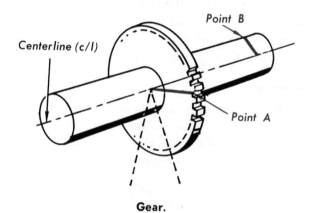

Gear.

If one end of a lever is moved 12 feet with a force of 1 lb (12 ft-lb total work), and the other end is shorter so that it moves only 2 feet, the shorter end must move with a force of 6 lb (since 2 ft × 6 lb also = 12 ft-lb). The first (input) end has to move six times as far, but the second (output) end moves with six times as much force. If you reverse ends (put work in at the short end), the converse is true.

The Principle on Which Gears Operate

A gear is simply a lever. Consider a single gear fixed on a shaft. This assembly revolves around a central line called the *centerline,* which is actually a fulcrum. Any point on the circumference of the gear can be considered as one end of a lever; and any point on the shaft's surface is then the other end of the lever. If work is done to rotate the gear, the shaft will rotate and do an equal amount of work; but the point on its surface will travel a shorter distance per revolution and with more force.

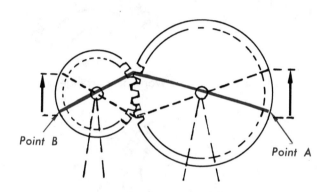

Two gears on separate shafts.

If a second smaller gear is also fixed on the same shaft, the preceding is then true of any point on the circumference of the smaller gear. When two gears are fixed on separate shafts so that one must revolve the other, you have two levers (instead of just one) operating together to produce the same result. By adding on gears and shafts in different arrangements, you can multiply the number of levers; but you will still always come out with the same result.

Gear Parts and Types

Since a gear and shaft rotate, the speed of their revolution is usually rated in terms of *rpm* (revolutions per minute); the force with which they rotate is called their *torque* (twisting force); and the distance that any point on a gear or shaft describes (travels through space) per revolution will be the circumference at this point (which, in turn, depends on the diameter). The

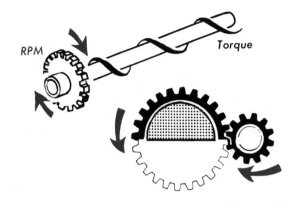

Gear ratio.

projections around a gear's circumference are its *teeth.*

For any two gears to mesh properly their teeth must all be identical, with alternatively disposed, identical spaces to fit the teeth. Since this is true, the quickest way to learn how many revolutions of the smaller gear will equal one revolution of the larger gear is to count their teeth. If the smaller gear has half as many teeth as the larger, it must revolve twice for each one revolution of the larger, etc. This relationship is called the *gear ratio.* In our example, the gear ratio is 2-to-1 (2:1).

NOTE: Gear teeth do not mesh at their tips. The point of contact between two teeth is approximately at the midpoint of eoch. Therefore, *not* the full gear diameter, but only the diameter as far out as the point of contact determines the amount of leverage obtained by a gear. This is called the gear's *pitch diameter.* The ratio between one gear's pitch diameter and another's is the same thing as the gear ratio.

Gears are named according to the type of teeth they have. The simplest is the *spur gear,*

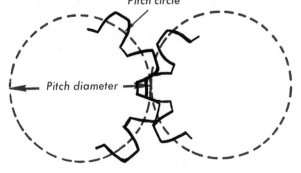

Gears properly meshed.

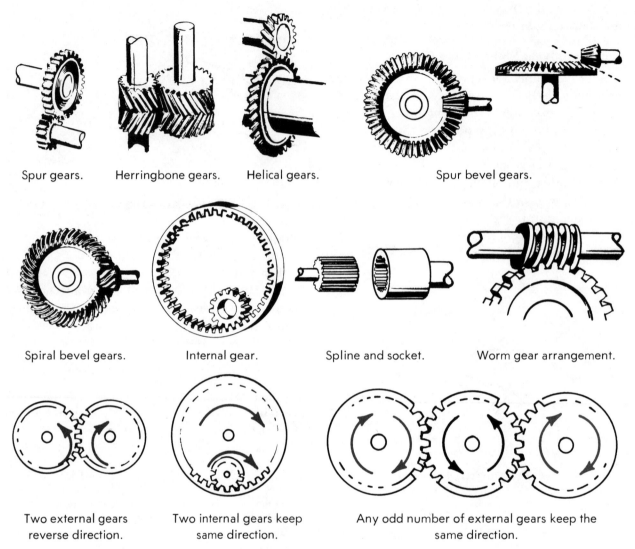

Spur gears.　　Herringbone gears.　　Helical gears.　　　　Spur bevel gears.

Spiral bevel gears.　　Internal gear.　　Spline and socket.　　Worm gear arrangement.

Two external gears
reverse direction.

Two internal gears keep
same direction.

Any odd number of external gears keep the
same direction.

Gear types and arrangements.

which has teeth cut straight across the gear's circumference. A *helical gear* has teeth cut across at angles and slightly curved (as if they were the threads of a flattened screw). Then there is the *herringbone gear,* which is like two helical gears welded side by side. All these types are used edge to edge on parallel shafts. They all operate exactly alike, except that the helical gear is quieter in operation than the spur. The herringbone gear is still quieter that the helical.

When two shafts are at an angle to each other, the gears used are called *bevel gears.* If the shafts are at right angles, they are 90° bevel gears, and the slope of each gear's face (outer edge) is 45°. The same relationship applies to other shaft angles. Again, if the teeth are straight across, these are called *spur bevel gears.* Another type is the *spiral bevel gear* (like the helical, but with teeth that are more curved).

When spacing requires that one gear rotate inside of another, this is called an *internal gear arrangement.* Any of the above types may be used. If a gear is "flattened out" to become a straight or slightly curved bar instead of a wheel it is called a *rack;* or, if it is only a pie-shaped piece of wheel, it is called a *sector gear.* Then, too, any small gear intended to be meshed with a larger one (or a rack) is called a *pinion* (or pinion gear). And, again, when we have two gears meshed together, we often call one the *drive* gear and the other is called the *driven* gear.

Whatever the names gears go by, the principle is the same. Substituting correct terminology in the lever formula given at the beginning, this principle is correctly stated as: $work = torque \times rpm$ (where "work" means the effort used in revolving the gear).

NOTE: Two types of mechanisms commonly referred to as gears do not belong in this group at all. One is the *worm gear* arrangement (which consists of a worm, or screw, arranged to drive a helical gear). This arrangement operates on a different application of the lever principle, so that the rule regarding gear ratio does not apply. The other is the *spline and socket* (often made to look like gears) used to join two shafts end to end.

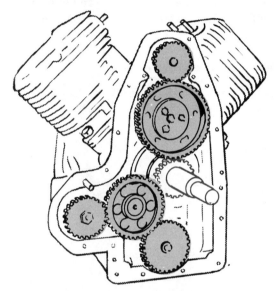

Gear train of one type of 2-cylinder engine.

How Gears Are Used in an Engine

As already noted, two or more meshed gears (called a *gear train*) can be used to increase torque, with corresponding decrease of rpm, or to decrease torque, with corresponding increase of rpm. They can also be used to alter the direction of rotation (see accompanying illustration), or simply to transmit power from place to place. (Refer to Chapter 10.)

Inside an engine, the torque applied to other moving parts is unimportant. But the rpm and (sometimes) the direction of rotation of another shaft with respect to the crankshaft, and the need simply to transmit power are important. The gears in an engine are used for these pur-

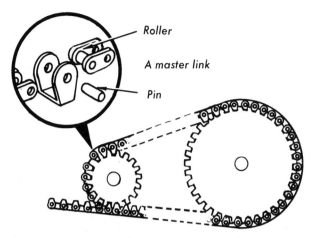

Sprockets and chains.

poses. Those used to control the rpm of another shaft or part are called *timing gears*. Since starting positions as well as rpm are important, timing gears always have timing marks of some kind to show which teeth are to be meshed.

Chains and Sprockets

Chains are sometimes used to accomplish the same purposes served by gears. The wheels with which a chain is meshed are called *sprockets*. The teeth of a sprocket are always cut to fit into the spaces formed by the chain *links* (each jointed section). To make a chain quiet, smooth, and longer lasting, the links usually are designed with rollers around the pins (crosspieces) where the sprocket teeth enter the chain. When so constructed, it is called a *roller chain*.

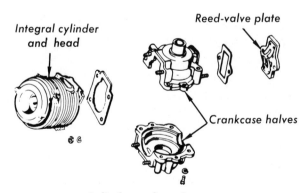

Cylinder and crankcase.

All the rules that apply to gears apply also to sprockets and chains. The gear ratio between two sprockets depends on the number of teeth in each.

Chains usually have at least one *master link* with a removable pin. This is so that the chain can be opened for removal or installation on its sprockets. Timing marks are also provided, when needed.

PRINCIPAL ENGINE STRUCTURAL ASSEMBLIES

Several typical assemblies of the principal engine structural parts are illustrated.

You will note that these vary considerably in arrangement and number of parts required. There are no fixed, nor even common practices regarding the designs of these parts; each manufacturer determines the best arrangement for his own purposes.

In the illustrations we have not attempted to show any parts other than those which combine to form the major portion of a cylinder-head, cylinder-block, and crankcase assembly. Actually, in each of the engines illustrated, many more attaching parts (accessories which will be

discussed later) are required to close all the holes and complete the structure. Assembled, however, the parts shown do enclose all the engine operating parts discussed so far.

TOLERANCES AND FITS

Today's engines require precision fitting of parts, often even to within one ten-thousandth (1/10,000) of an inch, and such parts are machined with extreme accuracy. In the manufacture and machining of such parts, however, it is not practicable for a manufacturer to discard every part which does not measure up "on the button." Then, too, in service, parts do wear; even if perfect at the start, they will become smaller in time by an amount measurable with a micrometer.

The Meaning of Tolerances

For these reasons, manufacturers establish *parts tolerances* on certain engine parts. Piston

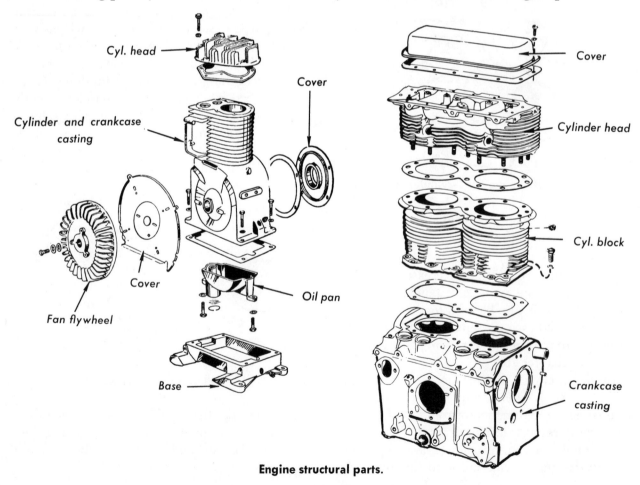

Engine structural parts.

rings; cylinder bores; valve seats, faces, and stems; shaft journals; bushings and bearings—these are typical parts for which tolerances (especially wear tolerances) are allowed. Tolerances may be listed in several ways.

One way is to give a wear tolerance, meaning how much a single part may be allowed to wear down in a certain dimension before requiring replacement or reconditioning. This is stated as a plus (+) or minus (−) amount (as the case may be) and may be written in parentheses following the standard dimension. For instance, the standard diameter of a cylinder bore may be 2.125 (2⅛) in.; but an otherwise perfect bore measuring up to six one-thousandths inch larger is still serviceable. This fact could be shown by the listing: "Cylinder Bore 2⅛ (+.006) in." (or the manufacturer might state specifically that there is an allowable oversize of .006 in. in the cylinder bore).

Another way is to give the mating tolerance between two parts which must fit together. This usually is stated as the amount of *clearance* required between them, or, as will be explained later, the amount of interference. For instance, the clearance between a ball bearing (inside diameter) and its shaft journal (outside diameter) might be stated as .005 in.

In the manufacturing of precision parts plus (+) and/or minus (−) tolerances are usually established to allow for slight oversizing and/or undersizing. The ball bearing previously mentioned may have a listed inside diameter (ID) of 1.125 (±.005) in., meaning that any bearing with an ID from 1.120 to 1.130 in. is acceptable. Its shaft journal may be listed as having an outside diameter (OD) of 1.120 (+.005) in., meaning that any size from 1.120 up to 1.125 is acceptable. Therefore, if the required clearance between these two is given as .005 in., parts will have to be hand-picked to obtain this clearance. Not just any bearing will fit properly on any journal, even though both are new.

Types of Fits

Quite often the manner in which two parts are to go together is stated in terms of a certain type of fit. There are four generally used classes of fits.

A *running fit* is the fit between two parts which must rotate or reciprocate (slide) without hindrance. The male part must be smaller than the female part by at least 0.001 inch for each inch of diameter. That is, there must be a *clearance*.

A *push fit* is that between two parts which are not quite free to rotate or reciprocate, but which can be assembled by hand. Male part should be 0.001 to 0.0001 inch smaller (a *clearance*).

A *driving fit* requires that the two parts be assembled by forcing them, as in an arbor press. This requires an *interference*. That is, the male part must be the same size as the female part, or up to 0.001 in. larger for each inch of diameter.

A *forced fit* requires assembly in a hydraulic press and an *interference* of 0.002 in. for each inch of diameter.

Parts like plain bearing liners and retainers which must remain stationary are assembled in one of these manners:

A *shrink fit* is accomplished by heating the female part (or chilling the male) so that when returned to normal temperature the female is actually stretched around the male.

A *spun fit* requires rotating of the female part to allow the male part to be spun-in while in a semimolten state.

Still more permanent fits than these are obtained, of course, by welding, brazing, or soldering, which chemically joins the two metals together.

Torque Tightening

In an engine, the firmness with which two parts are fastened together (with bolts, nuts, screws) is often quite important to the proper operation of the engine. Cylinder heads, for instance, must be tightened on very securely to prevent combustion loss.

For this reason it is common practice to state how tight the holding bolts, etc., are to be made—in terms of the number of foot-pounds of torque (twist) that is to be applied when tightening them. When torque tightness is specified, a *torque wrench* must be used. This is a special wrench that makes it possible to accurately measure the required foot-pounds of tightness.

NOTE: There is also danger from overtightening, which can result in warping or breaking of parts. Also, from uneven tightening caused by drawing down one side or corner of a part before the others are tightened at all. Generally, care must be taken to tighten all the bolts evenly, each in turn a little at a time, and not to exceed specified torque tightness.

A WORD ABOUT METALS

The first metals to be used for the structural parts, shafts, and other stress-bearing parts of an engine were iron and its derivative, steel. Fifty years ago, steel alloys, let alone all the other various alloys that we have today, were unknown. It was not uncommon for early engine manufacturers to have rooms full of broken parts which were returned by customers because of failure.

Today in the automotive field, over 40 different steel alloys alone are in use. An alloy is created by adding small amounts of another metal (sometimes as little as one or two parts in 1000) to change the properties of the original metal. Manganese, silicon, nickel, chromium, tungsten, molybdenum, and vanadium are the alloying metals used with steel. These have varying desirable effects according to the quantities added, other metals added, or even the method by which the additions are made. For instance, manganese added in amounts of 11% to 12% makes an extremely abrasion-resisting steel much used for parts subjected to wear. Iron (including steel) and the many alloys above are called *ferrous* metals, while all others are called *nonferrous* metals.

In the small-engine field, not only many of the steel alloys, but many of the nonferrous metals and their alloys also are largely used. Aluminum and its alloys, in particular, are highly favored because of their lighter (than steel) weight. Weight is not only important from the standpoint of overall engine bulk, it is also of great importance in the fabrication of moving parts (such as a piston) where inertia must be considered.

We can't go into the details here regarding the specific properties and advantages of all the various materials used by different small-engine manufacturers. The point is, however, that in each case a particular metal is chosen to do a very specific job. Some other metal (even though superior in many ways) may not be capable of doing the same job as well. Beware of substitute parts which may have all the appearances of being satisfactory, but which may not be right for the job.

CHAPTER 6

The Rotary Combustion Engine

THE ENGINE WE WILL DISCUSS

RC (*rotary combustion*) is a general classification which includes any positive-displacement (capable of compression), internal-combustion engine that develops mechanical power by rotation of the combustion-driven part. The combustion-propelled part may be a piston, a piston assembly, or a housing, such as a cylinder block, that contains the piston(s). Many mechanical designs have been proposed and a few have been developed for accomplishing rotary combustion. In this chapter, however, our discussion will be limited to the Wankel design.

The original Wankel engine design (referred to as DKM) achieved rotary combustion by using two rotors, one within the other, inside a stationary housing. The KKM-Design engine is somewhat different; this engine design uses a single rotor within a stationary housing. The latter (KKM-Design) incorporates the fundamental principles of all RC engines currently being developed and/or produced by industry. Since development of the first KKM-125 Model, a number of improved models have been built by Wankel and his NSU associates. These, only, are correctly referred to as NSU/Wankel engines. Other firms (notably, Toyo Kogyo in Japan, Daimler-Benz and Citroen in Europe, and Curtis-Wright and General Motors in the U.S.A.) have been and are developing the original KKM-Design for a number of specific uses. Each firm

has a different tradename for its product(s).

As a result of this widespread and intensive interest, the original design has and will be considerably improved, modified, expanded and otherwise altered for adaptation to a variety of commercial applications. It would not at this time be practical, nor is it the intent of this text, to cover all possible variations; we will discuss only the basic KKM-Design.

ADVANTAGES OF THIS ENGINE

This engine does *not* afford any distinctively different process for converting fuel energy into mechanical power. It operates on exactly the same internal-combustion principles as a four-stroke-cycle, reciprocating-piston engine. There are the same five events: intake, compression, ignition, power and exhaust. In fact, its rotor is actually a piston assembly, and the rotor housing assembly serves the purpose of a cylinder block.

Wankel.

Therefore, the advantages which have aroused so much interest in this design do not derive from a new concept in the use of fuel energy (such as jet propulsion); on the contrary, they result from an improved means of utilizing existing knowledge and the technological improvements acquired in furtherance of this knowledge.

NOTE: Although the KKM-Design rotary engine cycles in a manner similar to the reciprocating-piston engine, power is developed by rotation of a rotor rather than by a sequence of piston strokes. For this reason it is proper to refer to this rotary as a *four-phase-cycle* (or *four-phase*) engine to distinguish it from the four-stroke-cycle (or four-stroke) reciprocating-piston type. It is also called the Wankel or the Wankel RC Engine.

The greatest advantages of the Wankel engine are its weight and size. Compared with an equivalent horsepower, standard V8 automotive engine a recently developed automotive-type four-phase engine weighed less than half as much, and occupied approximately half as much space. Even better, the lbs-per-hp ratio for small horsepower four-phase engines is much less than half the ratio for comparable four-stroke engines.

Of almost equal importance, this rotary-type engine is more economical to maintain and operate. A comparison between the preceding two engines shows the V8 as having 1029 parts of which 388 move, while the rotary-type has only

Standard V-8.

633 parts with only 154 of them subject to the strain and wear of movement. Further, the latter uses ports instead of valves, thus eliminating the many valve-train parts which contribute much to the maintenance problems of the four-stroke engines.

NOTE: In the automotive-engine field, production-cost analysis favors the rotary type; but it appears that production costs of the two types are about equal for small-horsepower sizes. The RC engine has quick throttle response without knocking when operated on gasoline with an octane rating of 70 or even less. It shows great promise of adaptation to diesel and other lower-cost fuels.

In addition, indications are that four-phase engines can operate with less internal power loss, at much higher output-drive speeds, and more smoothly, quietly and pollution-free than four-stroke engines. These performance gains contribute to greater overall efficiency in the development of mechanical power.

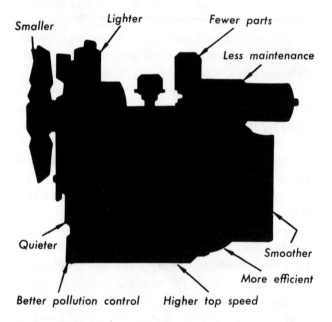

Internal power loss is reduced in a number of ways. With fewer moving parts there is less friction. Substitution of no-load port-opening operations for heavy-load valve-opening ones saves considerable power. The rotating parts are subject to much lower inertia loads than reciprocating parts which, in consequence of the strains put upon them, must be strongly built, thus adding to the dead weight which a four-stroke engine must accelerate and decelerate during operation.

Moreover, movement of the mixture and of the burned gases is one-directional and less torturous; there is no need to reverse the direction of gas movement to effect exhaust, as in a four-stroke engine, nor to move the mixture and exhaust gases through winding passageways into and out of the working chamber. Depending on the port locations, dwell periods can be made shorter, and effective use of combustion can be extended to longer duration than in a four-stroke engine; this would result in a reduction of unburned fuel loss out of the exhaust. In a four-stroke engine some breakaway-torque power is wasted unfreezing the rings each time a piston starts from top- or bottom-dead-center, especially after the rings have worn grooves in the cylinder walls at these positions. There is no comparable loss in the operation of a rotor because the seals (equivalents of rings) do not halt and reverse their movements.

Finally, even though momentum loss due to continuous reversal of a four-stroke piston is negligible, no equal loss occurs in a rotary engine in which rotor momentum is unidirectional.

Higher output-shaft rpm is practical for several reasons. The four-phase engine has a higher volumetric efficiency; that is, the working chamber can be filled with a proportionately larger amount of charge at full-open throttle, due to the more direct and less interrupted flow of intake mixture and to the more thorough inertia-aided exhaust of the previously burned gases. Lubrication problems, especially at higher speeds, are reduced because the rotor moves slowly in relation to a piston (for each output-shaft revolution the rotor makes only one-third revolution; a piston completes one revolution for each crankshaft revolution).

This relatively slow rotor movement also reduces bearing strain, as do two additional factors: (1) Every bearing in a rotary engine is under constant positive (forward) torque due to unidirectional rotor rotation; there are no bearings subjected to negative torque such as applied to crankpin and wrist pin bearings each time a piston is thrust "backwards" for compres-

sion and exhaust. (2) The "throw" of the rotor, which travels a slightly eccentric course, is much smaller than the crank throw of a piston. In addition, there is no strain on the rotor seals (which travel only in one direction) comparable to the flutter produced in piston rings at very high engine speeds.

Most important, a rotor is subjected only to centrifugal and slight inertia forces. The momentum variations to which a rotor is subjected are indeed negligible compared to those of the reciprocating parts of a four-stroke engine. In fact, there are no destructive momentum factors comparable to those which exist for pistons and valves, which tend to disintegrate if the speed limit of a four-stroke engine is exceeded.

Superior smoothness, quietness and pollution-freedom are assured by additional favorable factors. There is less vibration; in fact, a two-rotor, four-phase engine operates with exceeding smoothness and quietness by comparison with a four-stroke engine in which all straight-moving parts, however well balanced, will at some operating speed set up noisy and annoying vibrations. Power output is more uniformly obtained with the rotary engine because each power stroke lasts for about two-thirds of a cycle instead of only one-quarter or less, as in a four-stroke engine.

Although it is as yet questionable whether the exhaust from a rotary engine will, in future, be

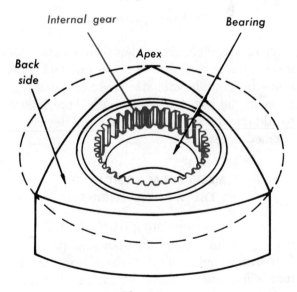

The rotor.

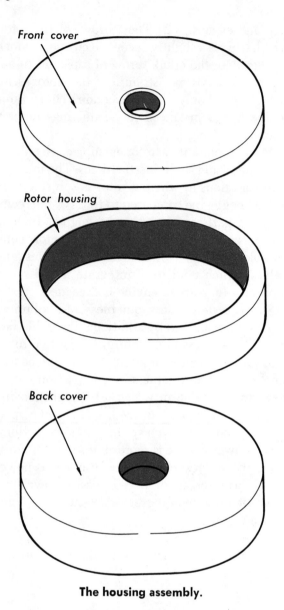

Front cover

Rotor housing

Back cover

The housing assembly.

chambers for the rotor; a main (or *eccentric*) *shaft* that serves the purpose of a crankshaft; a spur (*stationary*) *gear*; and an *internal ring gear*. A camshaft and valves are unnecessary. All accessory systems—fuel, ignition, exhaust, lubrication and cooling—can be furnished by using parts like, or similar in principle to, those in use with four-stroke engines.

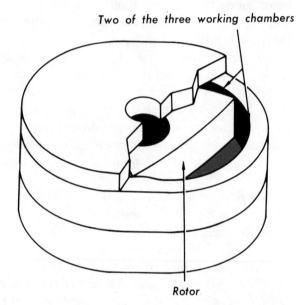

Two of the three working chambers

Rotor

The rotor has two flat, parallel sides and three identical, slightly convex-curved faces which form an equilateral triangle. It also has a large diameter bore that is concentric with a circle drawn through the three apexes (tips) of the triangle. Installed side by side in the bore are a bearing and the internal ring gear. We shall refer to the rotor side at which the bearing is installed as the rotor front; the ring gear side is then the rotor back.

Three pieces make up the housing assembly: a center piece (*rotor housing*) and two (*front and back*) covers. The rotor housing forms the perimeter of the assembly, and its inside face is shaped to be in constant contact with all three rotor apexes as the rotor rotates. The inner faces of the front and back covers make constant contact with the respective rotor sides. Three fully enclosed spaces (*chambers*) are thus formed, one opposite each rotor face at whatever angle of rotation the rotor may be positioned.

Positioned crosswise at the center of the housing, the *main* (or *eccentric*) shaft is held by two

more or less pollutant ridden than exhaust from a comparable four-stroke engine, one fact is certain—the considerable size and weight advantages of the rotary make it much more feasible to add on whatever pollution-control devices are deemed necessary.

MECHANICAL DESIGN OF THE WANKEL ENGINE

There are five operating parts which differ in basic principle from corresponding parts of a four-stroke engine: a *rotor* which serves as a piston; a *housing assembly* which, serving like a cylinder block and head, provides three working

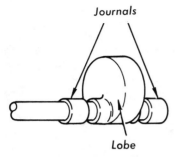

The eccentric shaft.

bearings, one of which is in the front cover through which the shaft projects. The second shaft bearing is located in the hub of the stationary gear, which is attached to the back cover. Therefore the shaft and stationary gear are concentric. The shaft is engaged with the rotor by a circular *lobe* that is positioned eccentrically on the shaft and rotates inside of the rotor bearing. Sizes of the lobe, internal gear, and stationary gear are such that the two gears are in mesh throughout rotation of the shaft.

Because of the eccentric lobe, rotation of the shaft would travel the rotor axis in a circular orbit around the shaft axis. Moreover, because the two gears are constantly meshed, this planetary rotor motion would simultaneously revolve the internal gear (and rotor) around the stationary gear and, therefore, around the shaft (which is concentric with this gear). Conversely, if the rotor is made to revolve, it (together with the internal gear) will simultaneously rotate around the shaft and orbit around the shaft center. For

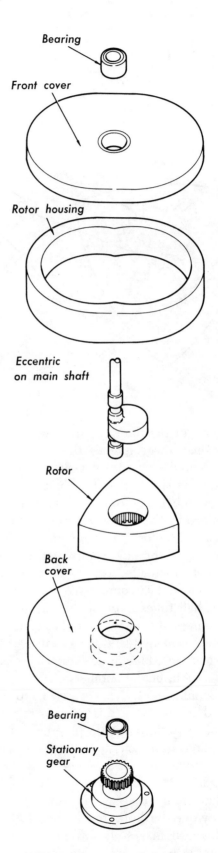

Assembly of the principal engine parts.

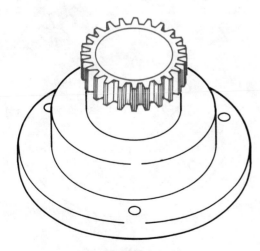

The stationary gear.

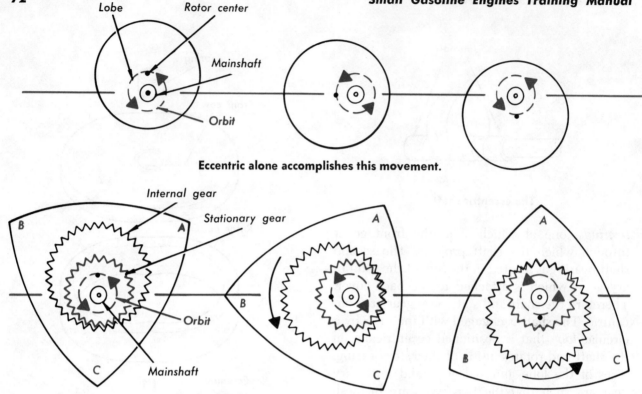

Eccentric alone accomplishes this movement.

Rotor revolves about stationary gear.

each orbit of the rotor the shaft will make one revolution. However, since the internal gear is larger than the stationary gear, the rotor will revolve more slowly than the shaft.

The ratio between rotor and shaft rotations is determined by the relative sizes of the two gears, which are therefore called *phasing gears*. This ratio is one of the key factors that makes rotor movement suitable for its purpose. The internal-gear radius and circumference are exactly one and one-half times those of the stationary gear, and the internal gear therefore has one and one-half times as many teeth. For instance, if the internal gear has 54 teeth the stationary gear must have 36 teeth; or the numbers could be 18 and 12, or 36 and 24, etc. This is a gear ratio of 3:2.

Stated another way, this 3:2 gear ratio means that only two-thirds of the internal-gear teeth are required to match all of the stationary-gear teeth, leaving an excess of one-third. (For instance, $36 - 24 = 12$, which is ⅓ of 36.) It follows that the internal gear (and rotor) will advance by one-third of a revolution (120°) each time this gear makes one full orbit around the stationary gear. As already said, for each orbit of

the rotor (and internal gear) the shaft makes one (360°) revolution. Therefore, the shaft rotates 360° for each 120° rotation of the rotor. During each full (360°) revolution of the rotor the shaft will complete three revolutions, and the rotor will complete three orbits around the stationary gear and the shaft.

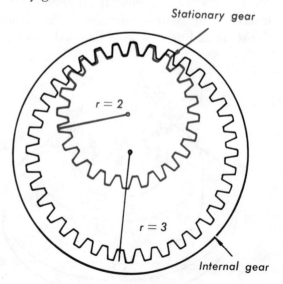

A gear ratio of 3:2.

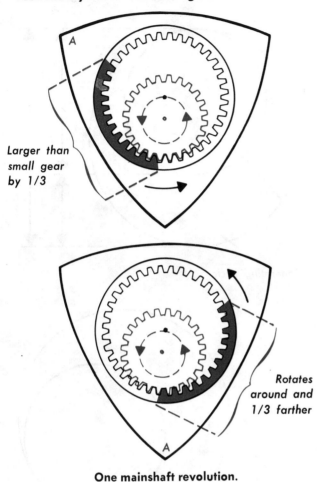

Larger than
small gear
by 1/3

Rotates
around and
1/3 farther

One mainshaft revolution.

As we have said, the rotor both orbits around the shaft and rotates—two motions are involved. These are the same two motions performed by a (*generating*) circle that rolls, without slipping,

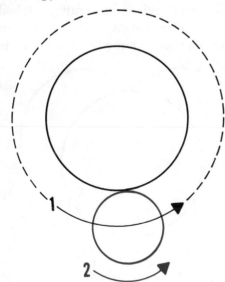

The same two motions.

around the circumference of another (*base*) circle. For the moment let us imagine that the stationary gear is a base circle and that a generating circle of largest possible diameter occupies the space between this gear and the internal gear. As illustrated, this requires the generating circle to be exactly half the size (diameter and circumference) of the base circle. The ratio between the base circle and generating circle is 2:1.

Because of this ratio the generating circle will revolve exactly twice while rolling once around the base circle. The center of the generating circle, as it rolls, will describe a circle, because this center will always be the same distance ($r = \frac{1}{2}$) from the base-circle circumference. However, a point on the generating-circle circumference will move eccentrically, just like any one of the rotor apexes moves. The point will be alternately closer and farther with respect to the base-circle circumference. It will move closer during any half revolution of the generating circle that carries it toward the base circle circumference; it will move farther by an equal distance during the following half revolution of the generating circle. Therefore, during the two full revolutions required to travel the generating circle around the base circle it will describe a two-lobed figure somewhat like an 8, ending up at the exact spot from which it started.

This figure is called a *two-lobe epitrochoid*. By definition, an epitrochoid (also called a prolate cycloid) is the line described by a point on a radius (not at center) of a circle rolled, without slip, around the circumference of another circle. Since the two circles can be of any desired relative sizes, there is an unlimited number of possible epitrochoidal configurations. For instance, shapes similar to three- and four-leaf clovers can be obtained; also, lines which do not define shapes because the point does not return to its starting place at the end of an orbit. To obtain a shape (a closed figure) the base-circle diameter must be a whole-number multiple (2, 3, 4, etc.) of the generating-circle diameter, so that the generating circle will complete an exact number of revolutions per orbit. To obtain the two-lobe epitrochoid, the ratio must be 2:1, as mentioned preceding. It is for this reason that the 3:2 gear

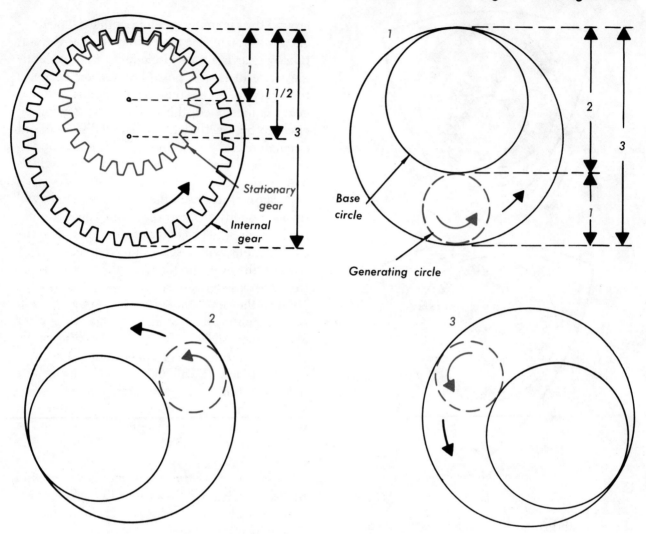

Stationary gear

Internal gear

Base circle

Generating circle

Generating circle moves like a rotor apex.

ratio—which assures a 2:1 ratio generating movement for the rotor—is critical to engine operation.

Actually, our imaginary generating circle does not lie between the two gears as we first pictured it for convenience. As a matter of fact, the rotor

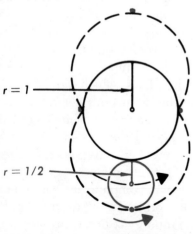

$r = 1$

$r = 1/2$

A 2:1 ratio.

A 3:1 ratio.

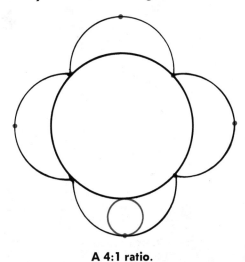

A 4:1 ratio.

revolves around the eccentric lobe. It rotates backward two-thirds of a revolution for each full orbit of the lobe, thus gaining only one-third revolution per shaft revolution. Since this is true, we have to imagine a base circle of the same diameter as the lobe and a generating circle of half this diameter rolling around this base circle. A rotor apex (any one of the three) can now be imagined as a chosen point on the generating circle, and the distance between this point and the center of the generating circle is the *rotor eccentricity*.

A two-lobe epitrochoid has two axes—the *minor axis* through the narrowest part, and the *major axis* through the widest part. If the se-

lected point is on the circumference of the generating circle, the minor axis will be the shortest possible and the major axis the longest possible. Moving the point in toward the generating-circle center lengthens the minor axis while also shortening the major axis. Rotor eccentricity (i.e.: location of rotor apex with respect to the center of the generating circle) is carefully determined to provide a suitable minor axis length, as will be explained later. (Keep in mind that if eccentricity were zero the resulting shape, as already explained, would simply be a circle.)

To return from imagining to reality, there is no generating circle; there are only the engine parts we have previously mentioned. The function of the generating circle is served by the eccentricity of the eccentric lobe which, as it is

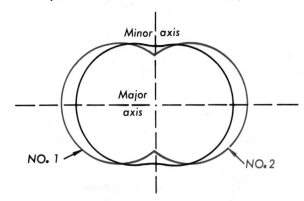

Housing is more nearly like No. 2.

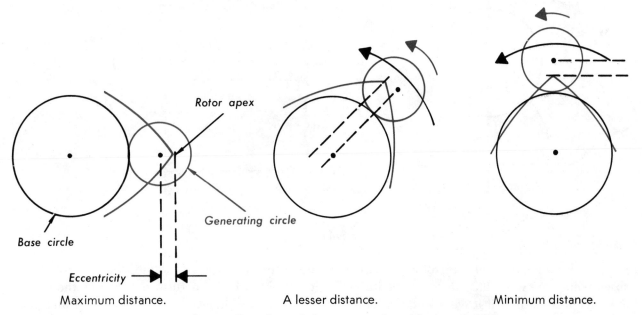

Maximum distance. A lesser distance. Minimum distance.

Apex distance from base-circle center varies with rotor position.

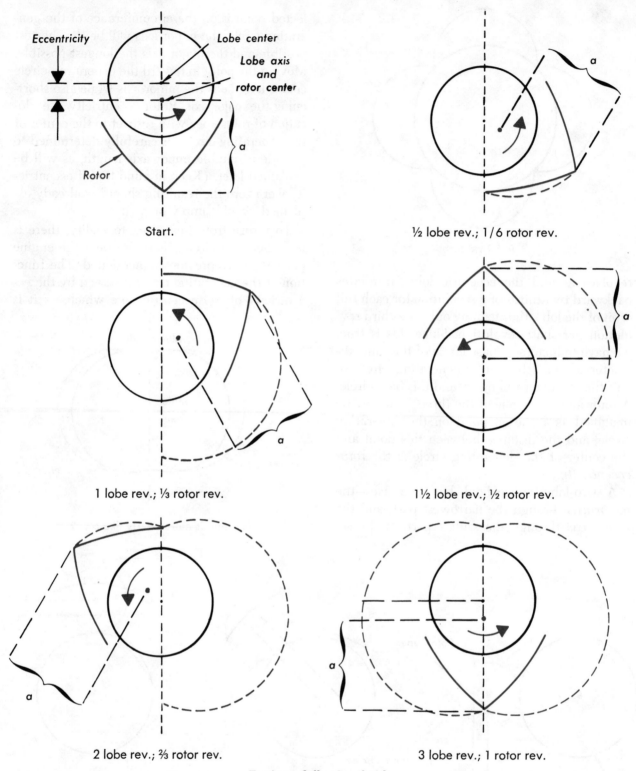

Start.

½ lobe rev.; 1/6 rotor rev.

1 lobe rev.; ⅓ rotor rev.

1½ lobe rev.; ½ rotor rev.

2 lobe rev.; ⅔ rotor rev.

3 lobe rev.; 1 rotor rev.

Tracing a full epitrochoid.

revolved by revolution of the rotor, causes a chosen point (apex) on the rotor to move in the same eccentric path as if it were a chosen point on a generating circle. In short, the amount of eccentricity of the lobe establishes the rotor eccentricity which, in turn, determines the exact epitrochoidal housing shape traced by an apex of the revolving and orbiting rotor.

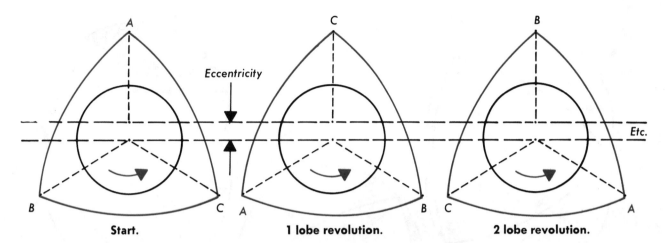

Start.   1 lobe revolution.   2 lobe revolution.

Having apexes that form an equilateral triangle, the rotor will trace the same housing shape with each of its three apexes. It will do so because of the 3:2 gear ratio which causes each apex to advance exactly one-third along the traced housing shape for each revolution of the eccentric lobe. Since the apexes are equidistant from the rotor center and equally spaced, each succeeding lobe revolution positions one after another of the apexes at the same selected point on the epitrochoidal shape, wherever the selected point may be. This would not be so if the triangle were not equilateral, or if there were 2, 4, 5 or any other number of apexes. Because it is so, all three rotor apexes are in contact with the rotor housing at all times.

It follows that the 3:2 gear ratio is responsible both for the housing shape and the rotor shape. Or, to state this fact more understandably, a two-lobe epitrochoidal rotor housing requires an equilateral-triangular rotor and a 3:2 gear ratio. In addition, the amount of eccentricity of the eccentric lobe determines the flatness of the epitrochoid at its minor axis.

HOW THE FIVE EVENTS ARE ACCOMPLISHED

As previously discussed, in the Wankel engine a rotor replaces the pistons of a four-stroke engine. A housing, which encloses the rotor so as to provide working chambers, replaces the cylinder block and head. Instead of a crankshaft there is a mainshaft having an eccentric lobe rather than a crankpin. And valves are eliminated by the use of ports, much like those in a two-stroke engine.

There are three working chambers, one between each pair of rotor apexes. As the rotor revolves these chambers also revolve; and because the rotor orbits the mainshaft as it revolves, the volumes of these chambers vary in a repeated sequence comparable to the combustion-chamber volumetric variations in a four-stroke engine. There is, however, a significant difference—each rotary-engine chamber completes a *full cycle* of volumetric changes during *one* rotor revolution, whereas a piston must make two "round trips" (each starting and ending at top-dead-center) to complete a full cycle of combustion-chamber changes.

NOTE: A rotary engine may be designed to rotate in either direction. Our illustrations show counterclockwise rotation.

The five events occur during each full cycle, in this sequence and in approximately this relationship to chamber volumetric changes:

1. *Intake*—while the chamber is expanding.
2. *Compression*—while the chamber is contracting.
3. *Ignition*—at or about the time of full compression.
4. *Power stroke*—which expands the chamber.
5. *Exhaust*—while the chamber is again contracting.

Because there are three equal rotor faces and three working chambers, three *independent* cycles—each separated in timing from the others

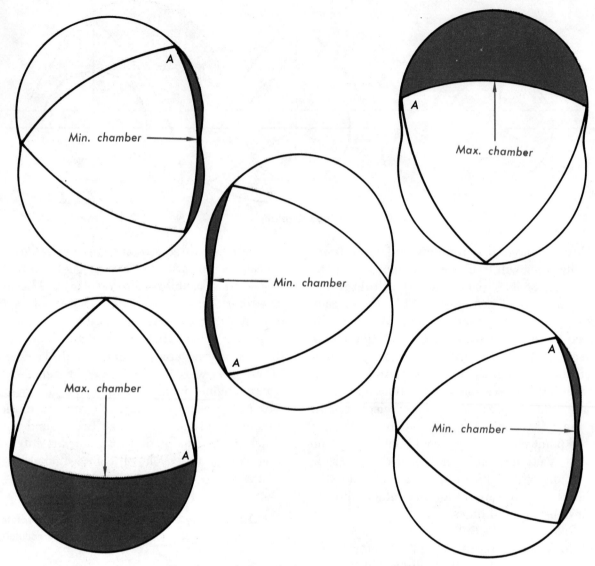

Chamber variations during one rotor revolution.

by a third of a cycle—are produced with each revolution of the rotor. The mechanical arrangement can, therefore, be likened to that of a three-piston, four-stroke engine, in which three pistons operate simultaneously. However, the comparison ends here. From a power-rating standpoint a single-rotor Wankel engine is the equivalent of an equal-displacement, two-cylinder, four-stroke engine. The Wankel produces one power phase per mainshaft revolution (three per revolution of the rotor, which revolves at one-third mainshaft speed); the reciprocating-piston engine produces one power stroke per crankshaft revolution (one-half cycle per revolution for each of the two pistons).

PERFORMANCE FACTORS

As with a reciprocating-piston engine, certain basic proportions and mechanical relationships determine the performance of a Wankel engine. Some of these concern only engine designers; some should be understood by those who will service the engines. We will briefly discuss the latter.

Displacement and Compression Ratio

In a four-stroke engine displacement is the cubic-inch volume of the cylinder space vacated by a piston in moving downward from TDC to BDC. Total engine displacement is the sum of

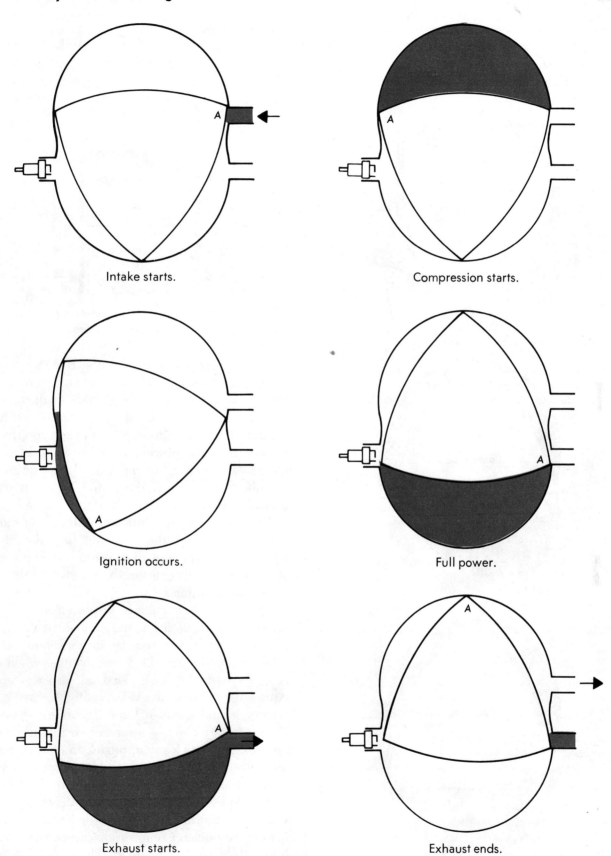

Intake starts.

Compression starts.

Ignition occurs.

Full power.

Exhaust starts.

Exhaust ends.

The five events for one rotor face.

the individual piston displacements and is often used in rating the maximum potential engine power because it determines the maximum rate at which the engine can, without supercharging, consume fuel. Compression ratio is the maximum volume contained between the cylinder head and piston at BDC divided by the minimum volume so contained with the piston at TDC.

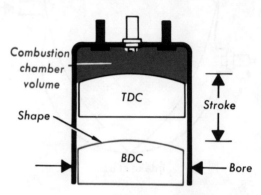

Four-stroke engine displacement and compression-ratio factors.

Since compression aids combustion, up to a certain point, this also is a factor that determines the power output of the engine. Both displacement and compression ratio are relative to the cylinder bore, the shape of the piston top, and the length of the piston stroke; the volume contained within the cylinder-head combustion chamber (into which the piston never enters) also affects the compression ratio.

Two crankshaft revolutions (720°) occur during completion of the 5 events (a thermodynamic cycle) for any one piston. Each power stroke occurs during a fourth of a cycle, 180° of crankshaft rotation.

For a Wankel engine displacement (also called *swept volume*) is the cubic-inch volume of a chamber when it is at its maximum size minus the volume of this same chamber when it is at its minimum size. That is, displacement equals the volume of space vacated between a rotor side and the rotor housing while the side rotates from its closest approach to the housing minor axis to its closest approach to the housing major axis. Total (single rotor) engine displacement is the sum of the three chambers' displacements; but there is considerable debate as to whether or not this is a compatible factor for rating engine power. Compression ratio is the maximum chamber volume divided by the minimum chamber volume.

Power phases per rotor revolution.

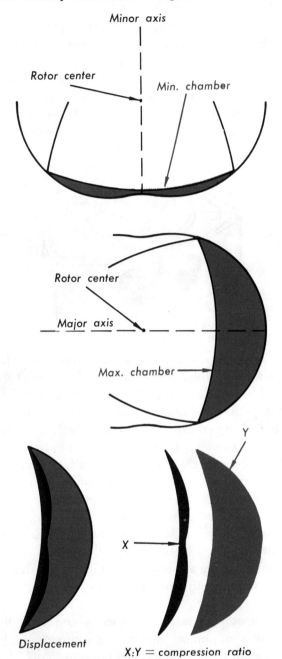

Displacement

X:Y = compression ratio

length of the rotor radius and the amount of rotor eccentricity, which can also be stated as R ÷ e.

NOTE: Rotor radius is the distance from rotor center to the top of a rotor apex. Rotor eccentricity is the distance between the mainshaft center and the rotor center.

Three mainshaft revolutions (1080°) occur during completion of the 5 events for any one rotor face. Each power phase occurs while the mainshaft is rotating 270°. The difference between 270° and 180° is the principal reason for the debate regarding use of displacement for comparative power rating. It should also be noted that the rotor turns 90° from minimum chamber to maximum chamber and rotates 360° to complete all 5 events.

When considering compression ratio an important fact must be kept in mind. This ratio is a dividend (maximum ÷ minimum). Therefore, if a certain ratio is desired, any increase of the minimum must be accompanied by an increase of the maximum equal to this increase × the dividend. For instance, if the maximum is 8 units and the minimum 1 unit the dividend is 8 ÷ 1 = 8. Increasing the minimum by 1 unit requires increasing the maximum by 8 × 1 = 8 units (16:2 = 8:1). Simply adding the same increase to the maximum would alter the ratio (9:2 does not = 8:1).

Rotor Face Design, Thickness, and Radius

As previously noted, the rotor must have apexes that form an equilateral triangle; but this requirement could be met by a rotor having straight, concave-curved, or convex-curved faces. A rotor with convex-curved faces is preferred for several reasons: (1) This shape provides relatively more area around the center for the internal gear and the eccentric-lobe bearing than straight or concave-curved faces would allow, and a curved face (either type) has more surface area for the dispersion of heat than a straight face. (2) The convex-curved faces form the widest apexes, which are the strongest, have the most internal space for cooling, and to the greatest degree minimize heat transfer from

In this case also, compression ratio is a power factor but, since the continuous movement of the chamber and gases affects combustion, limitations are different from those of four-stroke engines (lower octane gasoline can be used). Both displacement and compression ratio are affected by four factors: (1) *Rotor face design,* which is equivalent to piston-top design. (2) *Rotor thickness,* which determines the distance between the housing covers. (3) The *rotor radius,* which determines the distance between apexes (4) The *R/e ratio,* which is the relationship between the

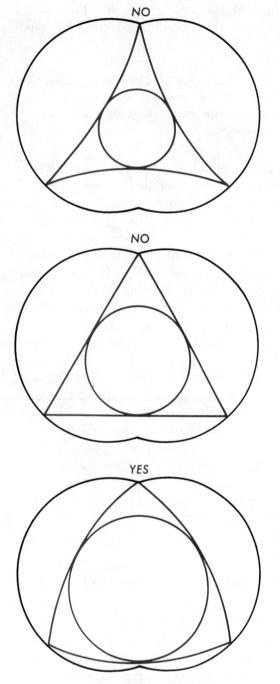

NO

NO

YES

Three possible rotor shapes.

chamber to chamber. (3) Most important, although this curvature reduces displacement, it permits attainment of the highest possible compression ratio. It does so because a convex curvature most nearly parallels the housing epitrochoidal curves, and thus produces the narrowest possible space between rotor and housing when a chamber is at minimum volume.

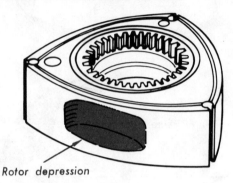

Rotor depression

Under proper circumstances (*refer to "The R/e Factor," following*) the rotor face curvature can be made to match the housing contour so closely that the minimum chamber is virtually reduced to two extremely narrow, feather-edged pockets connected together only by the slim clearance space between the rotor face and the minor-axis crown. Because maximum volume is geometrically greater than minimum volume when calculating compression ratio, the smallest possible minimum volume is the most desirable basis on which to begin. For this reason, rotor faces generally are designed to match the housing contour at the minor axis as closely as possible.

Although an extremely narrow and virtually two-part minimum chamber is quite desirable for exhaust and induction (*refer to "Port Designs," following*), such a chamber would greatly hamper the start of combustion. To overcome this disadvantage (without altering the face cur-

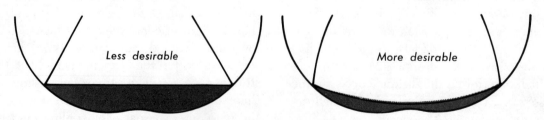

Less desirable   *More desirable*

Matching rotor faces to housing contour.

Effect of rotor-face depression.

vature) a depression is formed in each rotor face. These depressions increase minimum-chamber volume, and thereby decrease maximum attainable compression ratio. In fact, the sizing of these depressions is one of the factors employed for ultimate compression-ratio determination. Moreover, the size and shape are also important for two other reasons. By increasing the surface-to-volume ratio within each chamber they aid heat dispersion. By breaking into the smooth plane of each face contour they improve turbulence of the mixture within a chamber.

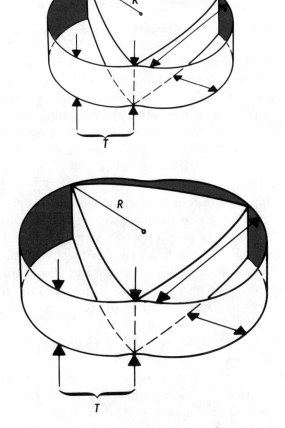

Rotor radius (R) and thickness (T) determine engine size.

Overall rotor size is determined by the rotor thickness and by the length of the rotor radius. The volume of a chamber for any particular rotor position equals the thickness × median width of the chamber. Chamber thickness varies directly with rotor thickness; the length and width values vary indirectly with the length of the rotor radius, which not only determines the distance between two rotor apexes but also affects the distance between a rotor face and the housing for all rotor positions. Therefore, varying either rotor thickness or radius will vary chamber sizes and, consequently, overall engine size and displacement. Moreover, varying thickness will *not* alter the compression ratio, but varying the radius will.

The R/e Ratio and Compression Ratio

Maximum-chamber volume adjacent to a rotor face is attained when the rotor apex opposite this face is at the housing major axis. Consequently, increasing the major-axis length, by increasing the rotor eccentricity, will increase the maximum-chamber volume. Whenever the major axis is lengthened, the minor axis is shortened. Moreover, minimum-chamber volume is attained when the rotor apex opposite the face is at the housing minor axis. Therefore it would appear that any increase of rotor eccentricity should be accompanied by a maximum-chamber volume increase and a corresponding minimum-chamber volume decrease.

On the contrary, however, increasing e also increases the minimum-chamber volume. It does because the rotor size and/or face curvatures must be reduced to allow clearance at the minor-axis crowns, and because these crowns become more pointed as e is increased, thus forming deeper and larger "pockets" at each side. Actually, since rotor face curvature is planned to match housing curvature at the minor axis, any alteration of e requires a change in rotor shape.

It follows that any increase of e alone enlarges both the maximum—and the minimum—chamber volumes, and vice versa. Increase of the maximum will be greater than the increase of the minimum, but will not be sufficiently greater as required to maintain any compression ratio already established by the relationship between e,

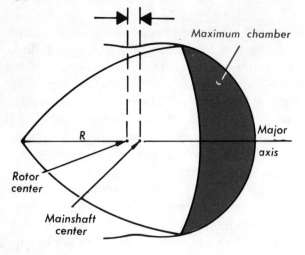

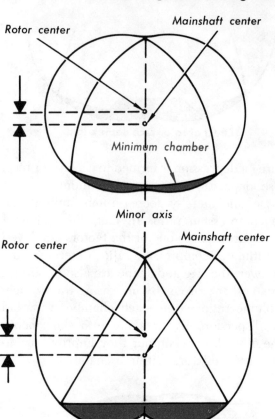

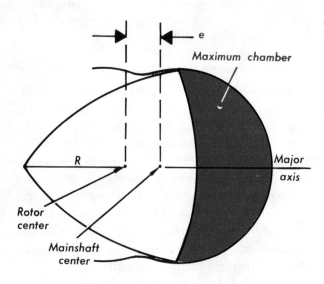

Increasing e increases maximum chamber.

Increasing e also increases minimum chamber.

R, and the rotor design. Consequently, the R/e ratio is just as important as rotor design in determining the compression ratio. A higher R/e ratio produces a greater maximum obtainable compression ratio. For instance, a radius of 23 units with an eccentricity of 2 units gives a very high R/e ratio of 11.5 and a maximum compression ratio of 30:1, while an R of 23 units and an e of approximately 5.9 units gives an R/e of 3.9 and a maximum compression ratio of only 10:1.

The R/e Ratio and the Sliding Angle

Whenever the rotor is positioned with an apex exactly at the minor or the major axis of the rotor housing, the rotor radius to this apex is at 90° to the housing tangent at this point. However, due to the fact that the rotor orbits as it rotates, at all

other positions of the apex the radius will *not* be at 90° to the housing tangent. The difference between its actual angle and 90° is called the *sliding angle* (also referred to as the *leaning angle*, the *Q angle* or the *angle of obliquity*). The angle varies at different apex positions and is greatest when an apex is midway between a minor- and major-axis intersection.

This variable angle is very important to engine operation because there must be a seal, similar in function to a piston ring, at each apex to separate one chamber from another. Like a piston ring, each apex seal must make as flat and unvarying a contact (with the housing) as possible in order to retain the forces of compression and combustion.

If there were no eccentricity all apexes would, at all times, contact the resulting circular housing at 90°. Increasing the eccentricity increases all values of the sliding angle. For instance, with the preceding R/e ratio of 11.5 the angle is only 15°, but with the 3.9 R/e ratio it is increased to 50°.

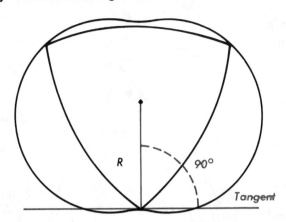
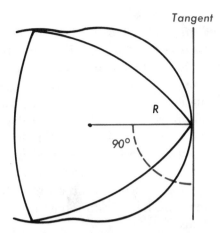

No sliding angle at these rotor positions.

It follows that, for this reason also (as well as in the interest of compression ratio), the degree of eccentricity—in short, the R/e ratio—is critical to engine performance. Too much eccentricity could create a sliding angle at which the apex seals would not function properly. This is the reason that rotor-face depressions, rather than increased eccentricity, are used for limiting maximum compression ratio as required.

SEALING

One of the most crucial problems relative to commercial production of the Wankel engine has been the development of adequate gas-leakage sealing. In general, rotary-engine seals must function in the same manner as the piston rings of a reciprocating-piston engine, but the gas seals (equivalents of compression rings) must do so under much more severe operating conditions. In order to prevent gas leakage from the chambers two separate paths must be blocked by these gas seals: (1) The path from chamber to chamber between each apex and the rotor-housing epitrochoidal bore. (2) The two paths, one on each rotor side, between each rotor face and the rotor center portion. Oil seals also are required to block or control leakage of oil at each rotor side from the mainshaft and phasing gears. *Apex seals* and *side seals* perform the respective gas-sealing functions. These are mounted in grooves or slots in the rotor to make sliding contact with the housing.

The apex seals, which are generally mounted in radial slots at each apex, are the most troublesome for several reasons. As each apex traverses the periphery of the rotor-housing bore, the apex

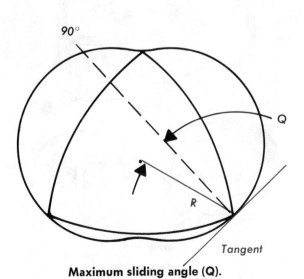

Maximum sliding angle (Q).

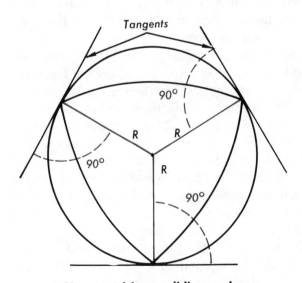

No eccentricity, no sliding angle.

seal varies in angle of contact from 90° to a more-or-less acute sliding angle, as mentioned preceding. Due to manufacturing limitations it is virtually impossible to mass produce housings that will, under all operating conditions, guarantee uniform-pressure contact with the the apex seals all the way around. Contact failure can result in *spitback* (firing from a combustion chamber back into the following compression chamber). Varying heat conditions at different housing areas and lubrication requirements also pose problems.

The sliding-angle problem is partially resolved by keeping the angle to the most practical minimum (not over 30°) and is further reduced by designing apex seals of substantial thickness with curved, rather than flat, contact surfaces. To solve the uniform-pressure problem: (1) The rotor-housing bore is made slightly oversize; that is, the (so-called) true epitrochoidal path of the rotor apexes is expanded by a small dimension (a) so that there is a constant clearance between each apex and the housing. (2) The apex seals are spring loaded to force them outward against the housing.

NOTE: The effective rotor radius (R) is increased by the addition of dimension a. Therefore the effect of this factor (a) on the R/e ratio must be taken into consideration.

A principal reason for side-seal problems is the required length of each seal. The manner in which each seal must sweep the housing side and the variance of pressure during the cycle phases are lesser problems. Side seals are generally located as close as practical to the respective rotor faces, are curved to parallel the faces, are inserted into grooves in the rotor sides, and are spring loaded to force them against the respective housing covers. Their surfaces are flat for maximum contact, and in some engines a double bank of seals is provided at each side.

Because the abutting ends of each adjoining pair of side seals must be interlocked, and because these side-seal ends meet the apex seals at 90° and the spring-loading of the seals tends to open the gap at this junction, special corner sealing also is required to block leakage past the rotor sides at the ends of each apex. The most

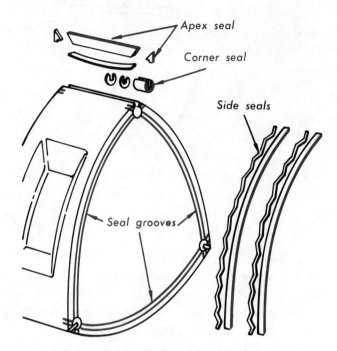

Typical apex and side seals.

common solution of this problem has been to use dowel-shaped, spring-loaded *link blocks* to seal all spaces between each apex seal and the two adjoining side-seal ends. Some manufacturers also employ a three-piece apex seal having a separate *corner seal* at each end so designed that the

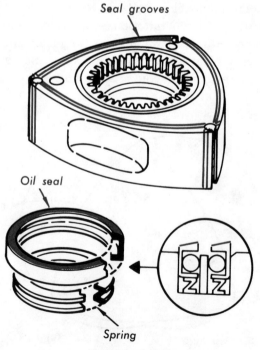

Typical oil seals.

spring load of the apex seal simultaneously thrusts the central part upward and the corner pieces sideways to ensure contacts at top and both ends.

Oil seals are circular and are installed in grooves at each rotor side, as close to the rotor center as possible. There may be a single or a concentric pair of oil seals at each side. They are flat topped and spring loaded so that, as the rotor both revolves and orbits, they serve to contain the lubricating oil within the phasing gear area and to scrape excess oil from the sides of the housing covers.

All of the seals have what are known as *primary* and *secondary* sealing areas. A primary area is that between the seal surface and the part against which it slides; a secondary area is one between the seal and the groove in which it rides. Except for the apex seals, the spring loadings are relied upon to preserve adequate primary-area contacts, and gas pressure serves principally to maintain the secondary-area contacts. At starting and during idling the apex-seal springs also serve the primary-area purpose; but as engine speed increases centrifugal force becomes the more dominant factor.

Because the apex sliding velocity varies considerably due to the orbiting movement (in one engine, for instance, variance is from 42½ to 108 ft./sec.) and because each apex moves alternately closer to and farther from the mainshaft axes, the centrifugal force acting upon an apex seal also varies considerably. This causes seal vibration and promotes the formation of *chatter marks* (scores parallel with the seal) in the rotor-housing bore. Extremely close fitting of seals in their grooves, special design apex-seal springs, adding oil to the gas mixture and various processes for improving the metal-to-metal interface to reduce friction are among the many solutions to this problem that have been tried with more or less success.

TIMING

The Considerations Involved

The combustion process of a Wankel engine differs somewhat from that of a reciprocating-piston engine. In the latter the combustion

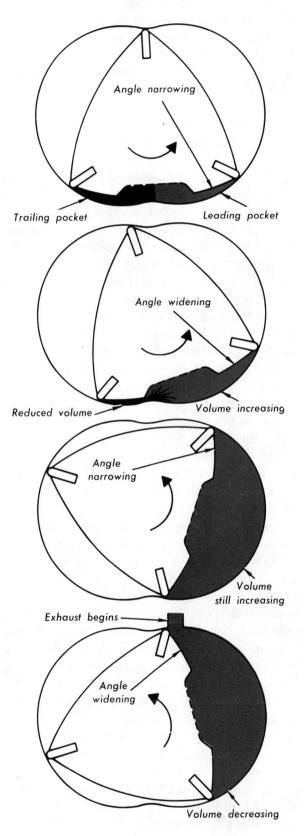

Combustion chamber variations.

chamber is a stationary area contained between the cylinder head, cylinder walls and piston crown, and the chamber enlarges uniformly as the piston is forced down by the expanding gases. In the rotary engine the combustion chamber travels with the rotor so that, during combustion, the mixture not only expands, it also moves forward (in the direction of rotor rotation) at a velocity determined by the engine rpm.

At the start of combustion the rotary chamber (then at or close to minimum volume) is more or less divided into two pockets by the minor-axis crown, and each pocket has a feather end where the respective rotor apexes contact the rotor-housing bore. During combustion the trailing-pocket gases are at first further compressed, as the trailing rotor apex approaches the minor-axis crown, and are then pushed into the trailing end of the leading pocket. The leading pocket as a whole is continuously enlarged, but the gases in it do not expand uniformly due partly to the pressure of gases swept in from the trailing pocket and partly to the changing sliding angles of the rotor apexes.

In any engine there are three major factors which affect the combustion process: the compression ratio, the mixture turbulence, and the manner in which the flame spreads following ignition.

The compression ratio necessarily varies with the amount of intake and will increase as the throttle is opened. Also, since the mixture burns and expands as the flame front spreads from the ignition point, unburned gases ahead of the flame front are subjected to additional compression during combustion. Higher compression results in quicker combustion, which means that more of the fuel heat energy is utilized during the very brief combustion period. More of the power is used internally by an engine to obtain greater compression, but this is more than offset, up to a point, by a greater gain in thermal efficiency and output power. For instance, tests have shown that pressure rise during normal combustion is 3.5 to 4 times the initial (compression) pressure of the gases. The efficiency increment diminishes rapidly, however, if combustion is abnormal.

Abnormal combustion can occur in any engine if the octane rating of the fuel is too low for the maximum compression ratio (compression ignition occurs ahead of timed ignition). It can also occur if: (1) There are hot spots in the combustion chamber (which can cause preignition pings or postignition rumbles of gas pockets in contact with the hot spots). (2) The mixture is too rich or too lean, or the fuel and air are not uniformly mixed by turbulence. (3) The flame front spreads too slowly or nonuniformly so that the gases farthest from the ignition point become overheated and overcompressed and ignite spontaneously before the smoothly burning flame can reach them (thus causing knock).

As previously noted, the Wankel engine affords superior tolerance with regards to fuel octane rating; compression ignition does not create a problem. Mixture richness or leanness is controlled in the same manner as in reciprocating-piston engines. Turbulence is controlled by the design of the rotor depressions, the intake manifold design, and the shape, size and (principally) the location of the intake port(s). The problem of hot spots is much the same as with reciprocating-piston engines and is dealt with chiefly by adequate cooling. Spreading of the flame front, alone, presents a problem unique to this type of engine.

This flame-front problem exists because of the chamber movement during combustion. In any expanding combustion chamber a movement, called a *transfer wave*, is created within the gases. Due to the traveling of the gases in a rotary chamber this transfer wave moves at exceedingly high speed. Moreover, as engine rpm increases the *gas transfer velocity* also increases. In fact, it has been determined that the flame front never overtakes the gas transfer wave at high engine rpms; some end gas remains unburned. And the problem can be further aggravated by quenching of the flame in the very narrow confines of the chamber's feather edges.

NOTE: For a rotor, BDC is the position in which an apex is exactly at the minor axis that lies between the exhaust and intake ports; TDC is the position in which this same apex is exactly at the opposite minor axis. Port openings and closings and ignition

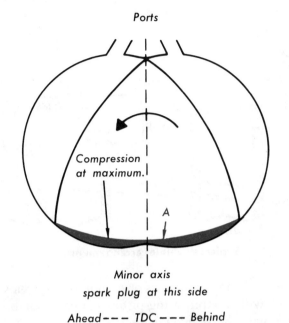

Ports

Compression at maximum.

A

Minor axis
spark plug at this side

Ahead --- TDC --- Behind

Behind --- BDC --- Ahead

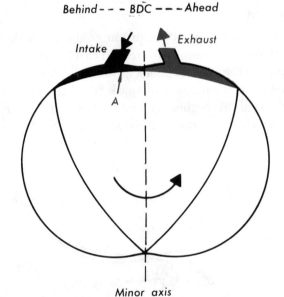

Intake   Exhaust

A

Minor axis
Spark plug at this side

Positions are for rotor face A.

timing are planned in degrees of rotor rotation with respect to these positions.

Port Designs

Various port types have been used, but the variations chiefly concern the intake (induction) rather than the exhaust port.

For the exhaust a single peripheral port, located close to the minor axis on the side from which the rotor apex approaches it, is preferred.

This port is generally oblong in shape so that a maximum-width opening becomes exposed as the rotor apex sweeps over it; and the length of this oblong and its exact positioning depend, first, upon the desired duration of fully confined combustion and second, upon the maximum "valve overlap" to be allowed. The port passages through the rotor housing and the exhaust manifold are designed to offer as little "elbow" restriction to exhaust-gas flow as possible, so that a minimum of combustion gases will be swept past the port and remain in the chamber during its (following) intake phase.

The inlet port may be located in the periphery of the rotor housing at the side of the minor axis opposite the exhaust port, or it may be located at an adjacent position in the end cover. In either case, exact location and shape are determined by the induction timing requirements and allowable "valve overlap." A peripheral port has been demonstrated to be most desirable for high-speed, high-output engines; the side port is favored for idling and low-speed, low-load operating conditions. With a side port it is practical to reduce the "valve overlap" to a minimum and yet obtain a satisfactory volumetric efficiency at higher speeds, provided port size and shape and the manifold are designed to handle the maximum possible volume (i.e., quick opening, late closing—with a minimum of "elbow" restrictions).

NOTE: In addition to single inlet ports of the types described, dual ports of various designs are also being used. One arrangement is a combination of a peripheral port with a side port (*combi-port system*); another is the use of dual side ports, one in each end cover. Both ports may be served by the same manifold, or separate manifolds may be used. In the latter case, one preheated (by exhaust gases) manifold, which is unobstructed at all times, leads to a side port, and ensures good starting and low-speed performance. The second, a peripheral-port manifold is closed or opened by a suitably controlled throttle valve, and serves only to furnish the additional amount of mixture needed under higher-speed/greater-load operating conditions.

Overlap, when using a peripheral exhaust port and a single side-inlet port, may be approximately 45% of 270° of mainshaft rotation, or with a single peripheral-inlet port may be approximately 74%. Therefore, use of a constant-open side port together with a throttle-controlled peripheral inlet port provides a desirable low-speed overlap down to 45% which is increased to about 74% when both ports are in use for high-speed operations.

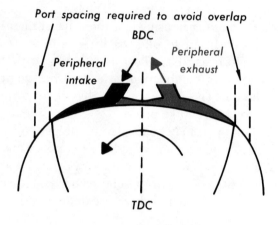

Typical single-inlet and exhaust-port arrangements.

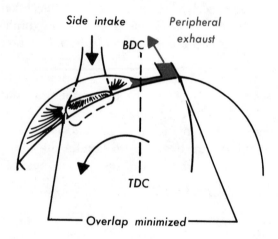

For use of heavy fuel a diesel-type nozzle (fuel injector) is substituted for the intake port(s) and is located adjacent to the spark plug at the rotor-housing TDC position. The nozzle is supplied from a high-pressure pump the operation of which is synchronized with the ignition timing. Compressed air is supplied through an inlet port. Vaporized fuel is burned as it is injected so that the flame front is practically stationary and almost any heat-value fuel can be used without adding to the size or weight of the engine. Rpm is controlled entirely by the amount of fuel injected. Either a side or peripheral air-inlet and exhaust ports may be used.

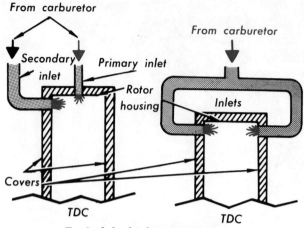

Typical dual-inlet arrangements.

Any of the carburetors or carburetion systems used with reciprocating-piston engines can be used with the Wankel engine.

Ignition

As with reciprocating-piston engines, Wankel engine spark timing with regard to TDC is a variable factor that depends principally upon engine rpm; the amount of spark advance must be increased at higher engine speeds. However, unlike the reciprocating-piston engine, the Wan-

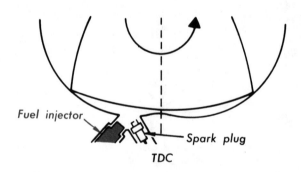

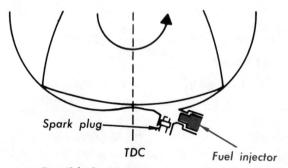

Possible fuel-injection locations.

kel engine presents a major problem with regards to spark-plug location and spark-plug selection. Location is important because of the gas transfer velocity; selection is important because of the excessive heat condition under which a plug must operate.

Tests have shown that locating the spark plug at or slightly ahead of TDC achieves better flame propagation with a greater variety of air-fuel mixtures under low-speed, low-load conditions. On the other hand, positioning the plug behind TDC provides the more rapid and thorough flame-front spread required for maximum burning of the mixture (and, therefore, more power) under high-speed, high-load conditions. To achieve both objectives—smooth combustion at idling and low speed together with maximum power at high speed—many engines are designed for two plugs.

One designer locates the first plug approximately 3° (of rotor rotation) ahead of TDC and the other approximately 2° behind TDC. The *transfer port* (opening between housing inner surface and plug chamber) for the first plug is much smaller in diameter than that for the second plug because of the much higher gas pressure in the trailing combustion-chamber pocket. This trailing-pocket plug fouls more quickly than the leading-pocket plug because quenching of the trailing-side flame front as the pocket is swept into the leading side tends to leave some pocket gases incompletely burned.

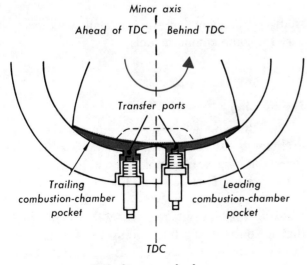

Use of two spark plugs.

Any of the spark-advance systems already developed can be used with a Wankel engine. Ignition timed to occur behind TDC produces more low-speed power, but also rougher operation. Ignition before TDC is more reliable with respect to a full range of air-fuel mixtures and is required for maximum power at higher speeds. In single-plug engines a spark-advance range from about 7° (of rotor rotation) after TDC to about 5° before TDC is generally used. Two-plug engines generally have two separate ignition systems with separate distributors and plug timing—at idling the ahead-of-TDC plug may fire at TDC or up to 2° before, while the behind-TDC plug may fire up to 7° behind TDC. The ahead-of-TDC plug may not be advanced for increased engine speed or may be advanced several degrees, while the behind-TDC plug generally is advanced to a maximum of about 5° before TDC.

Most RC engine manufacturers are using specially designed spark plugs that will resist fouling and preignition under the unique engine-operating conditions.

LUBRICATION

There are two distinct areas of a Wankel engine that require lubrication: (1) The phasing gears and eccentric lobe and the mainshaft bearings. (2) The rotor seals (particularly, the apex seals).

In most engines developed to date lubrication of the apex seals (and, to some extent, of the side seals) is obtained by mixing oil with the mixture—either by adding oil to the gasoline (2-2.5% of SAE 30 or 40 oil by volume) as in a 2-stroke reciprocating-piston engine, or by vaporizing oil through a metering pump and jet nozzle or wick into the gas-air mixture in about the same proportion. If used, a jet is located close to the carburetor in the intake manifold; the wick is located in or close to the induction port. Some developers, however, prefer to rely upon controlled oil leakage past the oil seals to lubricate the side and apex seals, rather than to induce oil into the gasoline or the mixture.

For lubrication of the mainshaft bearings, eccentric lobe, and phasing gears a more-or-less

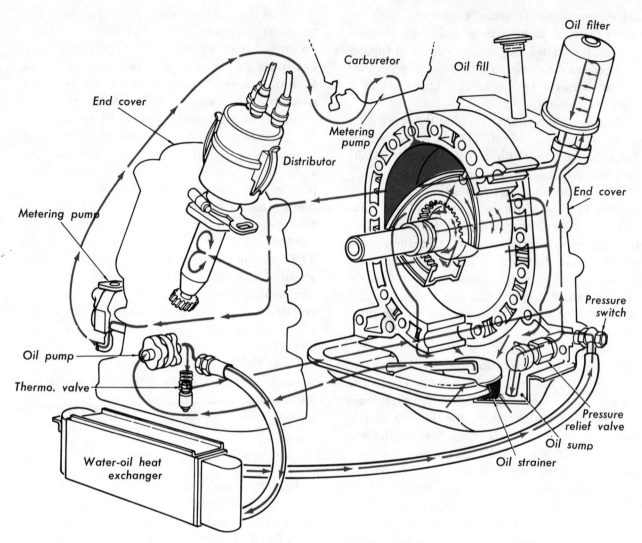

Lubrication system using oil to cool rotor and a metering pump to deliver oil to air-fuel mixture for seal lubrication.

conventional oil-sump, oil-filter, pump and cir- culating system is relied upon. Oil can be trans- ported from the pump and back to the sump through lines and fittings, or may be circulated in part through channels provided within the mainshaft. In some models the rotor is also hol- low and oil is squirted into the rotor to help cool it. Oil used for cooling purposes may or may not be precirculated through a special oil cooler.

In all cases oil consumption has proved to be relatively negligible. Moreover, oil fed directly through the gasoline or the mixture or oil al- lowed to leak past the oil seals is either con- sumed in the combustion process or ejected; none is returned to the oil sump. Therefore, pe-

riodic oil changes have been deemed unneces- sary for some engine models.

COOLING

The Housing

The RC engine presents a unique cooling problem because each operating phase for all three rotor faces is completed within the same housing area. Therefore, the housing has a "cool lobe" that is continuously swept by the incoming mixture from the intake port—and a "hot lobe" that is continuously undergoing the power and exhaust phases. The hot lobe is subjected to ex- cessive temperature rise, especially at higher en-

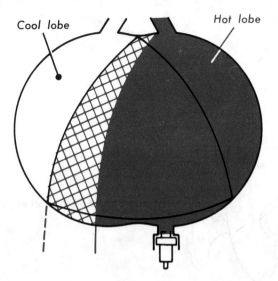

Cool lobe   Hot lobe

gine speeds, whereas the cool-lobe temperature never rises sufficiently to require cooling of this area. As a result of this uneven heat distribution there is a tendency for the housing and covers to become distorted, thus preventing proper functioning of the rotor seals. In addition, the repeated, sudden-combustion temperatures, if not otherwise absorbed, can destroy the oil film to cause excessive apex seal and housing-bore wear —and can result in thermal fatigue of the housing metal so that cracks occur around the spark-plug holes. In short, a Wankel engine is extremely sensitive to the proper functioning of its cooling system.

To more nearly equalize engine temperatures, the hot-lobe housing sector is provided with superior cooling and the cool-lobe sector in some models is heated by directing exhaust gas through channels in or around this housing-assembly portion. Both liquid- and air-cooled engines have been developed.

Liquid cooling is affected by directing the coolant through channels which are located so as to provide the greater wetted area where most needed and/or are sized to increase the flow of fluid past the hot-spot areas. A conventional water pump provides the circulation and, except for marine engines, a heat exchanger (radiator) with a fan is also used.

Air cooling depends principally upon closely crowded cooling fins machined around the periphery of the rotor housing and on the end covers. To provide more cooling for the hot-spot areas fins at these locations have larger surface areas and are more closely spaced to speed the passage of air between them. The cool-area fins are reduced in size and more widely spaced so that the air-pressure drop between them is much less. A fan driven from the mainshaft provides a flow of air relative to the engine speed, but the considerable pressure drops created by the narrow fin spacings are relied upon to cool the hot spots so that a high-volume, power-consuming fan is not required.

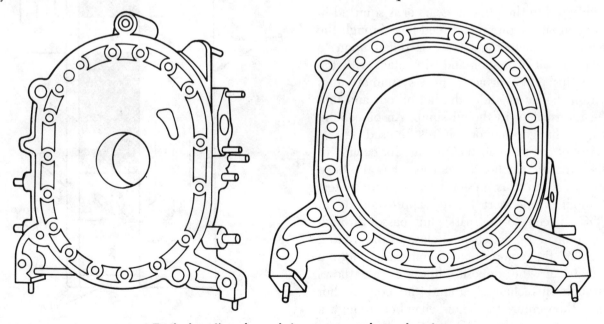

Typical cooling channels in a cover and rotor housing.

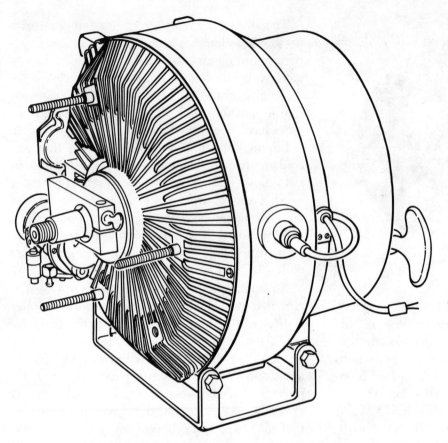

Typical air-cooled small engine.

The Rotor

As previously mentioned, general practice is to use the lubricating oil to cool the rotor. Oil either is sloshed into the hollow rotor from the supply fed to the phasing gears or is squirted in by an auxiliary pump and feed line through the mainshaft. Centrifugal force is relied upon to keep the oil in motion and in contact with the rotor tips' inner surfaces, and a system of vanes within the rotor guides the hot oil out and into the return flow to the oil sump. An oil-cooling heat exchanger may or may not be used.

Use of the fresh air-fuel mixture for cooling of the rotor and the housing hot lobe has also been proposed. In this design the mixture enters through a side port located approximately at TDC adjacent to the spark plug, passes through the rotor in a cross channel close to the trailing apex of the combustion chamber and into a channel at the opposite housing side, flows through this channel around the housing hot lobe, then enters the intake chamber through a side port near the BDC position. When com-

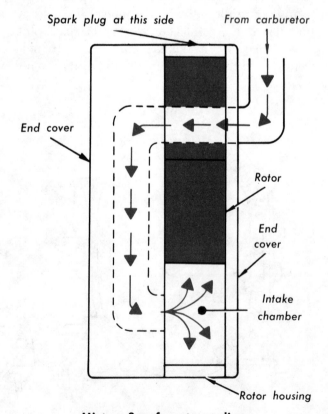

Mixture flow for rotor cooling.

bined with air cooling this arrangement provides for preheating of the mixture together with additional cooling for the spark-plug boss and the hot rotor apex at the trailing side of the combustion chamber.

MULTIROTOR ENGINES

As previously discussed, a single-rotor RC engine (such as we have been discussing) has the power equivalence of a two-cylinder, four-stroke engine. Each additional rotor theoretically adds "two-cylinders" of performance. Certain limitations are, however, imposed by the basic design; for instance, incorporation of more than three rotors creates major problems with regards to mainshaft design and a practical method of true-alignment assembly and reassembly after overhaul.

Because a single-rotor RC engine has the disadvantage of some roughness, twin-rotor designs are preferred for automotive use, even in the low-powered economy models. This roughness, however, is not a distinct vibration; rather, it is a phenomenon associated with the noise and feeling of the combustion process (much like the external effects produced by early valve-in-head reciprocating engines)—and practically disappears as the engine speed is increased beyond idling. The exhaust noise also changes with engine speed—from a distinct popping like that of a two-stroke engine to the purr of a four-stroke engine. It follows that the single-rotor engine disadvantage for automotive use has little or no

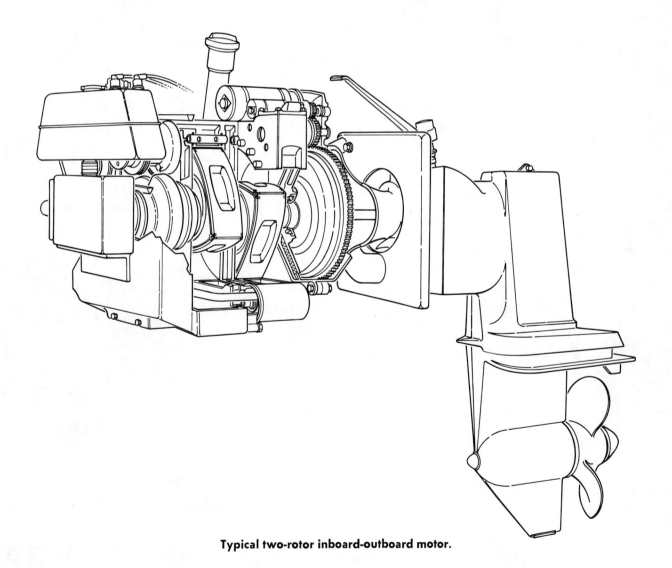

Typical two-rotor inboard-outboard motor.

bearing upon its use for a majority of small-engine applications. The economy and simplicity of single-rotor designs therefore make this type preferable for most of the applications covered by this text.

When two rotors are used, one side cover between the two serves both. Each rotor is fitted with its own phasing gears and mainshaft eccentric lobe, usually has its own distributor, and may be supplied from a separate carburetor or horn of a multihorn carburetor. The two rotors

are positioned on the mainshaft with their six apexes 60° apart, and the rotor power phases are alternated so that there is a power phase for each 180° of mainshaft rotation.

In a three-rotor engine the rotors are positioned with the nine apexes 40° apart, and there is a power phase for each 120° of mainshaft rotation. A four-rotor engine has twelve apexes 30° apart, with a power phase for each 90° of mainshaft rotation.

CHAPTER 7

The Fuel System

A WORD ABOUT GASOLINE

Gasoline is derived from petroleum (also called *crude oil*) which, fortunately, is an abundant liquid (second in quantity only to water) found mostly in underground streams and pools of saturated sand, gravel, and other rock formations. Deposited perhaps millions of years ago, it is the residue of vegetable and animal life once on the surface, then partially decayed, buried, and subjected to tremendous pressures and temperatures. Coal, diamonds, and some other precious stones are of the same origin.

What Petroleum Is

Petroleum is a mixture (Chapter 2) containing many hydrocarbons, compounds made up of different groupings of hydrogen and carbon atoms. If you will recall, it was said in Chapter 2 that oxygen atoms have an especially strong affinity for (ability to combine with) certain other types of atoms. Carbon atoms are those for which oxygen has the strongest affinity. Hydrogen atoms, too, are high on this list. Hence, petroleum molecules when subjected to heat will oxidize very rapidly and thoroughly. In short, they are highly explosive. (The combustion power of gasoline far exceeds that of dynamite or even nitroglycerine, quantity for quantity.)

If a liquid is vaporized (by boiling) any solids it contains will be left behind, so that when it is condensed back into liquid form it will be pure (free of foreign matter). Also, if two liquids that have different boiling points are mixed, they can be separated in the same way; that is, by boiling off the lighter liquid (one with the lower boiling point) first. When liquids are thus separated from each other and from any solids that were in them we call the process *fractional distillation*.

Now, the hydrocarbons that compose petroleum vary considerably in the numbers of carbon atoms and hydrogen atoms which make up their molecules. Some contain a great many atoms per molecule. These are said to have high *molecular weight* (meaning they are relatively massive). Such liquids require higher temperature to boil than those liquids having a lower molecu-

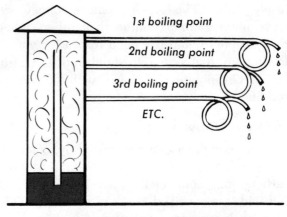

The principal "idea" on which fractional distillation is based.

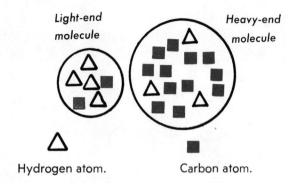

Typical of why gasoline has light and heavy ends.

lar weight. We say that the higher weight (*heavy-end*) liquids are less volatile than the light-end liquids.

Petroleum is subjected to fractional distillation. The most volatile hydrocarbons are boiled off first. This continues until only the least volatile liquids and foreign matter are left. In this way petroleum is separated (*refined*) into petroleum ether, gasoline, naphtha, benzine, kerosene, gas oils, lubricating oils, fuel oils, petroleum jelly, paraffin, and petroleum coke (to name the better-known products in the order in which they boil off).

But each product is not necessarily just one of the many liquid hydrocarbons that composed the original petroleum. On the contrary, gasoline (for instance) is still a mixture of hydrocarbon liquids boiled off at a certain temperature range during the distillation process. Therefore, gasoline has some parts which are more volatile than others. It has, as we say, some *light ends* and some *heavy ends*. The gasolines of different refiners will vary in the numbers of light and heavy ends. (This is why some gasolines will leave more *gummy residue* than others when they evaporate and is a good reason for *not* allowing any fuel system to stand until the gasoline in it evaporates and clogs it with the gummy residues.)

The Newer Gasolines

Even though gasolines vary slightly, commercial standards require that all the liquid hydrocarbons in a commercially acceptable product fall within a specified range of volatility. This, then, excludes many of the liquid hydrocarbons which are too heavy to volatilize at the specified

temperature. But the ever-increasing need for gasoline makes it desirable to obtain as high a yield of acceptable hydrocarbons as possible. Therefore, several new manufacturing processes have been developed to accomplish this.

One is called a *cracking process*. Some of the heavier hydrocarbons (normally found in kerosene) are cracked so that their molecules are broken up into smaller and lighter molecules. The latter are then volatile enough to be included with gasoline.

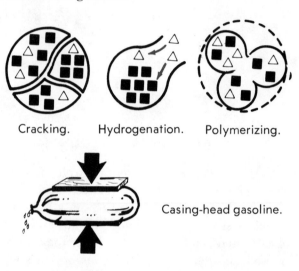

Cracking.   Hydrogenation.   Polymerizing.

Casing-head gasoline.

Methods of distillation.

Another process is *hydrogenation*. Molecules which have too many heavy carbon atoms in proportion to their number of the lighter hydrogen atoms are made lighter by adding more hydrogen atoms.

A third process is *polymerizing*. Some of the too-simple molecules (too few carbon atoms) are bunched together to form larger molecules having the proper molecular weight and volatility. Gasoline made by this process is often called *poly-gasoline*.

Still a fourth method is the production of *casing-head gasoline*. Natural gas, often found in the same pools with petroleum, is cooled and compressed. Its heavier ends are thereby extracted in a liquid form suitable as a gasoline or for blending with regular gasoline.

In addition, gasoline can also be manufactured out of low-grade coal and even from vegetation. Such gasolines are called *synthetic gasoline*.

Octane Rating of Gasolines

Two of the liquid hydrocarbons contained in commercial gasolines are called *heptane* and *iso-octane*. An excess of heptane in relation to the amount of iso-octane produces a fuel which, when used in a high-compression-ratio engine, tends to burn unevenly. This results in a more-or-less loud knocking noise accompanied by loss of power. On the other hand, pure iso-octane is knockless, even at high compression ratios. The proportion of iso-octane to heptane in a gasoline is a factor in determining how much the fuel can be compressed without knocking. Therefore, gasolines are rated according to the percentage of iso-octane present in relation to the percentage of heptane; this is called their *octane rating*. The higher the octane rating, the more the fuel can be compressed without knocking.

NOTE: Some so-called "premium" gasolines on the market have small quantities of tetraethyl lead and ethylene bromide mixed in. This has the effect of reducing the tendency of a low octane gasoline to knock (by making the heptane in the gasoline burn more uniformly).

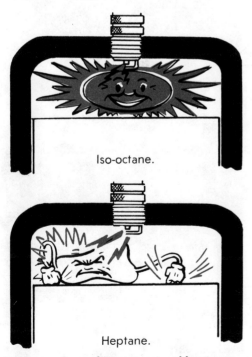

Iso-octane.

Heptane.

Comparison of iso-octane and heptane.

The compression ratio at which knocking becomes a serious nuisance is approximately 6 to 1.

For engines having a small compression ratio (which includes most small engines) the octane rating is not too important. Since practically all presently sold standard gasolines do have a suitable rating for compression ratios up to 6 to 1, most small engine users do *not* have to buy premium or extra-high octane fuels.

The Importance of Proper Gasolines

The ideal gasoline for any engine is one that will vaporize and readily mix with air under the required conditions—but one which will not vaporize too soon. As previously said, perfect combustion can occur only when the gasoline does vaporize and mix with the air in a manner that permits the oxygen atoms to combine quickly and completely with the carbon and hydrogen atoms of the gasoline, when the mixture is heated to the flash point of the gasoline. On the other hand, if the gasoline vaporizes at too low a temperature, it may form small clouds of gasoline vapor in the fuel line which will then block passage of the gasoline. In fact, such clouds can even shut off the supply of gasoline and result in what we call *vapor lock* (and a dead engine).

The ideal gasoline is also one that will form as nearly perfect as possible a combustible mixture with air. As will be explained, poor combustion can result from a gasoline having too many heavy ends.

From a practical standpoint, any good grade of commercial gasoline is satisfactory, if properly handled. By proper handling we mean not contaminated with dirt or other foreign matter, nor mixed with one of the less volatile fuels like kerosene. We also mean not allowed to evaporate so that the lighter ends are lost and too high a proportion of heavy ends remain. Gasoline that is not pure or that has stood too long while evaporating is *not* a proper fuel.

HOW THE GASOLINE-AIR MIXTURE AFFECTS COMBUSTION

Perfect combustion occurs when each carbon atom from the fuel combines with two oxygen atoms from the air to form the gas known as *carbon dioxide*. (CO_2). If there are too many

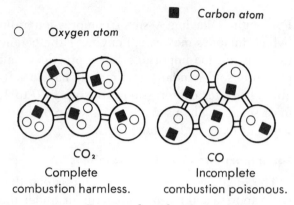

○ Oxygen atom

■ Carbon atom

CO₂
Complete
combustion harmless.

CO
Incomplete
combustion poisonous.

Degree of combustion.

carbon atoms for the amount of oxygen available—so that only one carbon atom combines with each oxygen atom—the resulting gas is called *carbon monoxide* (CO), and the combustion is incomplete. That is, CO gas is still a fuel. It can be further burned by continuing the combustion process to form CO_2. Also, of course, if there are way too many carbon atoms, some will not find any oxygen atoms to combine with.

NOTE: CO_2 is nonpoisonous and nonflammable; but CO *is both flammable and very poisonous* to inhale (will kill a man quickly). During even excellent combustion, some CO is formed, but in poor combustion a dangerous amount may be formed. Poor combustion also results in a residue of carbon. Some of this residue will blow off as black smoke, some as soot, and some will remain in the engine to form hard carbon deposits.

It follows that to avoid poor combustion, it is essential that sufficient oxygen be present. That is, the air-gasoline mixture must consist of the right proportions of each to ensure there being enough oxygen atoms for all the carbon atoms. The best mixture for average gasoline is approximately 15 parts of air to 1 part of gasoline, by weight. But gasoline weighs 600 times as much as air at sea level. Therefore, for each gallon of

gasoline we need approximately 1200 cu. ft of air at sea level—about the amount in a 10 × 14-foot, average-height room.

NOTE: Air contains 21% oxygen, 78% nitrogen, and 1% other gases. Gasoline, as previously said, is composed of carbon and hydrogen, plus some impurities. During combustion, the hydrogen and nitrogen are mostly released as harmless gases; some oxygen and hydrogen combine to form water vapor; and the other substances go into making other compounds, such as acids) which unfortunately add to the carbon deposit and have a corrosive effect on some engine metals.

THE COMPLETE JOB AND PARTS OF A FUEL SYSTEM

Summing up all the requirements for good combustion, these, then, are the tasks which a fuel system can perform:

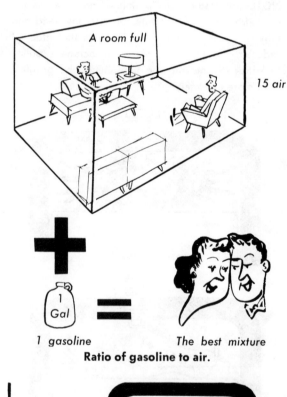

A room full

15 air

1 gasoline

The best mixture

Ratio of gasoline to air.

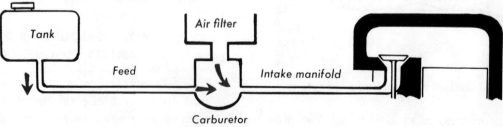

Tank

Air filter

Feed

Intake manifold

Carburetor

Basic parts of a fuel system.

1. Store an adequate supply of gasoline in a manner to prevent its evaporation and to prevent it from igniting.
2. Vaporize the gasoline as needed, and mix it with the proper proportion of clean air in a manner to prevent vapor lock and to ensure as thorough mixing as possible.
3. Convey this mixture to the intake valve ready to be sucked into the cylinder. At the same time, permit the operator to control the amount of mixture being fed to the engine.

A modern fuel system requires the following parts to accomplish these tasks:

1. An enclosed *fuel tank* for gasoline storage.
2. Either a forced, vacuum, or gravity-type *feed system* to transport liquid gasoline to where it is mixed with the air.
3. An *air filter* (*cleaner*) capable of taking in and cleansing the large volume of air needed.
4. A *carburetor* to meter the gasoline (measure it out in proper ratio) and to vaporize it into the cleansed air so that the gasoline and air become mixed.
5. An *intake manifold* (of whatever length and design is required) to carry the gas-air mixture to the cylinder(s) and, if necessary, to keep it warmed during the journey so that the gasoline will not condense back into (unmixed) liquid state.

NOTE: Automotive engines have also been equipped with special silencers (to eliminate the whistle of the in-rushing air) and antiflashback devices (to protect the fuel system should combustion occur while the intake valve is open so that burning gases flash back through the fuel system). With small engines, however, these are either not required or, where found necessary, the functions of such devices are incorporated into the designs of the other parts.

THE PRINCIPLE ON WHICH ANY FUEL SYSTEM OPERATES

We have already explained the vacuum principle of intake (Chapter 3). Thus, due to atmospheric pressure, air from the surrounding atmosphere will flow, with considerable force, into any vacuum to fill it, and equalize the pressure inside with the pressure outside. To state this in another way: Since the outside pressure at sea level is approximately 15 psi, whenever a vacuum or partial vacuum is created so that the air pressure inside the vacuum is less than 15 psi, air will be pushed from the outside to the inside until the inside pressure is raised to 15 psi. We state this more briefly by saying that *air always flows from a high-pressure area into a low-pressure area.*

NOTE: This is what causes wind to blow—sometimes with such force as to be a tornado. Air heated by the sun over a desert or flat area (where the heat becomes concentrated) expands and becomes lighter than surrounding air so that it rises (to float on top). Then the surrounding air rushes in to fill the low pressure (partial vacuum) area left by the rising air. Other factors (such as moisture) also contribute to this.

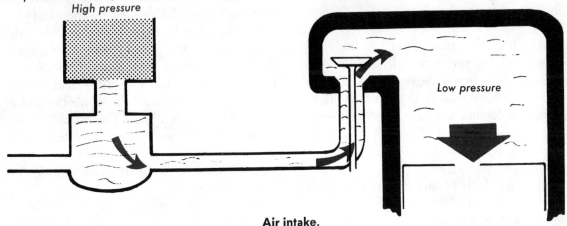

Air intake.

Liquid Flow

In dealing with the flow of a liquid from one place to another, several factors must be considered—the pressures developed within the body of liquid; the pressure of the air resting on the liquid's surface; and the force of gravity.

As explained in Chapter 2, atmospheric pressure is due to the weight of the air on top. In the case of air (or any gas), the weight compresses the air beneath so that it is under a pressure that is directly proportional to its degree of compression. Moreover, we can further compress any gas in a compressor to correspondingly increase its pressure, or we can decompress it (by creating a partial vacuum) to reduce its pressure. And a gas will always flow from a higher pressure area to a lower pressure area—and will always expand indefinitely—if allowed to do so. In fact, all the gases of earth would expand and flow outward away from earth if it were not for the pull of gravity.

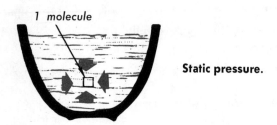

Static pressure.

A liquid differs from a gas in one major respect—it will not expand to fill a vacuum, and it is not easily compressed (in fact, for all practical purposes, cannot be compressed). This is why we can use any liquid, *by hydraulic process,* to transmit force with the same effectiveness as in transmitting force through a solid (like a lever).

However, like the atmosphere, a body of liquid does develop pressure in its depths due to the weight of the liquid above. Each molecule is subjected to a force relative to its depth which, since liquid molecules are free to slide, could propel it in any direction. The deeper you go into the ocean, for instance, the greater will be the pressure surrounding you.

In a contained body of liquid this pressure is simply what we call *static pressure,* meaning that it does not cause any flow of the liquid.

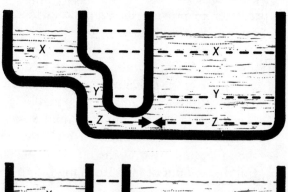

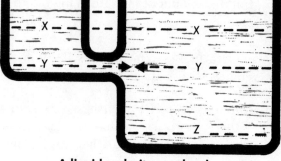

A liquid seeks its own level.

Each layer of molecules from top to bottom is already trapped by the molecules around it. To express this fact we say that any contained body of liquid seeks its own level.

The level a liquid seeks refers only to the top surface; the bottom contour does not matter. Also, it does not matter how many separate pools of liquid surface there may be; if the pools are in any way connected below the liquid surface they all together comprise one body of liquid. For instance, each pair of connected tanks illustrated holds a single body of liquid, and the liquid surfaces in each pair are at the same level. This is so because at any depth (X, Y, or Z) in either tank all the molecules are subjected to the identical force determined by this depth.

On the other hand, if its container is suddenly enlarged so that a liquid has room to spread out, the static pressure becomes a *dynamic pressure* (one which produces results) and forces the molecules to move until the whole body of liquid again finds its own level. This would occur even if the container were in a complete vacuum. In short, the pressure caused solely by the liquid's own weight can cause it to flow under proper circumstances.

Atmospheric pressure on the surface of a liquid also can cause it to flow if the pressure

1—Plug just pulled

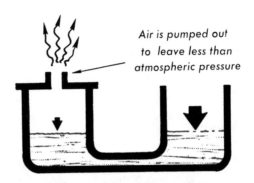

Air is pumped out to leave less than atmospheric pressure

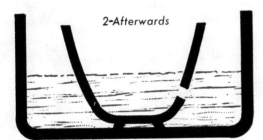

2-Afterwards

Dynamic pressure.

Liquid relevels so that both remain at atmospheric pressure

(weight) is different above different areas of the liquid's surface. If a pair of tanks are both exposed at the top to the same atmospheric pressure, the pressure has no effect. Capping one tank in the pair still would make no difference since the trapped air remains at the same pressure it was.

However, after capping one tank, if you should use a suction pump to remove some of the air and create a partial vacuum, the air pressure

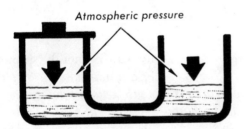

Atmospheric pressure

in the capped tank would then be less than that on the liquid in the exposed tank. Enough liquid would flow from the exposed tank into the capped tank to compress the trapped air and again equalize its pressure with the atmospheric pressure on the exposed tank.

NOTE: If all the air is pumped out of the capped tank, liquid will flow into it until the pressure (weight) of the liquid alone will equal the combined liquid and atmospheric pressure on the ex-

posed tank. This is the principle of a barometer. Mercury is pushed up into the vacuum tube at left until the pressures at each side of line A are the same. Thereafter, any change in atmospheric pressure (due to changing density and vapor of the air) will cause the mercury level to rise or fall, to re-equalize the pressures.

In the third place, gravity—which is responsible for both of the preceding liquid-moving forces—can move a liquid in a manner not associated with either of the preceding forces. Free-falling water (like rain, for instance) has no internal pressure, nor is it falling due to any atmospheric pressure on it. It falls solely because of the gravity pull on it.

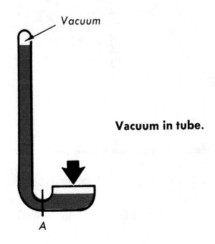

Vacuum

Vacuum in tube.

A

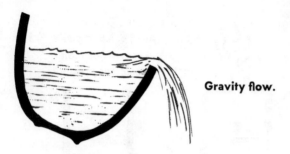

Gravity flow.

Practically speaking, liquid generally flows in response to a combination of two or all three of the preceding forces.

To demonstrate this combined effect, the accompanying illustration shows a tank with a hose outlet. When held as shown (view 1), the air and static pressure at the bottom of the hose are the same, per square inch, as they are at the tank opening into the hose. No liquid will flow. Even if the hose end is curved back down (as indicated by the dotted lines), the air trapped in the hose still exerts its same pressure and blocks the flow. To start a flow we can do several things:

1. We can straighten the hose out with its end below the liquid level in the tank. The now lower liquid pressure in the hose, plus the downhill run from the surface in the tank to the hose end, will start a flow.
2. We can leave the hose curved (dotted lines) and lower the air-bubble part—even though the pressures remain unchanged—until the downhill run from the tank to the air bubble starts a flow and pushes the air out.

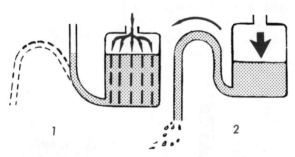

Effect of gravity on water flowing through hose.

3. We can leave the hose exactly as shown in the dotted lines and eliminate the air by forcefully filling the hose with liquid. This is called *priming*. With the air pressure

removed, the inequality of pressures will cause the water to start flowing, uphill past where the air was. Once past this point, it will flow by gravity downhill out the hose end. Since the path through the hose is now a solid stream of liquid, so that any stoppage of flow in the uphill portion of the hose would necessarily leave a vacuum (a low-pressure area), the liquid must continue to flow until the surface of the tank is lowered to the height of the hose end. This pressure-started but gravity-finished type of flow is called syphoning (view 2).

NOTE: The denser (heavier) a liquid is, the greater will be the effect of gravity on it, both in generating static presssure and in causing gravity flow. Also, the bigger the drop (the longer gravity has to work on the molecules of a liquid), the greater the velocity of the liquid will become toward the end of the drop. For gravity is a continuing, constant force (which means that for every second it is applied, it will add another increment of momentum). In short, a falling object *accelerates* (gains velocity) in direct proportion to the distance (time) it has fallen (less, of course, any loss due to friction).

Gravity also works on gases just as it does on liquids. But since a gas is so much less dense than liquid, the direct effect of gravity flow in a gas is so negligible that it can be disregarded.

The Venturi Principle Applied to Liquids

It is a law of physics (Bernoulli's principle) that: *If the speed of a liquid flowing through a confined area is increased at any point, the pressure of the liquid decreases at that point.* "Pressure" in this case means the sum total of forces causing the liquid to push outward against its

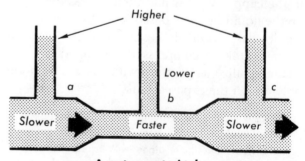

A water venturi tube.

container. This law can be demonstrated by the use of a *liquid venturi tube,* as illustrated. When liquid is forced through the horizontal tube so as to fill all of the tubes throughout, the liquid level in the central vertical tube (which is connected at the *throat,* or restricted portion, of the horizontal tube) will be lower than the levels of liquid in the two other vertical tubes. Obviously, something must cause this. Let us examine the cause.

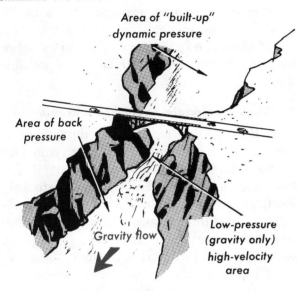

The flow of a river.

Consider a river. If the bottom were level, end to end, at whatever depth rain might pile up the water at one spot between the banks, the static pressure of the water would be equal in all directions. With the banks to hold it at the sides but with the two ends open, the water would spread out in two directions until it found its level. The only reason a river does always flow downstream is that it is flowing downhill, by gravity. And so long as its flow is unrestricted, the equalized static pressures within the water add nothing to its flow.

If a narrow canyon restricts the river's flow, the same amount of water that is flowing in the river above must also pass through this canyon; otherwise the excess water left upstream would pile up continuously to form an ever-enlarging lake. Obviously, then, the water must speed up through the canyon. But the force of gravity alone—which has caused its flow up until now—

cannot suddenly increase the water's velocity. Gravity is a constant force and its pull on the molecules does not vary. Therefore, a new force is needed to literally boot each molecule as it enters the canyon and hurry it through.

This new force comes from the development of a dynamic pressure in the water above the canyon. The water upstream does pile up and broaden out until its increased depth and bulk are sufficient to create the higher pressure and additional force required to increase the velocity through the canyon.

If there were nothing but open sea below the canyon, the water would actually squirt through to settle in the sea. It is more likely, however, that the river will continue to meet all manners of other restrictions (narrowing banks, rapids, debris, etc.). At each obstruction the water must back up, then speed up. The whole progress of the river is one of developing high- and low-pressure areas to aid the gravity flow. Downstream from the canyon or any restriction, the pressure again builds up due to the *back-pressure* created by whatever restriction lies downstream.

In our horizontal venturi tube, gravity plays no part, nor can the water in the wider areas increase its depth. Therefore, dynamic pressure (as we have described it) also does not exist. The only force available is the hydraulic pressure which causes the water to flow in the first place. But the liquid still has to speed up in the venturi throat and must be given extra momentum at the entrance to the throat. Consequently, the throat has the same effect as the canyon; it holds back the liquid until the pressure above accumulates and becomes enough greater than the pressure in the throat to accomplish the speed-up.

NOTE: Again, if there were no container downstream from the throat, the liquid would simply "squirt out" and there would be no pressure build-up here. If, however, the venturi is submerged in liquid, as shown, the back-pressure of the surrounding liquid opposing the flow will cause the pressure in the wide area below the throat to almost equal the pressure in the wide area ahead of the throat. Or, if the venturi is simply one obstruction of many in a long length of pipe, the pipe's friction effect plus the effort of the other obstructions will build up back pressures the same as in a river. Actually, in a long pipe, as friction uses up the original hydraulic force, the average pressure will drop. It can even drop to nothing. In such case, due to the inadequacy of the original force, the liquid would only flow so far, then the flow would stop altogether. Whatever the circumstance, however, so long as the liquid does flow, the pressure will always be greatest above a venturi throat, lowest in the throat, then almost as great again just below the throat. Instruments for measuring water flow and velocity are based on this principle.

The Venturi Principle Applied to Gases

A venturi works exactly the same with a gas as with a liquid, with this one exception—a gas is easily compressible, whereas a liquid is not. Consequently, as a gas approaches a venturi throat, it will partly speed up as a liquid does, but will partly become compressed and literally squeeze more molecules through. Compression

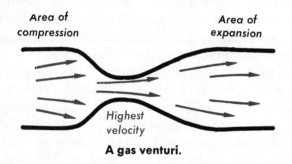

Area of compression

Area of expansion

Highest velocity

A gas venturi.

increases the pressure in a gas. Therefore, that portion which is compressed will add to the pressure in the venturi throat. The overall pressure differential between the area above the throat and the area in the throat will not be as great as it is in a liquid.

Actually, this compression factor introduces a new element to consider—the time element. Until there is time for all the gas above the throat to become so highly compressed that it will build up sufficient compression pressure to increase the velocity as required to equalize the flow, the flow actually remains unequal. There will be less gas flowing through the venturi than will be flowing through the area ahead of it. Gas will accumulate (under continuing compression) above the throat. Also, the compressed gas will expand to refill the area below the throat. During the interval in which less gas

passes the throat, the expansion of the gas necessarily exceeds the amount by which it has been compressed. And expansion reduces the pressure. As a result, the pressure in the area just below the throat will be *even lower than the pressure in the throat* during such an interval.

Once this interval is ended, however, and the compression above the throat builds up to equalize the flow, the expansion factor will just equal the compression factor. All the pressures will be just the same as those explained for a liquid. In short, when compression reaches the point where there is no need for further compression, the gas will flow exactly like an uncompressible liquid.

The Venturi Principle Applied to a Carburetor

Air does not flow continuously through a carburetor. It flows in starts and stops, as called for by the intake stroke(s) of the piston(s). The time interval of each flow period is extremely short. Moreover, the air is not being hydraulically pumped through against atmospheric pressure at the other end. On the contrary, it is being sucked through by a partial vacuum. These factors all influence the design of a carburetor venturi and make it somewhat different from the liquid venturi previously illustrated.

Due to the short interval of air flow, the lowest pressure point in a carburetor venturi is in the area *just beyond the throat*, where expansion is taking place. This is accentuated by the fact of the partial vacuum beyond. The throat, itself, does not serve the purpose of a low-pressure area; it serves only to create the conditions for this low-pressure area beyond.

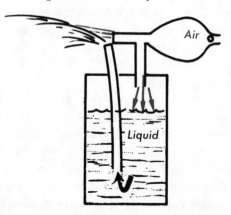

Air

Liquid

An atomizer uses friction drag.

Friction Drag

Let us go back to our river and the canyon. Anyone familiar with such a spot knows that a breeze of some proportion always blows over the water there. This is due to *friction drag.* The molecules of the water flowing downstream bump and carry along with them molecules of the air above. Wherever the water velocity is sufficient this action is sufficient to carry along an appreciable amount of the air. As the lower layer of air is carried away, more air blows in to replace it. In fact, this phenomenon is one of the contributing factors to the birth of hurricanes in areas above vast and rapidly moving ocean currents.

The principle of friction drag can be reversed. Fast-moving air will carry away molecules of an adjacent liquid. An atomizer, for instance, depends partly on this principle. By an increased pressure on its surface, the liquid in the bottle is made to flow up and bubble out the nozzle; and by simultaneusly squirting a fast-moving airstream over the nozzle end, this bubbling liquid is picked up, molecule by molecule, and atomized.

Friction drag is used in a carburetor, too, to atomize the gasoline. The liquid gasoline nozzle is in the venturi, facing into the lowest-pressure area and just where the highest velocity airstream will squirt over it and pick up molecules of the gasoline.

THE APPLICATION OF THESE PRINCIPLES TO THE FUEL SUPPLY

Carburetion means the vaporization of the liquid gasoline and the mixing of this vapor with a proper ratio of air. Both vaporization and mixing are done in a carburetor, which is designed to meter (measure in a continuous flow) the gasoline and air, and to vaporize the gasoline into the air for mixing. The mixing process then continues as the air and gasoline vapors flow and tumble through the intake manifold and on through the intake valve into the cylinder.

A simplified diagram of the carburetor function is shown in the accompanying illustration. For the purpose of this explanation we will assume that liquid gasoline is available under atmospheric pressure, as shown at "a," that air is free to enter at "b," and that the engine is operating at its normal running speed.

When the piston moves down for intake, it creates a partial vacuum at "e" and, since the valve is open, also at "d" in the intake manifold (which is why the intake suction usually is referred to as *manifold vacuum*). Air under high atmospheric pressure at "b" rushes through the venturi to refill the vacuum at the low-pressure areas "d" and "e." Because the piston is moving rapidly, the air flows from "b" to "e" at a considerable velocity. Moreover, in passing through the venturi throat at "c" it moves at still greater velocity, and its pressure at "d" remains much

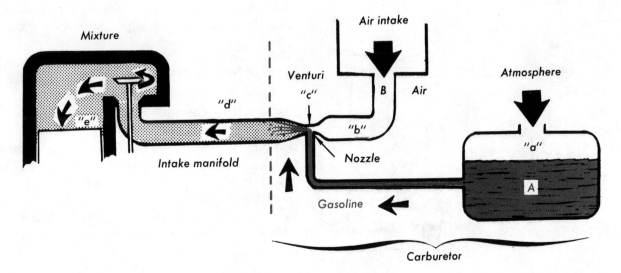

Use of a venturi in carburetion.

lower than atmospheric pressure. In addition, it creates a considerable friction drag at "c" on the gasoline in the nozzle at this point.

As the pressure on the gasoline surface at "a" is high atmospheric pressure, while the air pressure at "d" (as already stated) is much below atmospheric pressure, the gasoline tends to flow from "a" toward "d" (just as the air flows from "b" toward "e"). Aided by the friction drag at "c," it is literally sucked out of the nozzle to spray into the air stream at "d." Thus, gasoline is made to flow from the available source "A" into the airstream that is flowing toward the cylinder(s), and to mix with the air. The tumbling action of the gasoline and air molecules, that re-

sults from the violence of the forces which cause each to flow, completes the mixing. By the time they reach the cylinder "c", these molecules are thoroughly tumbled to form a combustible mixture.

NOTE: Later, we will discuss exactly how the gasoline-air ratio is maintained at various engine speeds. Basically, of course, this is determined by the nozzle size in relation to the other factors involved in causing the gasoline to flow.

In the following few pages, and before going on with the story of carburetion, we will discuss how the gasoline supply is made available at such a point as "a," and how air is made available at "b."

SYSTEMS COMPONENTS

FUEL TANKS AND FEED SYSTEMS

A fuel tank may be of any construction that will safely store the liquid gasoline and keep it clean and enclosed. Customary features are: (1) a *filter screen*, fitted in the inlet under the tank cap, to strain dirt out of the gasoline as it is poured in, or fitted around the tank outlet to strain the gasoline as it leaves the tank; (2) a *drain cock* for cleaning and/or a *fuel shut-off*

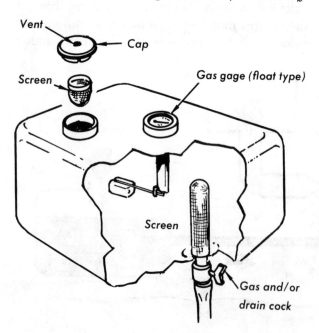

Vent
Cap
Screen
Gas gage (float type)
Screen
Gas and/or drain cock

Types of fuel tank accessories.

cock to control fuel flow from the tank to the carburetor (sometimes one cock serves both purposes); (3) a *gasoline gauge* built into the tank or cap to register the amount of gasoline inside.

No one tank need have all the preceding. This depends on the manufacturer. One requirement of all tanks, however, is an *air inlet* (or *vent*) at top (usually a pinhole in the cap or spaces under it when it is mounted) so that air can flow in to replace the gasoline used. Without such an inlet a gravity feed system won't work at all; and a forced feed system would create a vacuum in the tank which could result in collapse of the tank as a result of the (then) superior outside atmospheric pressure.

Types of Feed Systems

There are four different methods used to flow gasoline from the fuel tank to the carburetor; these are referred to as a *gravity system*, a *suction-feed system*, a *standard-type vacuum system*, and a *forced-feed system*.

In a *gravity system* the fuel tank must at all times be at a higher level than the carburetor. Fuel then flows downhill to the carburetor, which is designed to admit the fuel when and as needed. Since the carburetor in any system depending on gravity feed alone is always vented to the atmosphere (just as the tank is

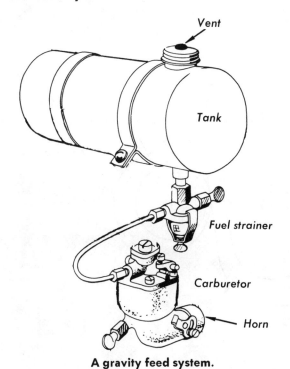

A gravity feed system.

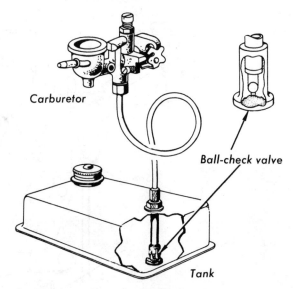

A suction feed system.

vented), a gravity feed line has to be lower, throughout its length, than the tank bottom, to avoid having any air pockets in the line. This is the type of system used with almost all small engines, because of its better economy and simplicity.

The type of vacuum system that is called *suction-feed* uses the manifold vacuum to suck gasoline all the way from the tank through the carburetor and into the airstream. In such a system the carburetor must be close to the tank (often is mounted right on top) as the amount of lift that can be created in this manner is limited. A *ball-check* (or *disk-type*) *valve* (see illustration) is sometimes employed in the fuel line. By dropping closed each time the suction ends, it prevents that gasoline remaining in the carburetor from dropping back into the tank. It thus ensures a ready supply of gasoline in the carburetor at the start of each intake operation.

A standard-type *vacuum system* employs the manifold vacuum to suck air out of an auxiliary tank located above the carburetor level. This creates a low-pressure area in the so-called vacuum tank so that fuel flows uphill from the fuel tank to this vacuum tank, then downhill from the vacuum tank to the carburetor. During intake, manifold pressure will be anywhere from

about 3 psi less than atmospheric pressure (on wide-open throttle) to about 12 psi less (on closed throttle). This pressure differential is sufficient to keep the vacuum tank filled, yet is not enough to hold the weight of the gasoline up in the vacuum tank against the pull of gravity through the syphon operating line to the carburetor. A vent hole in the tank top admits enough air to ensure the gravity flow and to prevent overfilling of the tank.

NOTE: The preceding system is seldom used on a small engine.

The fourth, or *forced-feed system*, employs a *fuel pump*. This pump feeds enough fuel from the tank to the carburetor to keep the supply adequate; but is so designed that it cannot overfeed.

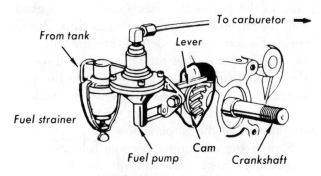

Fuel-pump operation.

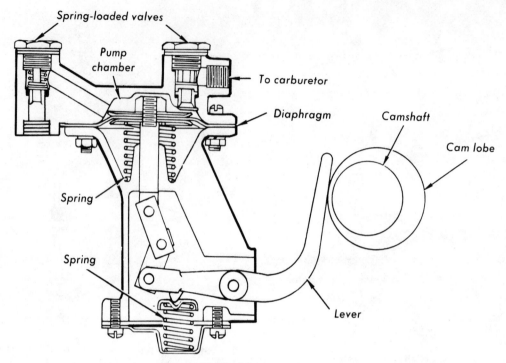

A typical fuel pump.

The Fuel Pump

Quite often (especially in two-stroke-cycle engines) the fuel-pump function is incorporated in the carburetor. This type of pump will be discussed later, together with carburetors. When a separate pump is used, it is of the type illustrated. The pump is operated by a lever arm that rides, under spring tension, on a cam on the crankshaft.

The large view (top of page) shows the principle of pump operation. Reciprocal movement of the lever pulls the diaphragm down against the spring below it, then allows the spring to push the diaphragm back up. In moving down, the diaphragm creates a low-pressure area in the pump chamber. Gasoline is sucked into the chamber from the fuel tank through the spring-loaded valve at left. When moving up, the diaphragm pushes the gasoline out through the spring-loaded valve at the right, to the carburetor. As the two valves are spring-loaded, each will remain closed except when forced open by pressure of the gasoline flow.

Should the carburetor become full, its valve (explained later) will shut off the line from pump to carburetor. Hence, the diaphragm spring will not be able to move the diaphragm up. The pump will be inoperative until the carburetor valve is again open, even though the lever arm continues to reciprocate, due to the action of its own separate spring.

The diaphragm is made of leakproof, flexible material and is mounted to prevent leakage of gasoline past it. It is usually fastened to the plunger by means of fiber discs, which grip the central portion around the plunger. There are many other lever arrangements besides the one illustrated. All operate in much the same manner.

Use of a Fuel Strainer

With gravity and forced-feed systems, but not with suction systems, *a fuel* (or *gasoline*)

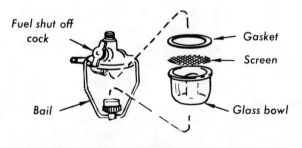

A fuel strainer.

strainer is often used instead of a strainer in the fuel tank. When a pump is used, the strainer and pump are often located together or even built into one integral unit.

This strainer consists of a glass bowl with an internal wire-screen strainer through which the fuel must flow on the way to the carburetor. The passage out is at the top. During operation, gasoline remains in the bowl long enough for heavier moisture, dirt, and other foreign matter to settle to the bottom of the bowl where it can easily be seen. And the screen prevents passage of such into the outgoing line. The bowl and screen are removable for cleaning.

The fuel shutoff cock may be located on the strainer top, instead of at the bottom of the fuel tank.

Use of a Priming Pump

In addition to the fuel pump, some fuel systems also employ a separate *priming pump* designed for manual filling of the fuel line and carburetor prior to starting the engine. Such a pump is required whenever the fuel tank is located some distance away from the fuel pump, as is the case with outboard motors. Without

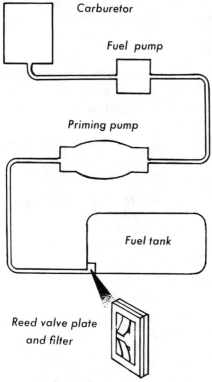

A typical priming-pump setup.

this feature, starting would be extremely difficult, if not impossible, since the feeble reciprocation of the fuel pump at the very slow starting speed would be insufficient to readily suck the air out of the fuel line and suck gasoline up to the fuel pump. Due to the expansion of air, the air that is sucked out of a line does not leave an immediate vacuum such as is created when a nonexpanding liquid is sucked out; instead, it merely reduces the overall pressure in the line. Hence, it takes a good many more sucking strokes to obtain the percentage of vacuum necessary for the liquid behind the air to flow all the way up to the pump.

A typical priming pump consists of a neoprene bulb located in the fuel line, between the tank and fuel pump (accompanying illustration). At the tank end there is a spring-loaded valve which will permit flow only in the direction of the fuel pump. This may be a simple reed valve (Chapter 4) or a ball-check type such as described for a suction-feed system. The bulb is relatively large.

When manually squeezed, the bulb expels a considerable volume of air out through the fuel pump and carburetor. Then, when the elastic bulb expands, it is necessarily filled with gasoline from the tank. Another squeeze (or two) then forces gasoline up through the fuel pump and into the carburetor, ready for starting of the engine.

NOTE: After the engine is started, the primed fuel pump keeps the gasoline flowing through the bulb from the tank to the carburetor. Further use of the priming pump would simply flood the engine by forcing too much gasoline through the line. Therefore this pump should *never* be used once the engine is running.

Fuel System Maintenance

The most important things to remember about any fuel supply system are: the gasoline flows by gravity and/or suction; all the lines, screens, and other openings through which it flows are precisely calibrated (sized) to permit just the right amount to flow at various engine speeds; and any artificial restriction may upset the flow and the mixture ratio (if it does not completely stop the flow). The same is true of the air supply.

Consequently, all lines must be kept fully open; screens or filters must be clean and open for free passage of gasoline; dirt and foreign matter must be kept out of the fuel tank, lines, pumps, etc., and out of the air passages. This means that the filters must not only be clean, but they must be in good shape to do the jobs intended.

In addition, all connections must be air-tight. The often very slight pressure differentials which cause the gasoline to flow can be drastically altered and destroyed by even a minute air leak into the system at the wrong place. Wherever gasoline can leak out there is a possibility that air will leak in.

Individual parts of the various components usually are made available by manufacturers for replacement purposes. Such items as screens, fuel-pump diaphragms, springs, gasoline lines, and gaskets must always be replaced (rather than repaired) if damaged. In some cases, such components as fuel pumps or filter assemblies are offered complete, on a trade-in basis. For the most part, the low cost of replacement items makes any lengthy repair service on individual parts and/or components undesirable.

AIR CLEANERS

The *air filter* (or *cleaner,* as it is usually called) is a simple but very important device. Conditions under which most engines are used stir up dust. Quite often the air is laden with pollen, seeds, and other suspended matter. If

allowed to get into an engine fuel system, such foreign matter would quickly clog it; nor would it do the engine, itself, any good either! Indeed, the air cleaner prevents rapid deterioration of the engine from grime and dirt. The most important thing about a cleaner is that it must work properly, yet freely admit all the air that the engine calls for. A clogged filter can rob the engine of air and reduce its efficiency, all the way down to zero.

Cleaners are of two general types: the *wet* (or *oil*) *type,* and the *dry type.* In the wet type the air is passed through either an oil bath or a filter element saturated with oil; in the dry type it is passed through a porous element. With both types air enters into the filter bowl (either through holes in the sides or bottom, or between the bowl and a cap), passes through the element, and then moves down a centrally located tube and out to the carburetor horn (air intake pipe).

Elements are made of many different materials: moss, felt, paper fibers, aluminum-foil mesh, etc. Most types can be cleaned (generally in gasoline) if not allowed to become too dirty or gummy with caked oil. Naturally, for the free passage of air, they must be kept clean or replaced. In the oil-bath type, oil must be kept fresh and at proper level.

Whenever oil is used—and especially in the oil-bath type—care must be taken to avoid spilling oil into the carburetor horn. Oil *cannot* be compressed by a piston stroke. Should enough of it seep into a cylinder, it would block the piston upstroke, possibly with considerable damage to the engine.

THE INTAKE MANIFOLD

In an engine with two cylinders or more, the intake manifold serves a dual purpose. First, it distributes the mixture from the carburetor to the various cylinders. Second, it allows the mixture to become more thoroughly mixed by reason of the turbulence resulting as the mixture passes through it. To serve these two purposes, an intake manifold has to be carefully engineered. It is *not* merely an assembly of pipes; the pipes have to be sized correctly to ensure

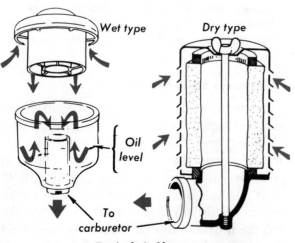

Wet type   Dry type

Oil level

To carburetor

Typical air filters.

even distribution of the mixture to all cylinders. The curves, too, must be plotted so as to cause the desired turbulence without, in any way, slowing down the passage of the mixture.

When lengthy intake manifolds are required, the problem of overcooling the mixture during its passage must often be contended with. A fast-moving airstream, especially if laden with moisture such as the gasoline vapors, tends to lose its heat and become cooler. Under extreme conditions, automobile intake manifolds will frost up as a result of rapid dissipation of heat out of the mixture and away from the pipe. Overcooling will result in condensation of the gasoline vapors back into droplets. This, of course, spoils the mixture for good combustion. To overcome this tendency, large manifolds are quite often heated by passing the hot exhaust gases around them, through encircling pipes. Or they may be located where engine or exhaust-pipe heat will keep them warm.

NOTE: Large engines are generally so equipped. Then, to avoid overheating (which could pre-ignite the mixture), they are fitted with bypass valves controlled by bimetallic thermostatic (heat-activated) elements which direct the exhaust gases over or away from the intake manifold, as required. Small engines, however, are never so elaborate.

In a one-cylinder engine the intake manifold loses much of its importance. It may even be omitted altogether. Actually, in a two-stroke-cycle engine there is no need for it as its one remaining function, completing the mixture of the gasoline and air, is more than adequately accomplished by the threshing about of the mixture in the crankcase.

THROTTLE VALVES AND GOVERNORS

The Throttle and Linkages

The *throttle valve* is the part of a carburetor which controls the amount of mixture allowed to pass to the cylinder(s). It is a flat disk (butterfly) on a rod so arranged that it can be rotated to close the passage into the intake manifold (*closed throttle*) or to open the passage wide (*full throttle*). When closed it does not fully block the passage; enough mixture will

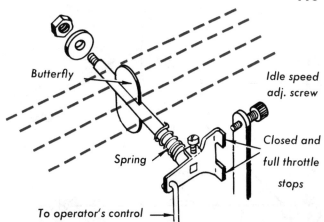

Typical throttle valve and control lever.

flow around the edges or through a small hole which may be provided in the butterfly to keep the engine idling. However, in any position other than full throttle, it does prevent the maximum amount of mixture from entering the cylinder(s)—thereby reducing engine speed.

NOTE: The balance between the force of combustion, on the one hand, and the force required to overcome friction and accomplish compression, on the other hand, is such that at any throttle setting an engine quickly levels off at a certain rpm. Adding a load will, of course, further slow it down.

Since the throttle is the primary (if not the only) control used during the operation of most engines, its linkage and operation are quite important to the satisfactory use of an engine. Many different types of throttle levers and linkages have been devised to make it convenient for an operator to control the throttle valve.

The lever may or may not be separable from its shaft. It usually is shaped to provide two stop positions (accompanying illustration): one at the full throttle setting; the other at the closed throttle setting. The closed setting is usually variable by means of an *idle-speed adjusting screw*.

Mounting of the shaft is usually such that the butterfly is necessarily centered in its opening. The shaft, lever, and butterfly are always held in the closed-throttle position, either by a coil spring (as illustrated), or by a spring somewhere in the linkage to the lever.

A throttle lever may be desiged for direct use by the operator, or for indirect use through a

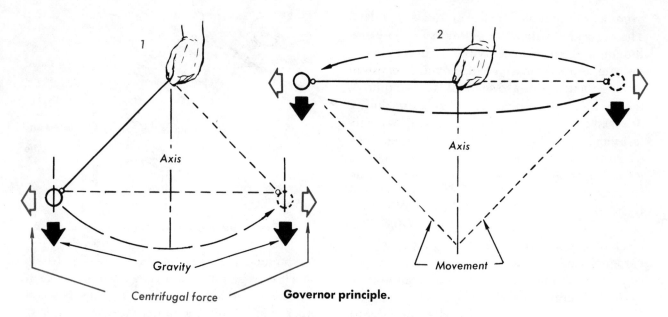

Governor principle.

remote-control lever and cable assembly. Then, again, the operator may not use the throttle at all, as explained in the following.

> NOTE: Slide-type throttles (used in motorcycle and similar carburetors, especially of foreign makes) will be covered at the end of this chapter.

Governors

A *governor is an automatic device* that is linked to a throttle lever and is actuated by the speed of the engine. If the engine is intended to provide the operator with a free selection of speeds from idling to maximum, any governor that is used will operate simply to take over the control and reduce the throttle setting *only when* a specified rpm is exceeded. On the other hand, if an engine is intended for a definite use (or uses) requiring certain fixed speeds, the governor will *always* be in control to the extent that, at each control setting made by the operator, the engine will run at a predetermined speed.

Quite often such engines as those used on rotary lawn mowers will be so governed that they run only at one speed. Then, again, two or more different speed settings may be provided, at each of which the engine is governed to run at a certain rpm. Still other engines (like outboard motors), which always operate under a known load sufficient to keep them from running away, do not require governors.

The primary purpose of a governor on a small engine is to prevent it from revving up to too high a speed when its load is suddenly reduced and the throttle is still opened to meet the load requirements. In some instances, governing is also done to protect the operator from a sudden surge of speed or from speeding and to open the throttle to maintain speed during periods of load increase.

> NOTE: There is considerable danger to engine parts from the effects of excessive speed. In small engines, extra power is built in to handle maximum anticipated loads; but, generally, the parts are not "beefed-up" to take the maximum speed which this extra power will produce under no-load conditions. It is calculated that the strain on engine parts increases by the square of the speed. That is, the stress on parts is quadrupled when the speed is doubled, etc.

There are two basic types of governors in use on small engines: The *mechanical governor* and the *pneumatic governor*. All governors operate in the same manner—they reduce engine speed by closing down the throttle setting or increase it by allowing the throttle to open.

The Mechanical Governor—As previously noted, inertia causes any object that is moving to tend to continue moving in a straight line, with a force equal to its mass times its velocity. When the object is revolved in a circle, its inertia will continuously tend to make it fly off at a tangent (straight line touching the circle circumference). The sum total of these tendencies (all around the circle) results in a constant force that pulls the object as far as possible away from the axis around which it revolves. We call this force *centrifugal force*. Like momentum, it is equal to the mass times the velocity of the object.

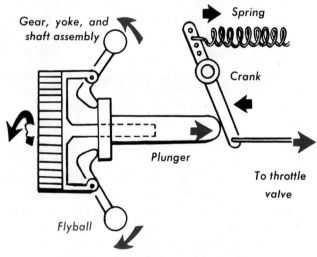

Principle of a mechanical governor.

If you tie a weight to a string and whirl it around, at first the force of gravity will hold the weight close to the ground. But as the centrifugal force increases, the weight will rise straight up until—with sufficient speed—it stands out horizontally, which is the greatest distance that the string will let the mass get away from the axis of rotation.

A mechanical governor is designed to be rotated on either the engine crankshaft or a shaft geared to the crankshaft. Its rpm is therefore equal or proportional to that of the engine. The governor itself consists of two or more weights (generally called *flyballs*) arranged so that centrifugal force will swing them outward on pivot arms. Instead of gravity, a *governor spring* (either inside the governor proper or in the linkage to it) opposes the centrifugal force and tends to keep the flyballs folded. With sufficient rpm, however, the centrifugal force of the flyballs overcomes this opposition; the flyballs move outward, and, in moving outward, they operate linkage connected with the throttle valve lever.

NOTE: One type of mechanical governor is illustrated. This employs a plunger and a crank linked to the throttle valve lever. The spring on the crank opposes unfolding of the flyballs. Moving the spring end to one or another of the holes provided will alter its leverage on the crank, to vary the amount of this opposing force and therefore the speed required to unfold the flyballs and operate the linkage.

Two other types of mechanical governors are shown in the following illustrations. These and the many more different-design mechanical governors in use all operate on the same basic principle.

In a mechanical governor the rpm at which the flyballs will unfold a certain distance is determined by: (1) the weight of the flyballs, and (2) the total force that opposes their unfolding. Consequently, governor operation can be altered by: (1) adding to or subtracting from the total flyball weight; (2) changing the tension or strength of any spring in the governor or in the linkages it operates; (3) repositioning any spring or link in a manner to alter the leverage somewhere within the governor or its linkage; and (4) tightening or loosening movable parts to vary the amount of their friction.

One or more of the first three methods of altering governor operation will be provided by the manufacturer, with exact directions for making these intended adjustments. If the flyballs are to be changed, a list of sizes and/or weights to be used to obtain specified governing speeds usually will be furnished. When spring or link positions are to be altered, quite often the choices are numbered, and a list is provided. Method 4 (changing friction) is never used, however, as results are too uncertain and not likely to last. For the same reason, governing should never be altered by bending links or altering any parts.

The Pneumatic Governor—This type is very simple. It consists of a *vane* (a flat piece like a metal flag on a pole) positioned to catch the breeze created by the flywheel, fan, or other revolving part of the engine proper. The breeze, which will blow with a force directly proportional to the rpm of the engine, lifts this vane to impart the motion required to operate the throt-

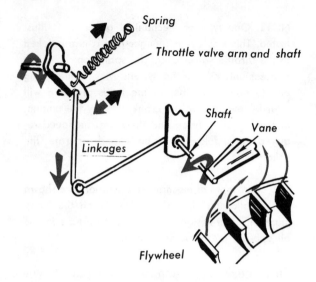

Principle of a pneumatic governor.

tle valve. A *governing spring* (somewhere in the linkage between the vane shaft and the throttle valve lever) opposes the breeze. It thus determines the amount of breeze force (engine rpm) necessary to lift the vane up and hold it in any particular governing position.

With this type of governor, the force acting on it is dependent on the breeze-creating action of the revolving part. It is not alterable, unless the part is altered or the breeze is blocked by some obstruction. Therefore, this force won't vary as long as the engine itself is in proper condition. However, the weight and contour of the vane, the tension and/or leverage of the governing spring or any other springs, and the leverages and/or tightness of the various links, are

all alterable factors that do determine the governor operation. And, as with a mechanical governor, the manufacturer will specify in each case which of these factors are to be altered (and how) to obtain desired adjustments.

Governor Linkages to Throttle—The accompanying linkage illustrations show several typical ways in which both types of governors are hooked up. All of these are simplified schematic drawings. Actual hookups will vary considerably in the number and shapes of links, number and types of springs, etc.

In the illustration "for direct operator control," only the plunger and crank of a mechanical governor like the one previously illustrated are shown. As pictured, the throttle is at idle position. Any movement of the operator's control will open the throttle and move the crank closer to the plunger. Operation of the governor will simultaneously move the plunger so that, at a predetermined rpm, the *gap* will be closed. At this point the governor will tend to oppose further opening of the throttle, but before it can actually affect the throttle setting, its centrifugal force—opposed by the pull of the *maximum speed adjust spring*—must overcome the force of the *throttle control spring*. Consequently, if the engine is under load, the throttle can be opened up to a degree determined by the positioning of the maximum speed adjust spring. However, at any time when the load is decreased and the engine revs up to the predetermined rpm fixed by the balance obtained by

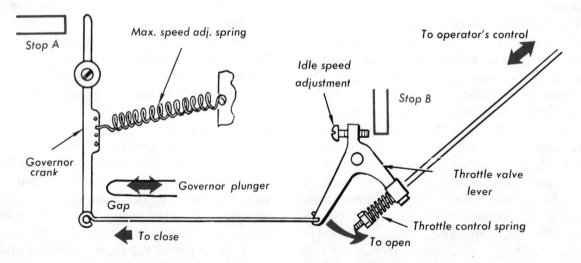

Linkage for direct operator control of all except maximum speed.

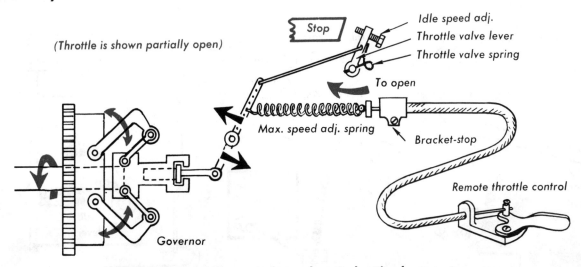

Linkage for governor control at each control setting by operator.

the two springs, the governor will take over and close the throttle down, until the rpm drops below the predetermined amount.

The next illustration shows another mechanical governor hooked up so that the governor plunger position will, at all times, directly determine the throttle opening. But the force opposing the governor is determined: (1) by the adjustable positioning of the *maximum speed adjust spring*, and (2) by the tension of this spring which, in turn, will be determined by how much it is stretched by operation of the *remote throttle control*. Consequently, at any control setting, the throttle will be opened and the governor will be held folded—unless and until the rpm exceeds the speed required to produce enough centrifugal force to overcome the spring and let the governor take control of the throttle.

The illustration, "linkage for governor control at just one running speed," shows a slightly different application of the arrangement just explained. Here, the governor (with its internal spring) holds the throttle fully open at all times, except when the rpm becomes sufficient to collapse this internal spring and allow the *throttle valve spring* to close down the throttle.

In the remaining three linkage illustrations we show other typical hookups for use with mechanical or pneumatic governors. In illustration "A" the governor will operate as in the last previous example, except that the operator is provided with a convenient *maximum speed adjust nut* with which to vary the *one* engine speed to meet his requirements. In illustration "B," the only control of the "one" running speed lies in the adjustments provided—unless the spring holder (X) is movable. (Quite often, X is a knob

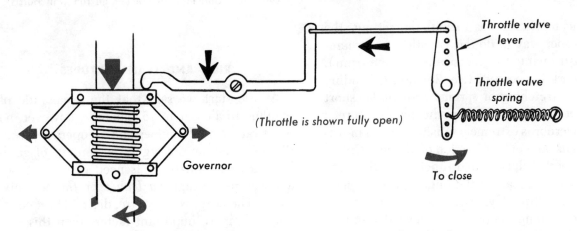

Linkage for governor control at just one running speed.

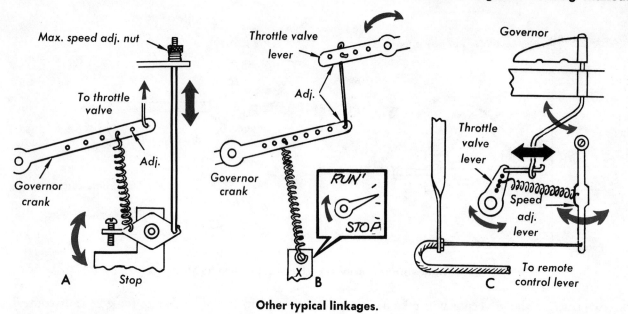

Other typical linkages.

to change spring tension, with two or more marked control settings, such as "Start," "Idle," "Run," etc.) View "C" shows use of a remote control with a pneumatic governor setup.

Governor Adjustments—As already said, whenever a manufacturer provides a governor he will also provide any adjustments that he deems necessary. He will also state how and under what conditions these adjustments are to be made. Usually, he will list desired engine rpm's for certain types of loads. He may then furnish a list of sized flyballs, numbered springs, or link positions (or the like) telling specifically what arrangement will result in what rpm. Or he may simply tell where to make the adjustment and then leave it up to the mechanic to "find" the proper setting to produce the desired rpm.

If the mechanic has to find the setting, the only proper way to do this is with a *tachometer* (an instrument for measuring engine rpm). Guesswork is poor practice. So, also, is bending of links, stretching of springs and similar short cuts not recommended by the manufacturers. Any governor is extremely sensitive. Total movement is usually only a fraction of an inch. Unless it is properly adjusted and functioning throughout, it can: (1) allow the engine to overspeed and wear out quickly, or (2) prevent the engine from developing full power when needed so that it will actually fail to pull its load. Indeed, we

cannot overstress the importance of correct governor operation!

Maintenance—Again, the manufacturer will state where to look for wear or damage that will affect governor operation. Repair is usually impracticable. Worn or damaged parts are best replaced. Any undesirable sloppiness of operation, due to wear or damage, may result in jerky engine performance due to hunting (erratic throttle movement as the governor tries to find its proper setting).

NOTE: We should mention here that hunting is not always the fault of the linkage. What is known as a surging governor condition also can be caused by improper fuel mixture, which results in some lean and some rich portions of mixture reaching the cylinder(s) and making the engine run with bursts of speed.

RICH AND LEAN MIXTURES

At the start we said that the usual ratio of gasoline to air is 1 to 15. There is, however, no exact ratio for all engines and all operating conditions. Sometimes a higher percentage of gasoline (a *richer mixture*) is needed; other times a lower percentage (a *leaner* or *thinner* mixture). The latter is, of course, desirable for economy, but more important factors than this dictate how rich or lean a mixture should be.

Starting

When cold, gasoline vaporizes more slowly than when warm. If cold enough when sprayed from the carburetor nozzle into the airstream, it will form droplets rather than a mist. With droplets, fewer carbon atoms are exposed to contact with the oxygen atoms and the mixing is incomplete. This condition could be so bad that the mixture would not even ignite in the cylinder(s).

To offset this condition a richer mixture is used. By greatly increasing the ratio of gasoline, the chances for proper oxygen-carbon contacts are multiplied. The opportunity for combustion to take place is much improved. The colder the gasoline and air are, the richer the mixture must be to ensure combustion. Consequently, for the starting of a cold engine (particularly on a cold day) a much richer mixture is usually required.

Running

When an engine is operating without a load at a given throttle setting, a leaner mixture will not only suffice for economy, it will actually run the engine better. The relatively higher engine speed for the throttle setting does two things. First, there is less time during compression to distribute heat evenly through the mixture. Second, there is less time following ignition for the mixture to burn. With too rich a mixture some of the gasoline in the mixture will go out the exhaust unburned.

Worse, still, since the unburned gasoline molecules are much heavier than the air molecules in the mixture, they will not be pushed out the exhaust as rapidly. They will lag behind and accumulate from cycle to cycle, until there is an appreciable amount of liquid gasoline in the cylinder. The engine will become increasingly sluggish. If continued, the ratio of gasoline to air inside the cylinder will become so great that there simply will not be enough oxygen molecules to produce a sufficient combustion force to keep the engine running. Should this happen, the engine will die from *flooding*.

On the other hand, when a load is placed on an engine without changing the throttle setting, the engine will at once slow down. Under load conditions, when more power is needed to maintain operation, the slow-down that occurs provides more time in which to heat the mixture during compression, and more time for complete combustion of a larger ratio of gasoline molecules. Thus, just when a richer mixture is required for additional combustion power, more time is provided for efficient burning of additional gasoline molecules.

All the preceding discussion is relative to speed. The faster the engine runs, the leaner the no-load mix should be, and, in turn, the less the under-load mix should be increased in richness. In addition to the fact that a slower speed makes it possible to burn a richer mixture, at slower speed the throttle is closed down. This means that less total mixture is admitted to the cylinder(s) and the amount of compression is reduced. In addition, air velocity and turbulence are reduced. All together, the factors add up to poorer mixing and poorer combustion.

Consequently, as an engine is throttled down from the running speed for which it was designed, less total gasoline and air are required; but an increase of gasoline ratio is called for. At a very slow speed, the mix should definitely be richer.

Too Rich or Too Lean

Too rich a mixture at any given time is, of course, wasteful of gasoline (unless, as at starting, the richness and waste are purposely allowed to accomplish a result). In addition, the raw gasoline introduced into the cylinder is harmful to the engine. If the engine is a four-stroke-cycle one, the gasoline will wash the lubricating oil from the cylinder walls, will drain past the rings into the crankcase and dilute the oil there. If it is a two-stroke-cycle engine (with oil mixed in the gasoline), the excess fluid will quickly foul and short out the spark plug and cause excessive carbon deposits. In either case, considerable damage to the engine can result in addition to the poor-running results already mentioned.

Too lean a mixture at any given time is almost as bad. It will cause the engine to heat up rapidly—possibly to overheat and burn up valves and other parts. Due to this superheating, precombustion, resulting in backfire through the fuel system is likely. There is a consequent loss of power and jerky engine operation.

A CARBURETOR CONTROLS THE
GASOLINE-AIR RATIO

Carburetor Requirements of Engines

In addition to controlling the amount of mixture through operation of the throttle valve every carburetor must, to the extent required, control the gasoline-air ratio and provide a richer or leaner mixture as the engine speed and load conditions vary. Different types and sizes of engines, and the different kinds of loads for which engines are designed, all add up to the fact that there are many kinds of carburetors offering a varied selection from the standpoint of efficiency. Automotive engines require extreme carburetion efficiency with provisions for instantly and accurately altering the ratio to meet immediate demands. Gasoline consumption and peak engine performance are very important. At the other end of the scale, a single-cylinder small engine designed to operate only occasionally (as on a lawn mower) and only under known load conditions, requires only a very simple carburetor. Gasoline consumption is inconsequential, and performance requirements (from the standpoint of acceleration and meeting emergency load conditions) are negligible.

> NOTE: Actually, a single automotive carburetor may cost nearly as much as an entire small engine of the type described. Initial cost of a small engine carburetor is therefore a big consideration in carburetor selection.

All carburetors have throttle valves, as previously discussed, with or without governor control. They all also have a *choke system.* Every type requires some means of making the gasoline available in the carburetor for mixing with the air; and most all carburetors have a separate *idling system* and *high-speed* (or *running-speed*) system, for correct metering of the gasoline at these two different speed ranges.

> NOTE: A suction-type carburetor (to be discussed later) is the one exception to the preceding discussion. It does not actually have separate idling and high-speed systems.

With the preceding comment, the general similarity of carburetors ends. One may have a very simple choke system, while another may have quite a complicated automatic type. The same can be said for the idling and high-speed systems. Some carburetors make provisions for special-engine power requirements, for economizing fuel, and/or for acceleration needs. Some have a very minimum of provisions even for ordinary idling and high-speed operations.

On the whole, very small, single-purpose engines solve the problem of economizing on the carburetor by using a simple carburetor designed for only two-speed operations, idling and running. To offset foreseeable needs for power to meet peak loads, the engine is run on a slightly richer-than-normal mixture at all times. This eliminates the need for complicated adjusting systems, etc. As the types of use for which an engine is designed become more demanding, however, even small-engine carburetors become more complex.

The Types of Carburetors

There are a good many different ways of classifying carburetors. Many of the terms formerly used (often for trademarking purposes) have lost their meaning and become more or less obsolete. For instance, we used to hear about "up-draft" and "down-draft" types (according to whether the manifold was above or below the carburetor). But most small-engine carburetors today, according to this classification, should be called "side-draft." Then, again, carburetors have been classified as "plain tube," "auxiliary air valve," "constant vacuum," and "air-fuel proportioning" types—depending on the manner of mixing air and gasoline. But these terms, too, have no particular meaning for us.

More realistically, small-engine carburetors fall into three distinct groups: the *float* (or *bowl,* or *wet*) *type;* the *diaphragm* (or *dry*) *type;* and the *suction type.* A float type has a bowl to hold gasoline with a float and needle valve to shut off the supply when the bowl is full. The diaphragm type has what amounts to a fuel pump built right into the carburetor with provision for storage of a small amount of gasoline above the diaphragm. And the suction type does not store gasoline at all. It simply passes gasoline right through it, into the airstream.

Plan for Telling the Story of Carburetion

Each carburetor type has its merits, which we will discuss when we discuss the types. In the meantime, we must first present another principle, followed by a general discussion of how all the principles apply. Next, we will take up the choke system, which is common to all carburetors. Last we will discuss the three separate types of carburetors and, in each case, the various systems used in them.

ONE MORE PRINCIPLE—THE AIRFOIL

Everyone who reads about airplanes is familiar with the term *airfoil*. An airplane wing is an airfoil; so are the rudders, flaps, elevators—even the fuselage to some extent. So, also, are the throttle valve and nozzle of a carburetor. Anything that projects into an airstream is an airfoil in the sense that the flowing air must flow around it.

An airfoil exemplifies a principle, actually the same principle of an air venturi. Take the airplane wing (illustrated) as an example. Due to the forward motion, it is rushing head-on into the air, so that wind blows straight from front to back. But due to its slight tilt, the straight blowing wind has to curl up over its leading edge and down under the wing around the trailing edge. More wind therefore has to pass over the top leading edge and down under the trail-

ing edge. The wind velocity at these points is increased, accompanied by a slight compression of the air to help all of it get by. Following these points the velocity returns to normal and the air re-expands to normal. Under the wing the slightly compressed air has a higher pressure; while over the wing the re-expanding air has a lower pressure. The pressure push as well as the glancing force of the rushing wind, both help to keep the airplane up.

The throttle valve is located between the venturi throat and the intake manifold end of the carburetor, in the widened passage portion. At full throttle it presents only the thin edge of its butterfly to the airstream; there is very slight deflection of the air around it. Equal amounts of air pass above and below it (or to each side, as the case may be). At any other setting, however, it becomes cocked in the airstream and sets up an airfoil action just like that of the airplane wing. As it approaches the closed throttle position, it becomes increasingly cocked, and the airfoil action increases. Hence, at all settings approaching closed throttle, there is a low-pressure area (accompanying illustration) behind the butterfly, and a high-pressure area along the face of the butterfly and at each edge.

If the opening of a passageway for gasoline is located in the venturi wall just behind where one butterfly edge will be at closed or slightly opened throttle, the greatly increased air velocity around the butterfly edge will tend to friction-drag gasoline out of the opening. This will result in a low pressure in this area and will tend to cause gasoline to flow into it from any higher-pressure area. In short, by adding a gasoline opening here, we have re-created the same activation that we have with a nozzle in the

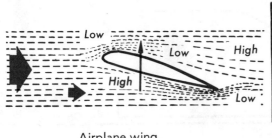

Airplane wing.

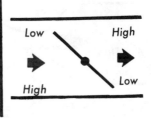

Throttle valve.

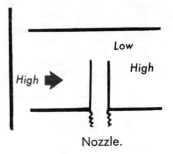

Nozzle.

Airfoils.

venturi. In this case, however, the flow of gasoline is activated by (and will increase in proportion to) the closing of the throttle valve. This, then, is the basis for an idling-speed system to satisfy gasoline-air ratio requirements at this engine speed.

Until now, we have talked simply in terms of a nozzle stuck out into the venturi as a means of providing the gasoline-air ratio requirements at running speeds. Actually, however, the nozzle projecting into the airstream (especially at this restricted, high-air-velocity area) also serves as an airfoil. The rushing air in front of it is additionally compressed a slight amount; behind it there is an even lower pressure area where its shape leaves a small pocket for the air to re-expand into. Actually, due to the air velocity at higher speeds, the pocket stretches backward

and becomes closer and closer (with increasing speed) to a perfect vacuum.

As speed increases, the increasing venturi velocity increases the friction drag on the gasoline in the nozzle, and the increasing pressure drop behind the nozzle tends to increase the gasoline flow. Here, then, we have the basis for a high- (or running-) speed system to satisfy gasoline-air ratio requirements at all speeds above idling speed.

HOW ENGINE SPEED AFFECTS CARBURETION

Before leaving the previous subject, let's review both of the two stated circumstances.

The airfoil action of the throttle butterfly increases as the engine speed approaches idling; it decreases and disappears as the throttle is

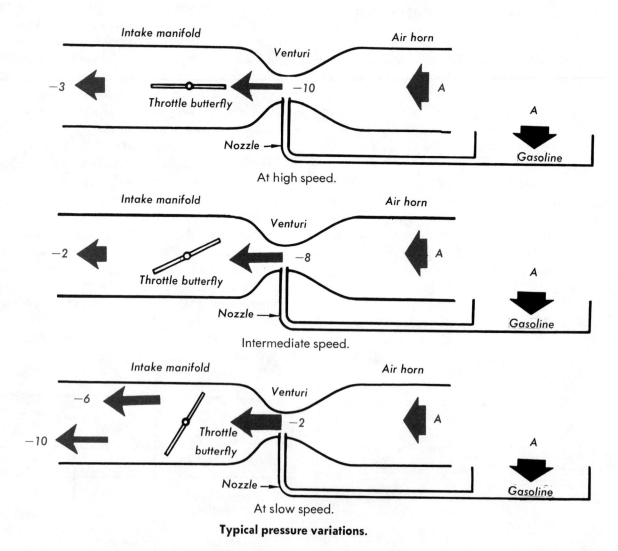

At high speed.

Intermediate speed.

At slow speed.

Typical pressure variations.

opened. The airfoil action of the nozzle increases as the engine speed increases; it decreases as the engine is throttled down. In fact due to the very low velocity of the air (even in the venturi) at slower speeds, the nozzle airfoil action also decreases to a practically negligible amount. One offsets the other by diminishing as the other increases, and vice versa. At some middle point, both during acceleration and during deceleration, the one action gives way to the other action, which then proceeds to supply the gasoline. Between the two actions, gasoline is always being sprayed into the airstream.

Pressures within the airstream vary not only from point to point throughout the carburetor air passage, but also in accordance with the engine speed. The accompanying illustration shows how pressures can vary from atmospheric pressure (A) in the air horn and on the surface of the stored gasoline to less than atmospheric pressure (−2, −10, etc.) in the venturi and intake manifold.

NOTE: The pressure differentials indicated are for an engine that demands a fairly large and constant volume of air (i.e., one with multiple cylinders). In a large engine, venturi air velocity can reach 200 or more miles per hour. A smaller engine requiring less total volume of air will not create such high air velocities. The pressure differentials will be accordingly less.

A small single-cylinder engine may create a top venturi velocity of no more than 20-30 miles per hour. Moreover, such an engine does not create a constant call for air. There is a lapse between intake strokes while the piston completes the other stroke(s). This factor also affects the pressure differentials.

At a slow engine speed the manifold pressure will be at its lowest, since the throttle valve restriction will prevent the cylinder(s) from ever being totally refilled. There will be very little pressure drop behind the venturi, but as already noted the pressure behind the throttle butterfly will be at its lowest.

At a high speed, the manifold pressure will be at its highest, since the partial vacuum in the cylinder(s) will be almost totally refilled at each intake stroke. There will be no pressure drop behind the open throttle butterfly, but the high velocity in the venturi will create a peak pressure drop behind the nozzle and in the passage behind the venturi.

At any speed, when a load is added the engine slows down but the throttle setting remains unchanged. The same volume of air is still available as there is no new restriction to air flow, but the engine's call for air is reduced. This will cause all of the pressures to rise—in the manifold, behind the nozzle and the venturi, and behind the throttle butterfly (if it is positioned to create an airfoil action). Conversely, at any speed, if a restriction is placed in the carburetor air horn to reduce the air available without reducing the engine's call for air, all the pressures will drop (which is what happens when the choke is closed).

At all speeds, the friction drag of the flowing air will be directly proportional to the velocity of the air at any given point in the system. Whereas the pressure differential is the prime factor involved in causing gasoline to flow out into the airstream, the friction drag is the prime factor that causes the gasoline droplets to be broken up and vaporized into a mist.

If the pressure were simply to be dropped without any air movement (say, by pumping air out to leave a partial vacuum), the gasoline would bubble out into the vacuum area in large droplets or a steady stream. It is the friction drag collisions of the molecules that breaks up the gasoline. The higher the air velocity, the finer the gasoline will be broken up for more perfect vaporization.

CHOKE SYSTEMS

The Classifications and Applications

The *choke* or *choke valve* is, like the throttle valve, a butterfly type. It is located in the carburetor air horn. There may be a simple external lever to operate it, or a remote-control lever or knob, or automatic controls. The valve is normally wide open to admit all the air possible into the carburetor, but is closed to restrict air flow when choking is required.

There are three purposes that a choke may be designed to serve. (1) To *temporarily* increase

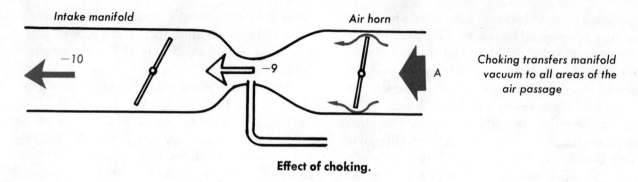

Intake manifold · Air horn · −10 · −9 · A · Choking transfers manifold vacuum to all areas of the air passage

Effect of choking.

the gasoline flow to produce the super-rich mixture that will provide easier starting. (2) To *permanently* adjust the mixture richness for all of a single operating period (after the engine is running) for an engine which will then run at one governed speed using this permanently adjusted mixture. (3) To *periodically* readjust the mixture richness in the proper manner when required by the engine's varying load conditions.

As previously said, any restriction in the air horn as provided by choking causes all the pressures in the carburetor to drop. This causes more gasoline to flow into the airstream, which, combined with the obvious reduction of air volume, enriches the mixture. And the degree of increase of richness is directly related to the amount by which the choke is closed. At full closing of the choke, all air would be shut off and only gasoline would be sucked into the cylinder(s). However, choke butterflys are designed to admit some air, even when fully closed —enough air to support starting combustion.

Choking to Start

Choking to start is nearly always essential in cold weather. On hot days, however, when the gasoline and air are at temperatures to more readily produce good vaporization, choking is less essential. It may not even be needed. This is especially true if the other carburetor systems are adjusted to produce a slightly richer-than-average mixture, which, as previously said, is true for many small engines. Overchoking such an engine will quickly result in flooding.

If a choke butterfly is designed to cut off practically all the air when fully closed, then right after the first "pop" of starting, the engine will instantly need more air to enable it to con-

tinue running. Under these circumstances, the choke must *at once* be opened enough to permit passage of the minimum air requirements. Furthermore, as the engine warms up, its heat, transmitted to the gasoline and air, will produce better vaporization. The need for continued choking therefore tapers off accordingly. Revving up an engine also diminishes the need for choking, as higher speeds require leaner mixtures. Consequently, as an engine warms and approaches its running speed, the degree of choking *must be further tapered off*. It must be ended if the carburetor is designed to operate at running speed with the choke fully open.

NOTE: For the good of an engine, if it is designed so that the operator can control the throttle, the engine should be run at a slow speed after starting until the engine warmth alone ends the need for choking. Revving it up to warm it and to quickly diminish the need for choking simply aggravates all the damage that can be wrought by unburned gasoline in the cylinder(s). If the engine has a lubrication system which won't function properly

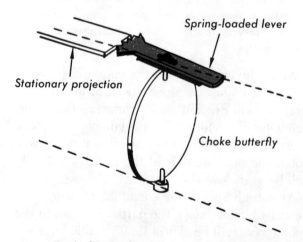

Spring-loaded lever · Stationary projection · Choke butterfly

Typical lever for operator adjustment.

until the engine and the oil are warm, the damage may be increased immeasurably.

Choking to Adjust Mixture

As was explained in connection with pressure differentials, at any engine speed an increase of load causes the air pressures in the carburetor to rise, thus reducing the flow of gasoline. However, at such a time, increased richness, rather than increased leanness, is required. In some carburetors this situation is met by devices or auxiliary systems related to the high-speed system. In very small engine carburetors, the situation is met, as already noted, by continually operating the engine on a slightly overrich mixture, i.e., overly rich enough to meet all anticipated load conditions. The desired overrichness may be obtained by adjustment of the high-speed system, or it may be obtained by choke adjustment, in which case it can be more readily reset by the operator to satisfy the conditions of a current load.

NOTE: Remember, it is much easier to calibrate a carburetor for an all-purpose one-ratio mixture when the engine has been built to serve just one specific purpose such as running a pusher-type lawn mower or a boat. No great load variations will be encountered. But carburetor calibration for varying conditions becomes increasingly complex as the types of uses and/or operating conditions are multiplied.

When a choke is intended for the purpose just discussed, its operating lever will be designed so that it can be locked into any one of several different settings. The usual lock consists of a stationary projection that engages with the notches in the end of the spring-loaded lever.

This type of choke is used for starting just like any other; but after the engine has warmed up at running speed, the choke, instead of being simply opened, is adjusted to whichever position seems to produce the required engine power. Different positions may have to be selected on different occasions, depending on the current load, weather, etc. In all cases the setting is according to the best judgment of the operator, who must listen to the engine to judge when it sounds best. Caution must be exercised to avoid overenriching and possible flooding. Usually, however, the limited number of selections provided will tend to make this easy.

Three Types of Chokes

Chokes may be manual, semiautomatic, or fully automatic. The first two types are most used on small engines.

The manual-type choke requires operator manipulation to control the extent of choking *from start to finish* of any choking operation. With the semiautomatic type the operator still has to manipulate the choke. However, an automatic valve that is part of the choke mechanism will function to partly open the air passage the instant that the engine starts. This automatic function guarantees enough air to keep the engine going, even though the operator may be slow in opening the choke. It only provides enough air, however, to allow the operator time to open the choke. He must still manually open the choke during warm-up, as already explained.

Manual Type—The manual type has a plain butterfly, shaft, and lever with which the carburetor air horn may be blocked or opened wide. There may be a spring (usually a coil encircling the shaft) to reopen the valve whenever it is released; or there may be none. When fully closed, the butterfly may close off practically all of the air flow or may be designed to admit passage of the minimum amount required to support starting combustion. When fully open, the valve will pass all the air needed at the engine's top speed.

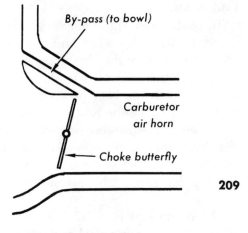

By-pass (to bowl)

Carburetor air horn

Choke butterfly

209

Typical manual choke.

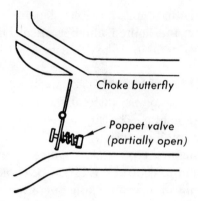

Typical semiautomatic choke.

Semiautomatic Type—A semiautomatic-type choke has a hole through the butterfly that is controlled by a small valve. This is a simple, spring-loaded valve which is set to open when the low pressure inside the carburetor drops to a certain level. Consequently, if the choke is fully closed after the engine starts (to increase suction and further reduce the venturi and manifold pressures) this valve will open. Its opening ensures rapid readjustment of the mixture ratio to meet the needs of the now-running engine.

NOTE: Our illustration shows a *poppet valve.* Spring-loaded, hinged *(flap-type) valves* and plain *reed valves* also may be used.

Either of the previously described choke types may have the lever-setting feature discussed under the preceding heading, "Choking to Adjust the Mixture."

Automatic Type—This type choke not only provides the choking needed to start, it also uses the choke to enrich the mixture to meet varying load needs.

To control starting, it employs a *thermostatic spring* to determine when and how much choking is required by the engine temperature. The spring (or thermostat) is made from a strip of brass fused to the back of a strip of steel and then coiled. Since the two metals have different coefficients of expansion, one will expand faster than the other as the spring is heated. Consequently, the spring will warp and uncoil or coil in a regular, dependable manner as it is heated or cooled.

The thermostat is usually enclosed in an insulated housing to shield it from the outside temperatures. It is then made to respond only to engine temperatures by conducting hot air into the housing from some part of the engine where the air will be heated in fixed relation to the engine temperature—usually from around the exhaust manifold. To keep this air flowing, the housing is also connected by hose (or tube) with some point in the intake manifold. Consequently, the vacuum there will "suck" air through the housing.

Careful calibration of the thermostat is required so that its opening and closing will correspond with the desired choke positions. Its shaft (which it rotates) is linked to the choke shaft in a manner to hold the choke closed when the engine is cold and to allow the choke to open as the engine warms up. It does *not* directly open the choke. If it were firmly attached to simply close or open the choke, it would be useful only for starting. Its connection with the choke shaft is through a *restraining lever* (or *link*) which merely opposes opening of the choke until the engine has been fully warmed.

Two other factors control choke opening, after the thermostat moves its restraining lever away. These are the velocity of the airstream entering the carburetor, and the engine rpm.

To use the airstream, the choke shaft is placed off-center in the butterfly, so that the portion of the butterfly that moves inward as the choke is opened is larger than the other portion. Consequently, whenever the engine picks up speed due to throttle valve opening and the suction increases, the added velocity with which the air is caused to flow will build up more pressure against the larger side of the butterfly. The choke will be forced open wider. In fact, the balance between the butterfly weight and the force of the airstream is such that the choke will open in direct relation to the engine's call for air.

However, this airstream alone is not allowed to control the choke. If it did, the choke would start to close every time the engine slowed, and the mixture might be overly enriched by choking. To prevent this, the airstream is made to share its control of the choke with a device that responds to the engine's rpm.

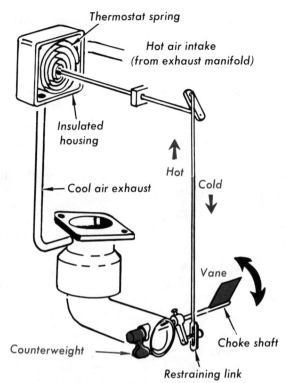

Vane- and counterweight-type automatic choke.

On the other hand, when the engine is throttled down to idling speed, the reduction of breeze force on the vane will not be as great as the reduction of the airstream force. In this case, the vane will prevent too much closing of the butterfly and allow free passage of all the air needed at the reduced speed.

Another more complicated and exact arrangement, seldom used on any but automotive engines, employs a small *piston* and cylinder set up to share control of the choke with the airstream.

At top, the piston is open to atmospheric pressure. Below the piston, the cylinder is connected by tube with the intake manifold at the point where the airfoil action of the throttle valve creates a lower-pressure area whenever this valve is not fully open. Consequently, as the engine is throttled down, the throttle position will determine the amount of suction tending to pull the piston down. The load condition of the engine will also help to determine this suction, since the manifold pressure varies as the load varies.

The piston is directly connected to operate the choke shaft, to open the choke as the piston is sucked down. Thus, both the piston and the airstream forces tend to oppose the force (a spring or weight) striving to close the choke. The balance is such that the choke will be fully open at the engine's top speed.

Now, if a load is applied with the throttle fully open, only the manifold pressure, which rises under these conditions, will tend to suck

In one arrangement (usually used on small engines) the "device" is a *vane* (like the pneumatic governor vane) placed where engine breeze will operate it in direct relation to engine rpm. This vane is on the choke shaft. At the other shaft end there is a *counterweight* which can be adjusted to balance both the airstream and the vane operations, so that the choke butterfly will be opened or closed as desired. This balance is such that the butterfly will be fully open when the engine is operating at its normal, no-load running speed, but will respond (by starting to close) to any slight lessening of the airstream and/or vane forces.

Now, should the engine speed be reduced by a heavy load, the breeze force on the vane will be lessened to allow the choke to partially close. However, since the throttle position hasn't been changed, the air volume requirement per stroke remains unchanged, and the airstream force will *not* be so greatly reduced by the engine slowdown. A new balance is therefore established which will allow just enough choking to provide the extra gasoline needed. And the instant the engine starts to regain speed, both the airstream and the vane will operate to reopen the choke.

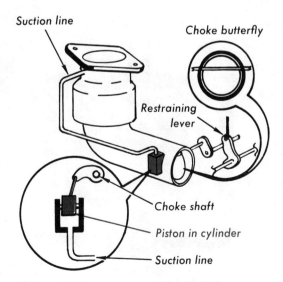

Typical piston-type automatic choke.

the piston down. There will be at this point less force trying to hold the choke open. Simultaneously, the airstream force is reduced by the reduced call for air as the load slows the engine down. With the lessening of these two forces the choke is allowed to close just enough to enrich the mixture as required. Later speeding up of the engine will increase both forces to reopen the choke.

On the other hand, if the slowdown is effected by closing of the throttle, although the airstream force is reduced considerably by the reduced call for air, the suction on the piston is increased by the lowered pressure behind the throttle valve. Consequently, the piston prevents the choke from closing more than the slight amount needed to help adjust the mixture for slower-speed operation. Applying a load at this point will again slightly raise the manifold pressure in order to let the choke close a little farther and to further enrich the mixture as required. Exceptionally good mixture control is achieved for all speed and load conditions.

FLOAT-TYPE CARBURETORS

FLOAT SYSTEMS

Two types of float systems are used: a non-adjustable type and an adjustable type.

A typical nonadjustable type is illustrated. The component parts are a *bowl* with removable cover, a hollow *float*, and a *float needle* centered in the float and guided so that the rise or fall of the float will cause the needle point to close or open the fuel passage into the carburetor. The *float tickler* shown may or may not be included. It provides a means of holding the float down to flood the bowl with fuel, if necessary for starting.

The second illustration shows a typical adjustable-type system. This is comprised of a *bowl*, a needle-type *float valve* to close or open the fuel passage into the bowl, and a hollow *float* with a pivoting *lever* arranged to operate the float valve as the float rises or falls. The fuel filter shown may or may not be included. Ball- and doughnut-shaped floats, the latter having other carburetor parts projecting through the center opening, are also used. In some instances there are two bowls and two floats. Then, again, the lever system and float-valve type and arrangement may be entirely different. However, all the adjustable-float type systems operate on the basic principle illustrated.

With both types of systems, the bowl is always open to atmospheric pressure. Sometimes this is through a small vent to the outside; more often it is through a connecting passage into the carburetor air horn, which ensures that only cleansed air will enter. Usually, a removable *drain plug* at the bowl bottom is provided for flushing. Also, the bowl either has a cover (first illustration) or is itself separable from the top part of the carburetor.

In the adjustable type, proper float-level adjustment is critically important. This is usually made by bending the lever. Remember, the depth of the fuel in the bowl, along with the atmospheric pressure on it, helps to govern the total pressure that will force fuel out of the bowl into the lower-pressure areas of the air stream. The fuel level must be properly maintained to keep the pressure correct for the calibrated operation of the other carburetor systems. Then,

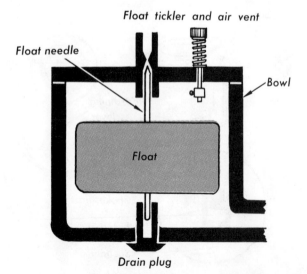

Float tickler and air vent

Float needle

Bowl

Float

Drain plug

Typical nonadjustable float system.

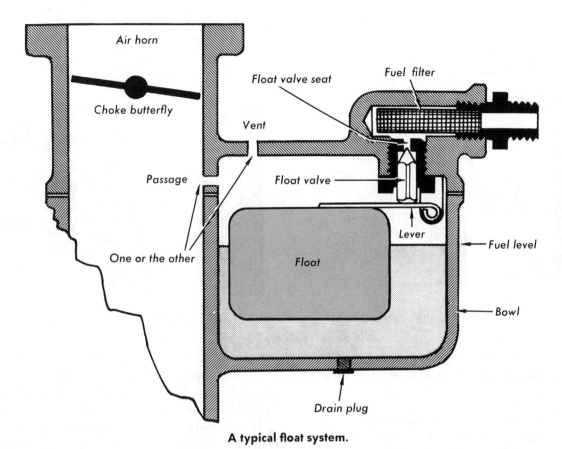

A typical float system.

too, erratic float and valve operation could flood or dry up the carburetor bowl. In all cases, manufacturers give very specific instructions pertaining to float adjustment.

All float-type carburetors will become inoperative if inverted. The fuel will run out into the airstream and the float valve will become jammed. Even if tilted too much, erratic operation will follow. The better types will permit tilting up to 45°, but poorer types should not be tilted more than 25° to 30°.

IDLING SYSTEMS

Operation and Principal Parts

As already noted, the *idling* (or *idle*) *system* is that part of a carburetor which uses the low-pressure area behind the closed, or partially closed, throttle butterfly to spray gasoline into the airstream during slow-speed operations. At such times the low velocity of air in the venturi makes the high-speed system totally or partially inoperative, depending on actual throttle set-

ting. And the idling system, in turn, becomes increasingly inoperative as the throttle is opened wider.

An idling system is comprised of a passage or series of passages in the carburetor, from the carburetor bowl to an opening into the air passage at a point just behind the throttle butterfly edge position when this valve is closed. The butterfly edge that moves outward toward the air horn as the valve opens is always the one. An orifice or *idle jet* that is very precisely drilled (calibrated) to pass just so much gasoline at such and such a pressure differential is placed somewhere in the passage. This meters the correct amont of gasoline through it. There is also an adjustable needle valve (*idle-adjusting screw*) designed for manual adjustment of the flow through the passage. This valve actually is a spring-loaded machine screw having a precisely ground needle point at the inner end, which screws down in a mating valve seat. Turning the screw clockwise closes the valve; counterclockwise opens it. The coil spring under the

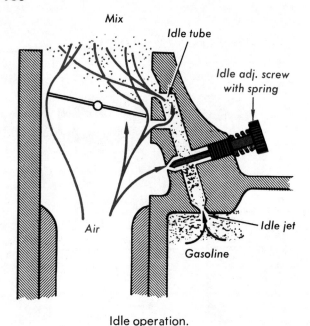

Idle operation.

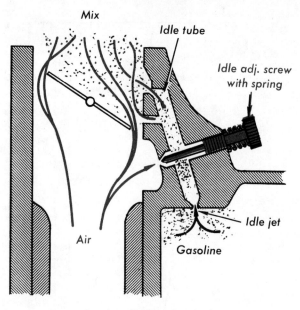

Intermediate operation.

Adjustable-air type system.

screw head holds the valve at any desired setting.

> NOTE: Do not confuse this idle adjusting screw with the idle-speed adjusting screw on the throttle valve. The first adjusts the flow; the latter adjusts the limit to which the throttle can be closed.

With different carburetors, the passageway arrangements and location of the idle adjusting screw will vary considerably. It is impracticable for us to cover all the many designs here. However, whatever the design, all these idling systems belong to one or the other of two general types. So we shall illustrate and describe the principles of these two types.

Adjustable-Air Type System

The two preceding illustrations, "Idle Operation" and "Intermediate Operation," show how one type of system works. In the first view, only the top opening (*idle orifice* or *port*) of the passageway (*idle tube*) is exposed to the low-pressure area behind the throttle butterfly. The other opening (*intermediate* or *secondary orifice* or *port*) is exposed to the high-pressure area in front of the butterfly. Some air will enter the idle tube through this intermediate orifice; more air will enter around the point of the *idle-adjusting screw*, depending on the setting of this screw.

Gasoline under atmospheric pressure in the bowl will flow up the idle tube, through the *idle jet*, which meters the flow. Above the screw it will mix with the air in the tube and start to vaporize. This mix will flow on out the idle orifice into the airstream where the high-velocity friction drag will help to carry it off and further the vaporization. Between the adjusted amount of air entering at the screw, the fixed amount entering through the intermediate orifice, and the fixed amount flowing in the airstream there will be exactly the proper total amount to mix

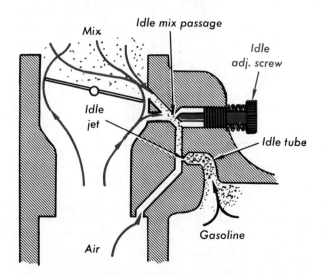

Adjustable-mixture type system.

with the metered flow of gasoline—to produce the mixture ratio desired for this throttle setting.

If the throttle is opened a bit wider, more gasoline is needed to combine with the increased air flow. At this setting (view 2), the intermediate orifice is now in the low-pressure area. Instead of air entering here, the orifice now provides an additional exit for the mixture in the idle tube, and an increase of suction to draw the mixture out. With the reduced amount of air and increased amount of gasoline now flowing out of the two orifices, the ratio requirement for this throttle setting is satisfied.

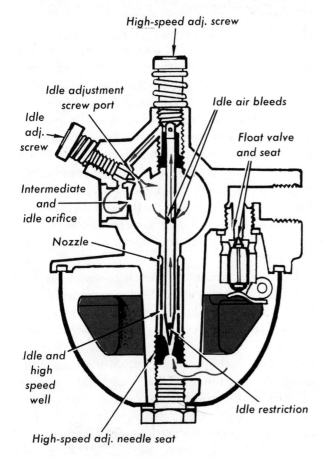

Fuel for idle system flows up through the hollow stem of high-speed adj. screw which shuts off fuel to both idle and high-speed systems.

Float and low-speed system.

Note that with the type of system described, the idle-adjusting screw adjusts only the amount of air admitted into the idle tube. This is the distinguishing characteristic of this system.

Adjustable-Mixture Type System

Our accompanying illustration shows how this type of system may be arranged. Note that it is like the previous one in all respects, with the exception that the idle adjusting screw is so located that it will adjust the mixture of air and gasoline flowing through the idle tube, rather than just the air. In the first arrangement, closing of the screw will increase suction on the gasoline while reducing the air; in this second arrangement, closing of the screw simultaneously reduces the suction and the air. The second arrangement therefore provides a smaller range of adjustment, but the adjustment is less sensitive and easier to make.

> NOTE: With an adjustable-air-type system, closing of the idle adjusting screw will enrich the mixture. With the adjustable-mixture-type system, closing of the idle adjusting screw will lean the mixture.

Other Variations

Some systems omit the intermediate orifice. In such case, the high-speed system has to take over as soon as the usefulness of the single idling-speed orifice diminishes to the extent of being unable to supply an adequate amount of gasoline.

With some arrangements the fuel for idling flows directly from the bowl through a separate passage; with others it flows first through a passage shared by the high-speed system. In the latter case the idle fuel flow can be cut off by closing the high-speed adjusting screw provided (see illustration, "Float and Low-Speed System").

In all previous illustrations the air that is introduced into the idle tube has been shown as being supplied from the carburetor airstream. This method ensures use of only cleansed air that has passed through the air filter, but it also has to take into account the varying air pressures within the airstream. A more constant source of air is the atmosphere. Some carburetors introduce *bleeder air* into the idle tube directly from the atmosphere. This is particularly true of outboard-motor carburetors, since cleansing is less important in the relatively dust-free air in which they operate.

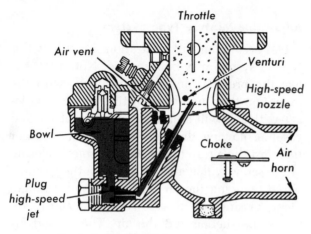

Typical plug-jet high-speed systems.

HIGH-SPEED SYSTEMS

As previously noted, the *high-speed* (or *power* or *main*) *system* is that part of a carburetor which uses the low-pressure area created by the venturi to spray gasoline into the airstream during all except the idling and low-speed operations. At such times the high velocity of air in the venturi makes this possible. At the same time the diminished (or ended) suction obtained from the throttle area has almost or entirely ended the functioning of the idling speed system.

Basically, the high-speed system is simply another passage or passages in the carburetor, leading from the bowl to the high-speed nozzle in the venturi. There is always a *high-speed jet* in the passage to meter the flow of gasoline through it. This may be: (1) a simple plug with a calibrated hole (a fixed jet); (2) an interchangeable series of such plugs which can be screwed out for substitution of one with a different hole size, to accommodate the carburetor operation to changing seasons or load conditions; or (3) a needle valve with adjusting screw similar to the idle-speed adjusting screw.

With the first type (used principally with multicylinder engines), other complementary systems built into the carburetor serve to vary the mixture to meet all operating conditions. This is also true of the second type, except that periodic variations are possible as already noted. The third (adjusting-screw) type is largely used for single-cylinder engines. It affords several

features. Whereas with the plug types the high-speed jet may or may not control gasoline flow into the idle tube as well as into the high-speed nozzle, with the adjusting screw type the screw always controls gasoline flow to both systems. Closing of this screw valve will shut down all gasoline flow from the bowl. Adjustment of the screw can be used also as a ready means of leaning or enriching the high-speed mixture to provide for current load conditions.

Needle-Valve Systems

A typical needle-valve arrangement is shown. Note that this valve controls *only* gasoline flow. Closing the valve (clockwise rotation) will lean, then shut off the fuel supply. Air is bled into the high-speed nozzle (for the same purpose that air is bled into the idling speed tube or passage), but it enters at a point beyond the needle valve and also beyond where the gasoline flows into the idle passage. As venturi pressure drops with increased engine speed, more air will be sucked through this bleed tube to relieve the increased suction on the gasoline and prevent overenrichment of the mixture. But none of the air entering here will have any effect on the idling system, as the liquid level in the high-speed nozzle will always be high enough to immerse the end of the idle passage.

In some carburetors this high-speed needle valve terminates in a simple spring-loaded screw head, intended only for occasional adjustment if necessary. The type illustrated, however, terminates in a handle fitting. If the handle used is one for manual operation—but without any calibrations—it is intended for occasional operator adjustment to vary mixture richness as required by various uses of the engine. If, however, the handle is a lever with marked settings, the valve is intended for constant operator use in operation of the engine.

When the valve is used as last described, the operator will have no control whatsoever over the throttle, which will be 100% governor controlled. To stop the engine, he will close off the gasoline supply to both idling and high-speed systems by shutting the needle valve. To start, he will open this valve to supply fuel to the idling system; then, when the engine warms up,

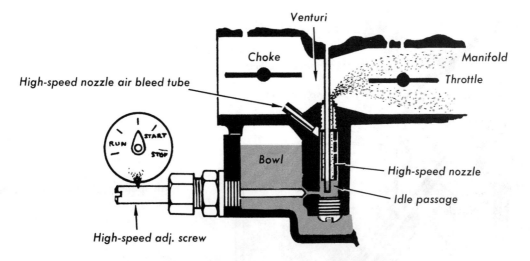

A typical high-speed system needle valve.

he will adjust the valve to produce the required richness for the high-speed system. Settings are usually marked STOP, START, and RUN (with several selections for the last).

NOTE: In all but the simplest carburetors designed for single-purpose and single-load condition engines, some one or more auxiliary systems are provided to vary the high-speed mixture to meet load and speed demands.

ECONOMIZER SYSTEMS

Systems of this kind operate to lean out the normal high-speed mixture whenever the reduction of load and/or speed makes a leaner mixture desirable. Two types of economizer systems are in general use. Both operate on a *back-suction* principle.

The Valveless Type

In the *valveless type* the air-vent passage from the carburetor air horn to the bowl passes around the venturi and joins a branch passageway that opens up into the airstream between the venturi and the throttle valve (see illustration, "High-Speed System With Economizer"). This branch passageway terminates in a carefully sized *economizer orifice,* and the air flow through it is also metered by a calibrated *economizer jet.* Hence the larger bowl vent will normally pass more air than the smaller jet will pass.

Moreover, how much (if any) air will flow out of the economizer orifice depends on the air pressure in the airstream around the throttle valve butterfly. Exceedingly precise calibration of all the factors is required.

Whenever the throttle is closed up to idling or intermediate-speed positions, the economizer orifice will be in the high-pressure area at the front of the butterfly—the pressure at the orifice opening will be practically as high as the pressure in the air horn, and very little air will be sucked out. The bowl vent will pass enough air to replenish the little sucked out and keep the air in the bowl under atmospheric pressure.

As the throttle is opened to (and beyond) the position shown, however, the low-pressure area behind the butterfly edge is moved down until the orifice opening is fully exposed to this low-pressure area. With increasing pressure drop at its end, more and more air will be sucked out through the economizer orifice. Consequently the limited amount of air passed by the bowl vent will become increasingly insufficient to meet the demand, and the air pressure in the bowl will drop accordingly. This pressure drop in the bowl will cause less gasoline to flow out of the high-speed nozzle, and lean the mixture whenever the throttle passes through this stage of opening.

If the throttle position at the peak of the economizing action (when the most air is being sucked out of the economizer orifice) is the max-

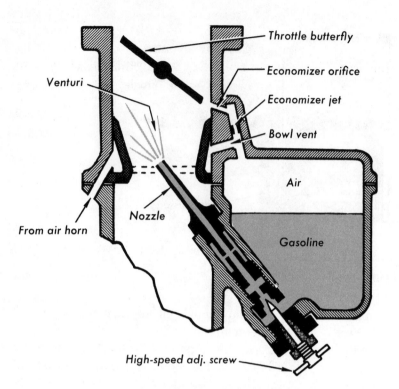

Venturi

From air horn

Nozzle

Throttle butterfly

Economizer orifice

Economizer jet

Bowl vent

Air

Gasoline

High-speed system with economizer.

High-speed adj. screw

imum opening required for top engine speed under the no-load condition, then the engine will be operating as desired. That is, it will be running at top no-load speed on the leanest possible mixture.

Now any increase of load requires a richer mixture. But whether the throttle is governor or operator controlled, it will be opened farther to meet increasing load conditions. And continued opening of the throttle will diminish its airfoil action and allow the pressure at the economizer orifice opening to build back up toward normal. This, of course, will allow the pressure in the bowl to build back up and push out the additional gasoline needed to enrich the mixture as required.

The Valve-Controlled Type

This type operates on exactly the same principle of creating a partial vacuum (back suction) in the carburetor bowl. However, instead of an economizer jet and orifice to calibrate the air flow, the throttle valve itself is directly used to control the size of the branch passageway (accompanying illustration). There is a slot in the throttle valve shaft which serves to join two separate parts of this branch passageway, whenever the throttle valve is opened, to place its butterfly in approximately the same position shown in the previous illustration. At a slower-speed setting of the throttle, this slot does not communicate with the right-hand portion of the passageway; and when the valve is opened wider, the slot will move out of communication with the left-hand portion.

The left-hand portion leads to an opening into the intake manifold side of the throttle valve, and the right-hand portion joins the bowl vent passageway. As a result, whenever the two portions are joined by the shaft slot, a partial vac-

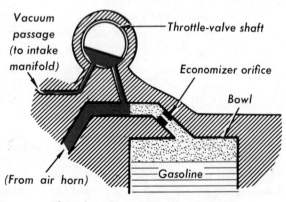

Vacuum passage (to intake manifold)

Throttle-valve shaft

Economizer orifice

Bowl

(From air horn)

Gasoline

Throttle valve economizer system.

uum will be created in the bowl. Also, during this period, the amount of vacuum will be varied in accordance with the amount of air that the slot will allow to pass. And, as with the first system, the arrangement is such that the peak vacuum in the bowl is reached at the engine's top-speed no-load throttle setting.

A METERING ROD SYSTEM

Carburetors using this system for control of the mixture ratio during high-speed operations do not have an adjustable high-speed screw; they use, instead, the plug-type fixed jet. The jet is called a *metering jet* (accompanying illustration). A *metering rod,* connected to the throttle lever, is arranged to move up and down through the center of the jet opening as the throttle is opened or closed. The end of this rod, in the jet, is either tapered or stepped down in diameter so that, as the rod is lowered, its increasing size will block off more and more of the jet opening.

Now, when the throttle is wide open, only the smallest-diameter portion of the rod end will be in the jet opening; all the gasoline required for top-speed top-load engine operation will be permitted to flow through the high-speed nozzle. As the throttle is closed, less and less gasoline will be permitted to flow. Thus, the mixture is leaned at the lower throttle settings used for lighter load and/or slower-speed operations. If

this same (metering) jet is used (by the design of the carburetor) to feed gasoline into the idling system, even at the lowest throttle setting enough gasoline will be permitted to flow to serve the idling system needs. But if the bowl connection to the idling system is through a separate jet, the rod may be designed to close the metering jet completely at the time the idling system comes into operation.

> NOTE: A slightly different arrangement of this same system is used in automotive carburetors. The metering rod is controlled by manifold vacuum as well as by the throttle setting. Addition of the vacuum control makes the system even more sensitive to rapid load changing conditions.

A POWER (ECONOMIZER) SYSTEM

This type of system serves much the same purpose as the economizer systems previously described. It differs in this respect, however: The simple, plug-type jet high-speed system used with it is calibrated to pass only the amount of gasoline required for top-speed no-load engine operation, and this power system is relied upon to enrich the mixture whenever necessary. Operation of the power system is more sensitive to load variations and permits a wider range of mixture ratios. But such a system is far more

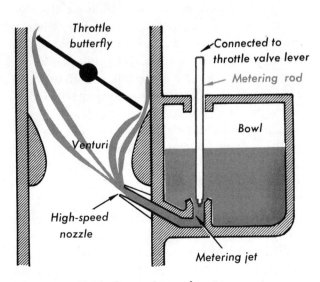

Typical metering-rod system.

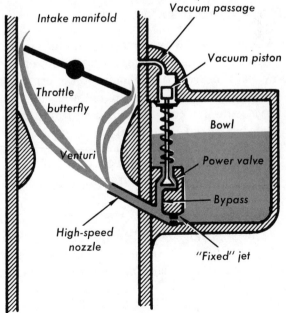

Typical power system.

complicated and expensive than the simple econ-omizer systems. It is used, therefore, only in larger-engine (primarily automotive) carbure-tors.

This system (accompanying illustration) uses a *bypass passage* around the fixed high-speed jet (between the bowl and the high-speed noz-zle). The bypass is controlled by a *power valve* that is opened or closed by a needle attached to a *vacuum piston.* The cylinder above the piston is connected, by a *vacuum passage,* with the intake manifold; and the underside of the piston is exposed to atmospheric pressure in the bowl. There is a coil spring around the needle which tends to hold it down (valve open), and upward movement of the piston will tend to close the valve.

Remember, manifold pressure is lowest at the slower speeds and rises as engine speed in-creases; it also rises whenever a load is applied to the engine. Consequently, the suction on the vacuum piston will be greatest at slower speeds, will lessen to allow the spring to open the power valve as the speed increases, and will also lessen to increase the valve opening whenever a load is applied. In short, gasoline will be allowed to flow through the bypass in direct relation to en-gine speed and/or load—to increase the mixture richness, as required.

ACCELERATING SYSTEMS

A system of this type may or may not be in-cluded in the same carburetor with one of the previously described economizer, metering, or power systems. It serves an entirely different purpose: the purpose of furnishing to the high-speed nozzle an additional spurt of gasoline during any period of acceleration in order to pick up speed or to pick up a load.

During such a period, the other carburetor systems cannot reset quickly enough to keep up with the opening throttle. The heavier gasoline molecules, having more inertia than the lighter air molecules, won't increase their rate of flow as instantaneously. The increase of air exceeds the increase of gasoline, and a leaning of the mixture occurs just when increased richness may be most needed.

If an engine is subjected to rapid changes of speed or load conditions, an acceleration system is essential to keep it from backfiring and pos-sibly stalling.

Two general types of systems are used. The simplest (most used with small engines) is an *accelerating-well system.* The other (most used in automotive engines) is an *accelerating-pump system.*

The Accelerating-Well System

This system provides excess fuel only when the throttle is suddenly opened from idling or near idling position. Should the engine be op-erating at a fairly fast speed and then be ac-celerated, no extra fuel will be provided. The system's use is therefore limited to engines with more or less constant operating speed.

Note the accompanying illustration. When-ever the idling system is in use, fuel will stand in the *accelerating well* at approximately the same level that it stands in the bowl (as shown). As soon as the high-speed system comes into use with opening of the throttle, gasoline is sucked through the *high-speed jet* from the bowl and through the *nozzle air bleeds* out of the ac-celerating well. Thus, during this initial period of transition from idling- or intermediate-speed operation to high-speed operation, the accelerat-ing well furnishes extra gasoline to enrich the mixture for the acceleration taking place.

Now, the air pressures on the gasoline surface are, of course, directly responsible for the flow of the gasoline out of the nozzle. But, whereas the bowl is vented, as previously explained, the accelerating well vent is partially blocked by the *nozzle air vent.* This vent will admit only the limited amount of air intended to be bled into the high-speed nozzle. Moreover, gasoline to re-fill the accelerating well must flow from the bowl up through the nozzle air bleeds. Consequently, once the well is emptied, provided the throttle remains open or continues opening, air will bleed into the high-speed nozzle through the nozzle air bleeds to lean the mixture back to operating ratio, and to prevent refilling of the well.

Whenever high-speed operation ends and the idling system again takes over, air will no longer be sucked down through the nozzle air vent,

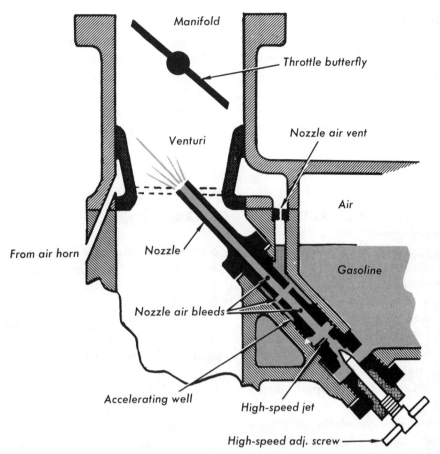

High-speed system with accelerating well.

through the nozzle air bleeds, and out the nozzle end. Gasoline will then flow back up through the bleeds, to refill the well for the next time it is needed.

The Accelerating-Pump System

This system is not illustrated as it is very unlikely that you will find such a system on any small engine. The principle of operation is as follows.

There is a separate passage from the bowl to the venturi. In this passage there is a piston-type pump with check valves arranged so that raising of the piston will draw fuel from the bowl into the pump cylinder. Lowering of the piston will then squirt this fuel out into the venturi. The piston is connected to the throttle lever through a spring arrangement. Whenever the throttle is opened—at whatever speed the engine may be operating—the pump spring will be compressed to an extent determined by the amount of throttle movement. In turn, the spring will then ex-

pand by forcing the piston downward, to squirt gasoline into the venturi. The rapidity of piston movement will be directly proportional at all times to the amount by which the spring is compressed; it will taper off as the spring expands. Closing of the throttle will always allow the spring to reexpand and then to pull the piston back up, ready for another acceleration. Careful calibration makes this an exceptionally sensitive system to care for all throttle-opening conditions.

IDLING AND HIGH-SPEED ADJUSTMENTS

With all the preceding types of carburetors it is necessary to warm up the engine prior to making adjustments. If any type of high-speed screw is provided for adjustment, it should be adjusted first to obtain optimum high-speed performance. Afterwards, the idling adjustments are made.

In making the idling adjustments it is generally preferable to first open the idle adjusting screw to a certain specified setting, then to adjust the idle speed adjusting screw, and finally, to return to the idle adjusting screw to make final adjustment. Sometimes it will be necessary to go through the two screw adjustments a second time to refine the adjustment. Occasionally, it will be necessary to test the acceleration from idling through intermediate, to high speed, and then make slight readjustments as required to smooth out the engine operation.

Every manufacturer furnishes a proper sequence of adjustments for his particular engine. These vary from engine to engine, often considerably. For proper engine performance it is absolutely essential that any one engine be adjusted according to instructions given. Following some general rule of thumb won't do the job properly.

DIAPHRAGM-TYPE CARBURETORS

COMPARISON WITH BOWL TYPES

Diaphragm-type carburetors have been developed for use on small engines which must operate in a variety of positions throughout changing angles to the horizontal, such as chain saw and lawn mower engines. Very acute angulation will not interfere with carburetor operation. The engine can even be run upside down for limited periods.

This type of carburetor employs the same type of idling system and adjustable high-speed screw type of high-speed system already described for the bowl-type carburetors. With the majority of such carburetors, no air is bled into the gasoline ahead of the airstream; liquid gasoline alone is sprayed out of the idle orifices and high-speed nozzle. And with carburetors of the type illustrated, both the idle adjusting screw and the high-speed adjusting screw are rotated clockwise to weaken the mixture. There is a diaphragm type, however (not illustrated), in which both screws do control air. The latter must be turned counterclockwise to lean the mixture.

There are no economizer, power, or accelerating systems employed with this type of carburetor. The screw-adjustable idling system (usually with an intermediate orifice also) and the screw-adjustable high-speed system between them are relied on to furnish all mixture requirements of the engine.

THE TWO TYPES

There are two distinct types of diaphragm carburetors: the type having its own fuel pump

attached, and the type without a fuel pump. The latter must be fed either by a gravity-type system or a system that incorporates a separate fuel pump.

The integral fuel-pump type can be used only with a two-stroke-cycle engine, as the pump is operated by the pressure pulsations in the crankcase of such an engine. The other type may be used with any small engine. In the accompanying illustration, we show a model which the manufacturer can adapt to serve either purpose. When furnished as a pumpless type, the carburetor section (upper portion in the illustration) alone is supplied. To convert this into an integral-pump type, the *bottom cover* of the carburetor section is removed. The fuel-pump (bottom) section is bolted onto the carburetor instead.

THE FUEL PUMP (WHEN USED)

This is a simple diaphragm-type fuel pump having *check valves,* as shown, in the fuel line. When the diaphragm is raised, gasoline is sucked through the check valve at right (which opens with the flow), while the suction holds the check valve at left closed. When the diaphragm is lowered, the process is reversed, and gasoline is pumped out the line to the left.

A hose or tube is fitted into the threaded opening of the *impulse channel* to connect the pump with the interior of the two-stroke-cycle engine crankcase. This channel opens into the small chamber above the fuel-pump diaphragm. Whenever the engine piston(s) is making an up stroke, the resulting partial vacuum in the

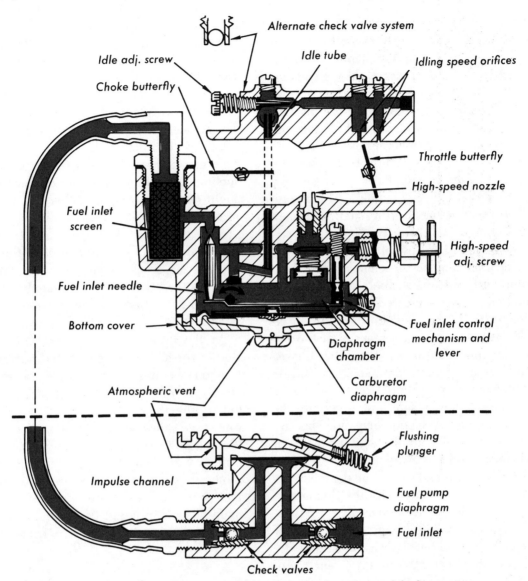

Idle adj. screw
Choke butterfly
Alternate check valve system
Idle tube
Idling speed orifices
Throttle butterfly
High-speed nozzle
Fuel inlet screen
High-speed adj. screw
Fuel inlet needle
Bottom cover
Fuel inlet control mechanism and lever
Diaphragm chamber
Carburetor diaphragm
Atmospheric vent
Flushing plunger
Impulse channel
Fuel pump diaphragm
Fuel inlet
Check valves

A typical diaphragm-type carburetor with or without integral fuel pump.

crankcase will then create a like partial vacuum above the diaphragm to suck it up. Conversely, whenever the piston(s) is making a downstroke, the resulting compression in the crankcase will push the diaphragm down. In short, the diaphragm will pump in harmony with the pressure pulsations in the crankcase.

If the carburetor is full, gasoline in the line and in the pump will hold both check valves closed. This will make the pump inoperative, even though the diaphragm continues to fluctuate.

The *flushing plunger* is provided to take the place of the removable plug which, of course, won't be there if the pump is attached. Either provides a means of flushing out any fuel, dirt, or foreign material that might accumulate in the vented air space below the *carburetor diaphragm*.

CARBURETOR OPERATION

Fuel enters the *diaphragm chamber* through the *fuel screen* and the *fuel-inlet needle seat* whenever the *carburetor diaphragm* is raised up to rock the *fuel-inlet control lever* and unseat the *fuel-inlet needle*. Whenever the diaphragm is lowered, a spring—held by an adjustable plunger (*fuel-inlet control mechanism*) and located at the end of the control lever's long arm—rocks the control lever back to reseat the fuel inlet needle, and stop the flow of fuel. The bottom side of the carburetor diaphragm is exposed to atmospheric pressure through an *atmospheric vent* (either in the bottom cover or in the fuel pump). The top side of this diaphragm is exposed to the manifold, both through the *idle tube* and its connecting passages to the *idling-speed orifices,* and through the connecting passage to the *high-speed nozzle.*

Whenever the engine is running to maintain a manifold vacuum, the resulting suction applied to the diaphragm will hold it sufficiently raised to permit an adequate flow of fuel into the diaphragm chamber. Since the chamber will thus be continuously filled with fuel, the suction on the diaphragm is not created by a partial air vacuum. Instead, this suction is transmitted through the liquid fuel, in a reversal of the hydraulic principle. That is, whatever the "degree of suction" at any instant, the diaphragm will be pushed up—by atmospheric pressure below— high enough to admit sufficient fuel into the chamber to supply the suction demand. This will prevent the chamber from being emptied of fuel, which would leave a vacuum in it. The fuel rapidly flowing into and out of the chamber has what we call a *minus pressure;* the diaphragm fluctuation is called *pressure sensing.*

When first starting the engine, this type of carburetor will be either dry or almost dry, depending on circumstances. In any event, there will be insufficient fuel in the carburetor to make the start. A special starting technique is required. The throttle valve *must* be cracked open a sufficient amount to expose one or both of the idling-speed orifices to the low-pressure area behind the butterfly edge (approximately as shown in the illustration). With the choke valve fully closed, all of the air called for by the slowly cranked engine must now come through the idle tube out of the diaphragm chamber. Whatever air there is in the chamber in place of fuel, its immediate expansion in response to the engine's suction will at once lower the pressure in the chamber sufficiently to raise the diaphragm and supply the fuel required.

In the carburetor illustrated, the idle tube and the high-speed nozzle have separate openings

into the diaphragm chamber (and this generally is the case). Also, note that the high-speed nozzle is blocked by a ball-check valve. This is always the case. During idling or intermediate-speed operation, the throttle butterfly will (as previously explained) create a low-pressure area to suck fuel out through one or both of the idling-speed orifices (the same as with a bowl-type carburetor). There will be practically no suction on the high-speed nozzle due to the very slight (if any) air velocity there. In fact, the air pressure at the nozzle tip will be somewhat greater than the minus pressure of the fuel in the chamber. If the nozzle were not blocked by the ball check (as shown), air would bleed back through the nozzle into the chamber to disrupt the fuel flow. Instead, the pressure differential holds the ball down to close the valve.

As the idling operation becomes ineffective due to further opening of the throttle, the suction on the high-speed nozzle increases sufficiently to raise the ball up, and suck the fuel out of the then-opened high-speed nozzle. At this stage, air cannot bleed back through the idling-speed orifices. The air pressure at these orifices will always be somewhat less than the minus pressure of the fuel in the chamber. In fact, since the idle tube and passages will remain filled with liquid fuel throughout all operating stages, some fuel will be friction dragged

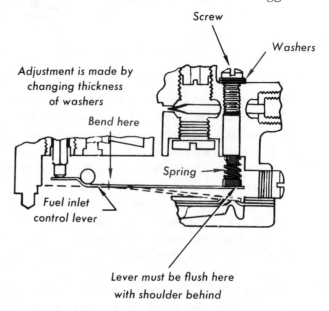

Typical lever and spring-tension adjustments.

out of the idling-speed orifices even after the throttle is full open and the high-speed nozzle is at the peak of its operation.

ADDITIONAL FEATURES

To accommodate the requirements of certain engines, the carburetor shown can be modified by removal of the plug shown at the right end of the horizontal passage leading to the two idling-speed orifices. When this is done, the additional through passage provided is connected directly to the intake manifold by an extension tube. Some additional fuel will then be sucked into the engine through the passage, during high-speed operation, to enrich the mixture in the engine. This modification is supplied by the manufacturer and must be calibrated to specific engine requirements.

Another modification offered by the manufacturer is shown on the illustration and labeled "Alternate Check Valve System." Ordinarily, the connection between the idle tube and the passage containing the idle adjusting screw is made through a plug having a single, fixed-size orifice for passage of the fuel. If the carburetor is inverted for any length of time, the additional help of gravity will flow an excess of fuel through the idle tube and out the idling-speed orifices to overenrich the mixture. Substitution of the ball-check plug (for the plain one) will prevent this. In normal upright position, the ball is at bottom—fuel enters at the sides and out through the two top openings, to supply the required amount. When inverted, the ball closes the center top opening, leaving only the side opening for fuel to flow through. The decrease of opening compensates for the increase of forces, to maintain the proper fuel flow.

ADJUSTMENTS

Remember that with this type of carburetor clockwise rotation of either the idle adjusting screw or of the high-speed adjusting screw will, with most makes, *lean* the mixture for the respective system.

Also, with this type of carburetor, the idling system functions to some extent, even when the high-speed system is functioning. Moreover, the high-speed adjusting screw does not affect the idling system. Consequently, it is usually best to first make the idle operation adjustments (idle-speed adjusting screw and idle adjusting screw), then make the high-speed adjustment. As with other carburetors, however, the manufacturer will always specify the desired sequence of adjustments. These instructions must be carried out exactly for proper operation of the carburetor.

Additional adjustment of the fuel-inlet control mechanism may or may not be required, depending on the manufacturer's specifications. Any such adjustment is usually made by proper bending of the fuel-inlet control lever and/or tensioning of the spring in the mechanism.

SUCTION-TYPE CARBURETORS

COMPARISON WITH OTHER TYPES

Used only for small single-purpose engines, this is the simplest type of carburetor constructed. Basically, it provides only a single idling-speed mixture together with a very limited range of operating-speed mixtures such as required by an engine which will operate only at one running speed and under calculated load variations. The direct suction of the engine lifts the gasoline from the tank through the carburetor (as previously explained under "Fuel Systems").

There are no provisions for fuel economizing, acceleration, or extra fuel for power, such as are found in bowl-type carburetors. Only one adjustment screw (called a *needle valve*) is provided. And the choke is, as often as not, a slide such as illustrated rather than a butterfly type. Also, there is no venturi, such as is used in the other carburetor types. The airfoil action of the throttle butterfly alone serves for both idling and running-speed operations. No air is ever bled into the gasoline prior to its being taken directly from the fuel tank and sprayed into the airstream.

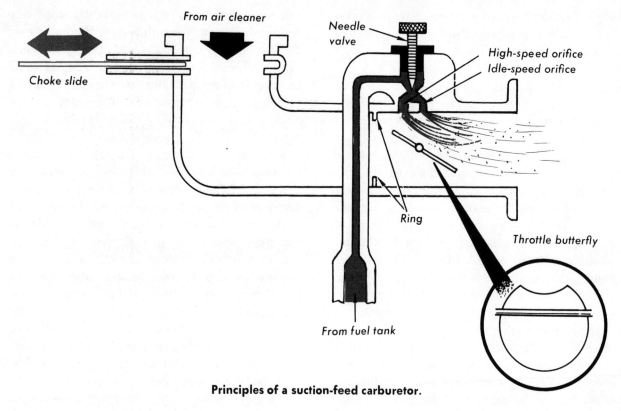

Principles of a suction-feed carburetor.

CARBURETOR OPERATION

Like the diaphragm-type carburetor, this carburetor must have the throttle valve cracked open, and the choke closed, for starting. Engine suction then creates a vacuum throughout the carburetor fuel passage into the fuel tank. Thereafter, atmospheric pressure in the tank will spray gasoline out the *idle-speed orifice* into the manifold. Immediately on starting, the choke must be opened to supply the required air.

With the throttle valve still in the starting position—or opened a very little farther—only the small idle-speed orifice will be exposed to the low-pressure area at the throttle butterfly edge. This orifice is calibrated to pass just the right amount of gasoline for idling operation.

Further opening of the throttle valve exposes also the larger high-speed orifice to the low-pressure area. Gasoline will now be sucked from both orifices and the total amount allowed to flow is governed by the adjustment of the needle valve. The actual amount flowing at any time will depend also on the degree of manifold vaccum, with slight variation depending on the exact butterfly position and consequent velocity

of the airstream around its edge by the two orifices. To keep the low-pressure area's suction from dropping off as the throttle is opened, a *ring* (or spiral) is generally placed (about as illustrated) ahead of the throttle butterfly. This ring keeps the passageway around the butterfly edge narrow—even at open throttle setting—so that the airstream velocity is slightly increased rather than reduced as the throttle is opened wide. The ring's added airfoil action is such that steady, unfaltering acceleration through the limited high-speed range is possible.

To increase the engine suction on the fuel, the air horn of the carburetor is usually somewhat restricted, ordinarily by narrowing of the passage around the slide choke. Also, the throttle butterfly usually has a flat or notch on the edge adjacent to the orifices, to increase the extent and duration of its airfoil action. For the same reason, the shaft may also be off center, as shown. Then, again, this type of carburetor is *always* adjusted for an overly rich mixture.

ADJUSTMENTS

Rotation of the single needle valve in a *clockwise* direction always *leans* the mixture. This valve is intended for high-speed mixture adjustment only; it must always be open more than enough to pass all the fuel that will pass through the idle-speed orifice.

Consider the amount of suction produced by the engine as one force, and the weight of the gasoline in the line and carburetor passage as an opposing force. It follows that the rate of fuel flow into the airstream during engine operation must depend, to some extent, on the level of the fuel in the tank. With a full tank, the column of fuel is shorter and lighter; with an almost empty tank, it is longer and heavier. As the column weight increases, the velocity of the flow (all else remaining unchanged) must decrease somewhat.

Consequently, it is *very important* with this type of carburetor, that the needle-valve adjustment be made *only when the tank is half full.* This will provide an adjustment that is a compromise between the two extremes of full and nearly empty and will ensure average carburetor operation as the fuel supply varies. The valve is generally set, under this condition, as rich as the engine can take it and still run evenly. Each manufacturer will, however, give the exact adjustment requirement for his engine.

A carburetor of this type does usually have an idle-speed adjusting screw. Idling speed is generally set quite high; but here again, you should follow the manufacturer's instructions.

FOREIGN-MAKE CARBURETORS

Carburetors of foreign make are used principally on imported motor scooters, motorbikes, and motorcycles, and these are adapted to such use by design differences which set them entirely apart (in both appearance and operation) from the carburetors previously discussed. Basically, however, their principles of carburetion are exactly the same. They function to mix gasoline with air, in ratios required by the varying speeds and load conditions of their engines.

These carburetors are classified here as *slide-throttle direct-proportioning* types. The chief differences in their construction are the use of a tubular slide in place of a throttle butterfly, and the simultaneous control (by the throttle cable) of a jet needle used to regulate gasoline flow. They also use a nonadjustable bowl with a fixed-float needle (refer to the section "Float System," under "Float-Type Carburetors"), and a compact wet-type air cleaner which incorporates choke provisions in its design. There is no venturi such as previously described; instead, there is a mixing chamber.

SLIDE-THROTTLE
DIRECT-PROPORTIONING CARBURETOR
Air Flow

The accompanying illustration schematically shows a typical carburetor's operation. Air enters through a *wet-type air cleaner*. If the carbu-

retor is equipped as illustrated, the amount of air entering can be controlled by an adjustable shutter cover on the face of the cleaner. This shutter then serves as a choke. If there is no shutter, one of the other types of chokes previously discussed will be provided.

The air flows through the *air horn* and to the *mixing chamber*. Here its flow is more or less restricted by the position of the *throttle slide,* which moves up or down in response to the throttle control (to which it is connected by a flexible cable). The farther down the throttle slide is, the narrower will be the air passage remaining below it. Thus, it serves not only to restrict airflow and perform a throttling action, it also serves as an airfoil. It is held down (throttle closed) into an adjustable idling-speed position by the coil spring shown.

Air is also bled into the idling system through an air vent leading from the air horn into the *idle tube.* The amount of air bled into the fuel through the passage is controlled by the *idle adjusting screw.* Therefore, this is an adjustable-air (only) idling system, and turning the screw clockwise will enrich the idle mixture.

Gasoline Flow

Gasoline enters the bowl through the float needle seat (at top) to cause the *float* to rise and regulate the flow into the bowl as previ-

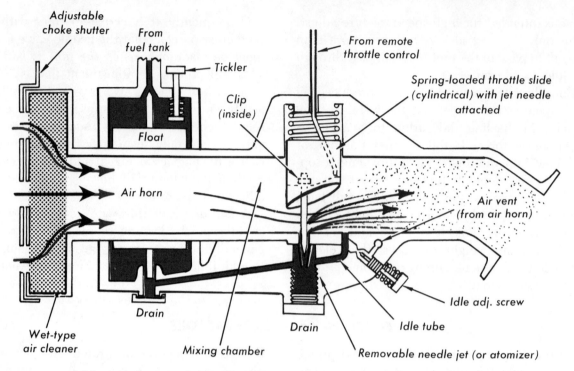

Schematic of a typical slide-throttle direct-proportioning (foreign) carburetor.

ously described. With the type of arrangement illustrated, the bottom end of the needle rests in, and partially blocks, the opening from the bowl into the carburetor passages. Rising of the float permits unrestricted flow of gasoline from the bowl into these passages. There is this advantage to the arrangement: At starting, the float can be held down by the *tickler* (bottom opening partially blocked) to completely fill the bowl to overflowing. When the tickler is then released, the excess gasoline in the bowl will rush out to flood the carburetor and thus aid choking.

When the float is up, gasoline flows out the bottom passage into the carburetor passageways. If the throttle slide is down (idling position), the very low pressure created by its airfoil effect at the end of the idle tube will cause gasoline to be sucked out through the idle tube into the airstream. In passing through the tube, it will be partially premixed with vented air, as already mentioned. During this idling stage, the jet needle fastened to the throttle slide will also be down. And the passage through the needle jet (or *atomizer*) will be closed.

As the throttle slide is raised to admit more air, it simultaneously raises the jet needle out of the needle jet. This admits first a small amount, then a correspondingly larger amount of gasoline, to flow through this passage into the airstream. The calibration of the needle and jet is such that just the right amount of gasoline will be sucked out into the airstream for each throttle setting. During the earlier stages of throttle opening, the idling system will also continue to spray gasoline into the airstream; but this system stops functioning when the throttle is approximately half opened. Due to the design of the mixing chamber and the slide, there will continuously be sufficient suction to flow the amount of gasoline needed for the throttle setting.

VARIATIONS

1. As already remarked, there may or may not be an adjustable choke shutter. If there is, it may have a cable control, or a simple thumb lever for operation.
2. There may or may not be a tickler. If there is none, the float needle will not have a bottom seat, and the passage from the bowl into the rest of the carburetor will be fully open.

3. The drains shown may or may not be included.

4. There may or may not be an idling system. If not, the jet needle positioning provides for all operating speeds.

5. When used with two-stroke-cycle engines, the wet-type air cleaner is self-moistening. Provision is made for the fuel mix, which contains oil, to dampen it while passing through the carburetor. As used with four-stroke-cycle engines, however, this cleaner must be kept moistened by the operator by occasional wetting in engine oil.

ADJUSTMENTS

Idle-speed setting generally is accomplished by means of a spring-loaded adjusting screw (not shown on our illustration) that is a part of the throttle rocker-arm assembly whereby the throttle slide cable is moved up and down. The arrangement is such that the screw can be adjusted to limit downward travel of the throttle slide.

When provided, the idle adjusting screw is used the same as with other carburetors. Remember, however, it also has considerable effect on higher-speed operations—up to approximately a quarter-open throttle. Thereafter its effect diminishes rapidly. Consequently, when adjusting it, the range of speeds which it does affect must be considered. The optimum adjustment throughout this speed range must be obtained. If this is impossible, look to the following adjustment for the remedy.

The only other adjustment provided is the repositioning of the jet needle with respect to the throttle slide. This needle is held in the slide by a clip (inside the hollow slide) that engages a groove in the needle. From three to five grooves may be provided. By changing to a groove that is higher on the needle end, weaker mixtures will be obtained throughout the range of higher-speed operations; while changing to a lower groove will enrich the mixtures. This adjustment requires partial disassembly of the carburetor.

In all cases, the manufacturer will provide specific adjustment instructions.

CARBURETOR MAINTENANCE

If we were to attempt to cover specific maintenance of each and every carburetor applicable to all small engines a volume as large as this one would be required to cover this subject alone. Consequently, all we can do here is to note a few generalities.

A carburetor is one of the simplest devices on an engine from the standpoint of maintenance. It is true that you would have to know much more about the science of carburetion than we have been able to discuss here, in order to design one, or even to rebuild one by machining new parts. Many, many man-hours of research and planning go into the design and calibration of each new model. But you will never find it either necessary or profitable to attempt to do the job of the designers. Carburetor maintenance need never be more than proper cleaning, proper replacement with authorized factory parts, and proper adjustment in accordance with specified instructions. Follow this rule, and carburetor maintenance becomes exceedingly simple.

There is one general rule applicable to all carburetors: *Neither fuel nor air must ever leak from any part of a carburetor; and every opening so intended must always pass exactly the amount of fuel or air intended by the manufacturer.* With this as a basis it is obvious that: (1) A carburetor must be thoroughly clean, throughout every part and every passage. (2) No orifice or passageway must ever be made larger or smaller, even by a microscopic amount (some are calibrated to the 1/10,000 of an inch or less). (3) All needle-valve points and seats and similar mating parts must be in tip-top shape. (4) All gaskets and mating flanges, etc., must fit snugly.

Each manufacturer will furnish cleaning, repair-and-replacement, and adjustment instructions. Cleaning usually is accomplished by soaking the carburetor in gasoline or other grease and grit solvent, followed by careful brushing and picking to remove all dirt, without damage to the finely drilled holes, etc. Repairs usually are limited to bending of parts for correct float or diaphragm lever setting. Most replacement kits are specified for minor or major overhaul purposes. These contain new gaskets, springs, etc., which may be required for replacement at the overhaul period specified. We have already covered adjustments.

Any carburetor is easy to maintain if you use the proper authorized parts and follow instructions. Any carburetor that is properly maintained is easy to adjust. If you follow all the rules of proper carburetor maintenance and engine performance is still faulty, look elsewhere for the trouble. Poor compression, bad valve action, insufficient ignition, or something as simple as dirty or improper fuel will very likely be the cause of trouble.

CHAPTER 8

The Ignition System

TYPES OF IGNITION SYSTEMS

There are two basic types of ignition systems used with small gasoline engines. The first is a *battery system* in which electrical energy, generated while the system is in operation (engine running), is stored in a storage (*wet-cell*) battery from whence it can be drawn whenever needed—whether the engine is running or not. The second type of system does *not* include a battery; current is generated and can be used only while the engine is running or is being motored by a starter.

A battery requires direct current (dc) for charging and, when charged, produces dc. On the other hand, a spark plug can be fired by either dc or ac; an electrical starter can be designed for operation on either dc or ac; and lights, a horn, etc., will operate on either type of current.

Any engine equipped with an electrical starter which does not operate on an outside source of power must have a battery. Electrical power for a battery system may be generated by a *generator*, which produces dc, or by an *alternator* or *magneto*, either of which produces ac. Whenever a generator is used (*a generator-battery system*) all components of the system operate on dc. If the system contains an alternator or a magneto the generated ac must be converted to a dc for charging of the battery; moreover, the battery dc is then used to operate the starter

and may be used for the other electrical components as well.

An engine equipped with a mechanical starter, or a starter operated by an outside source of power, does not require a battery. Most small two- and four-stroke-cycle engines are so designed and are equipped with a *magneto system*. Small rotary-combustion engines may also be so equipped.

Basically, a magneto is a type of alternator (and is sometimes so named), but it differs from an alternator (and a generator) in one major respect: it is very much simpler in design and the generating units are housed within the engine flywheel assembly instead of within a separate, engine-driven component. There are several types of magneto systems: (1) A *conventional system,* in which all parts of the ignition system are part of the magneto assembly; (2) An *energy transfer* (*ET*) *system,* in which the ignition coil is a separate unit; and (3) A *capacitor discharge* (*CD*) *system,* in which solid-state electrical ignition components (rather than mechanical components) are used.

NOTE: Because generators and alternators are similar in construction, battery systems using these components will be discussed, following, under "Battery Systems;" and because all magnetos are basically different from generators and alternators, these will be discussed under "Magneto Systems," and this section will include even those magnetos having provisions for battery recharging. Solid-state ignition components are discussed in a third "Solid-State Ignition Units" section.

BATTERY SYSTEMS

With any battery system dc must be used for charging of the battery, and the battery, in turn, produces dc. Now, any electrical current requires two pathways—one going out (from source to operating unit) and one going back (to source). In any system employing a battery one side (*terminal*) of the battery is grounded to the engine and/or framework of the vehicle in which the engine is mounted. Usually, it is the negative battery terminal that is grounded (though it can be the positive). The same (negative or positive) terminal of the generator—or one terminal of the alternator—together with one (either) terminal of each electrical component in the system are also grounded. Therefore, one side of every electrical circuit in the system passes through the engine (or framework), and to complete each circuit only one wire (for the other circuit side) is needed. It follows that any

battery system is a *single-wire system* (one wire only to each component) with the engine (or framework) taking the place of the second pathway. Both 6-volt and 12-volt battery systems are in use.

COMPONENTS OF A GENERATOR-BATTERY IGNITION SYSTEM

A simplified one-wire system is illustrated. Its principal components (as shown) are: (1) *generator*; (2) *battery*; (3) *ignition switch*; (4) *ignition coil*; (5) *distributor*, which incorporates the *breaker*—known as the *points*—and usually incorporates the *condenser* also; and (6) a *spark plug* for each engine cylinder. A *cutout* usually is provided. Its function is to prevent the battery from discharging back through the generator to operate it like a motor. However, a manual switch (controlled simultaneously with the ignition switch) could be substituted. Also, whenever the generator is to be operated by the engine over a wide speed range, a *voltage regulator* is required to keep the charging rate reasonably constant at all speeds. This regulator, when used, usually incorporates the cutoff fea-

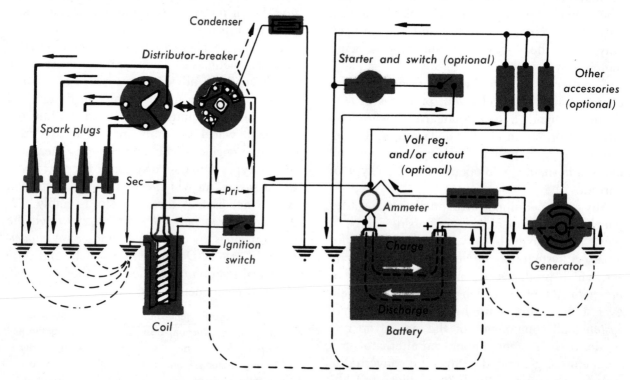

A simplified generator-battery system.

ture. An *ammeter* to indicate charging or discharging rate, whichever is greater, is more often than not included; and the *starter with switch* and *other accessories* (such as lights) are, of course, optional.

While actual connections from device to device and to ground will seldom be found to be exactly as illustrated, the diagram shows the basic principle involved. Ground connections are quite often made internally so that they are not apparent. One wire, when practicable, may carry currents of identical voltage and direction to several devices. Variations in generator and voltage regulator design account for other differences.

Whenever the wiring is complex, or where convenience can be served, the manufacturer generally furnishes the wiring in the form of one or more *wiring harnesses*. These are preassembled wire groups which, as a rule, have proper-type connectors already fastened to the various wire ends and marked in some manner to aid in making the proper connections. Markings may consist of plus and minus signs stamped on the connectors or printed on attached bands, or of color coding. Color coding is a system of coloring or shading the wrappings around the exposed wire ends to indicate where each wire is to be connected in the system. An accompanying diagram or guide to explain each color coding is necessary.

DC GENERATORS

Magnetism, like electricity, is a phenomenon we have learned a great deal about without knowing the actual reason for its existence. We know, for instance, that there are natural magnets, that some substances can become magnetized, and that there is a relationship between magnetism and electricity which can be put to good use in several ways.

A *coil* is a length of wire neatly wrapped into a compact assemblage of uniform, continuous loops. When an electric current flows through a coil it exhibits the qualities of a magnet, and its magnetic strength is directly proportional to the quantity of electricity in motion within the coil. That is, the magnetic strength basically

is determined by the amount of voltage applied (which affects the amount of current flow) and by the number of coil loops, since the longer the wire within the coil is, the greater will be the total amount of electricity simultaneously in motion through the coil. Wrapping a coil around a ferrous metal core (iron, steel, or steel alloy) also increases its magnetic strength. A device so designed is called an *electromagnet*.

Any magnet will attract ferrous metal and will also either attract or repel another magnet, depending on whether the unlike or the like poles of the two magnets are adjacent. Thus a magnet possesses a force capable of creating mechanical motion. Relays, solenoids, and electric motors are familiar appliances which employ electromagnets to create desired motions. There is an established relationship between magnetism and mechanical motion. Through magnetism electrical force can create motion; conversely, motion can create electrical force.

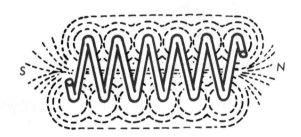

Magnetic fields around a coiled conductor.

A generator uses this converse relationship between motion and electrical force. In simplified explanation we say that an electromagnet is surrounded by a magnetic field in which energy is stored—the energy that is capable of creating motion. If a conductor (anything capable of carrying a current) is forcibly moved through the magnetic field this magnetic energy changes into electrical energy (an emf) within the conductor. Unless the conductor is part of a closed circuit, this emf simply is a potential (inactive) force; but when there is a closed circuit the emf creates a current through the circuit. Hence, mechanical motion created by some source of physical force—a gasoline engine, waterfall, etc.—can be converted into an emf which will cause an electric current.

Construction Features

A dc generator in an ignition system is driven by the engine, the source of physical force, and creates an emf. This emf exists whenever the generator is revolving. There is a current only, however, during those moments when the generator is part of a closed electrical circuit. Like a battery, a generator produces no current unless it is connected.

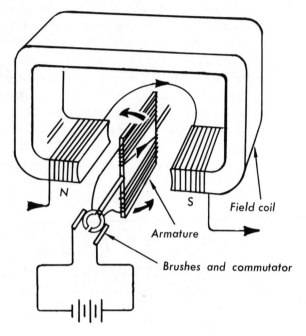

N   S   *Field coil*

Armature

Brushes and commutator

A simplified dc generator.

For construction convenience the coil, which is forcibly moved through a magnetic field by operation of the generator, is wound horizontally on rectangular-shaped cores that are positioned around the generator shaft to form an assembly called the *armature*. This armature is centrally located between the poles of a magnet so that, when it is revolved, as many as possible of the coil loops will be in motion within the magnetic field between the two poles. The magnet, which is an electromagnet, is called the *field coil*. This coil is energized by part (or all) of the current caused by the generator. The armature-coil ends are connected to the separate path of the segmented ring around the shaft at the end of the armature. This ring, the *commutator* is contacted as it revolves by two *brushes* so arranged that the armature coil can at all times be con-

nected through these brushes with a closed external circuit.

A simplified generator is illustrated to show the relationships between the previously mentioned parts. In this generator the two parts of the armature coil shown will be moving across the space between the two field-coil poles only during about 50% of each armature revolution. During the remander of each revolution they will be moving (as actually pictured) from one pole to the other. Maximum emf is created at that instant when the armature coils are closest to the poles; this emf shrinks to minimum when, as illustrated, they are farthest from the poles. Because of this a generator creates a fluctuating emf and causes a *pulsating* current that regularly rises and falls in value.

With the simplified generator illustrated, pulsation would be extreme. A commercial generator has, however, a number of armature-coil parts arranged like the blades of a paddlewheel around the armature. It also has four or more field-coil poles. Consequently, the moments of maximum emf occur much oftener than twice per revolution, and the moments of minimum emf are much less severe. The pulsation is greatly reduced.

In the simplified generator each armature-coil part moves down past the field-coil S pole and then up past the N pole. That is, it cuts across the field between the two poles in first one direction, and then in the other. In each direction the created emf is such as to cause current in a direction opposite from the other. Hence, the current through the armature coil alternates (is ac). However, the commutator has just two segments, one connected to each coil end. Each brush contacts one segment at a time. During each revolution each brush will contact one segment for half the revolution, then a reversal of segments takes place to correspond with the reversal of current through the armature. Consequently, current always flows out one brush and back through the other. In the external circuit the flow is dc.

When many armature coils are used, there are two such commutator segments to serve each coil as it cuts the flux lines of one of the field coils. These segments, called *bars*, are so ar-

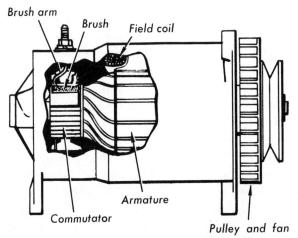

A typical dc generator.

ranged that current being produced in which-ever coil is then at its maximum current-producing capacity, will flow out the one brush and back into the coil through the opposite brush. The bars are of copper separated by mica, or similar insulating material called *separators,* and all these parts together are shaped like a cylinder around the end of the armature shaft.

Coil wires are soldered to the top inner edges of the bars. The brushes are made of carbon and slide inside *brush holders,* with *brush springs* behind them to keep them in firm contact with the revolving commutator. *Brush pigtail leads* (of flexible wire) connect with each brush to carry current to and from the armature.

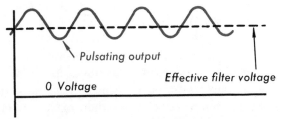

Typifying dc and ripple-filter effect.

The slight current pulsation of any dc generator is called *commutator ripple.* When undesirable it is eliminated by a *ripple filter* connected across the two brushes. The filter is either a condenser, or condenser and series-connected choke coil. It functions to oppose changes in the generated voltage by storing current during each "high tide," and then releasing it during each "low tide."

Various Types

The generator is wired so that the electromagnets of the field coil are excited (magnetized) by the current produced by the generator. When these field coils are connected in series with the armature the machine is called a *series generator;* if they are connected in parallel, the machine is a *shunt generator.* Some, called *combination generators,* have one part of the field coils in series and the other part in parallel. Any field coils that are in series must carry all of the generated current without offering too much resistance. They are wound with a few turns of heavy wire. Those connected in parallel must offer enough resistance to take only a fraction of the generated current. These are wound with many turns of fine wire.

As previously explained, any generator will produce more voltage and greater amperage through the same resistance when its armature is revolving faster, or when the size of its magnetic field(s) is increased. With a series generator, any increase of armature speed will send all of the increased current through the field coils to build the magnetic fields up to larger size, and to even further increase the output, and so on. Such a generator will therefore increase or decrease its voltage output tremendously with

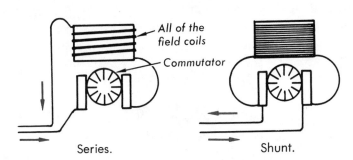

Series.   Shunt.   Combination.

Typifying three generator types.

variations of speed. And since the amount of resistance in the circuit it is supplying will determine how much current will flow at any particular voltage, this generator will increase its output voltage whenever its load calls for increased amperage.

With a shunt generator, only a fraction of the generated current acts on the field coils. Voltage variation with speed variation is much less. Also, an increased load demand for amperage, which flows in a parallel circuit, takes amperage away from the field coils instead of causing the increased amperage to flow through them as in the series generator. This, therefore, reduces instead of increases the generator output voltage. A combination generator has some of the advantages of each type. Voltage will increase moderately fast with increased speed and will remain practically constant under load variations.

Ignition-system generators are nearly always of the shunt type, especially if the engine with which they operate is one which will be run at greatly varying speeds (such as an automobile engine). The principle of shunt operation is put to another good use, as explained in the following section.

Third-Brush Generators

Even with a shunt generator, voltage output will vary somewhat with generator speed. Then, too, the generator's load will vary as accessories connected to the battery are turned on or off; for though its principal task is to recharge the battery, in so doing it also furnishes current to the system. And, again, it must produce more voltage than the battery in order to charge the

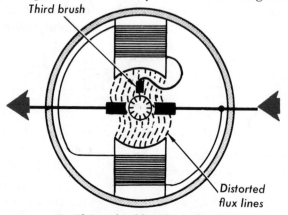

Typifying third-brush application.

battery, for it must force current backward through the battery, in opposition to the battery voltage. If it is constructed so that even during its slower-speed operations it will produce the required voltage, when it is run at top speed it could produce so much voltage, and consequent current, that it would burn itself out.

To overcome these problems the ordinary shunt generator has been modified for automotive and many other similar uses by the addition of a *third brush*. Instead of shunting the field across the two normal brushes, as is done in the ordinary shunt generator, the field is shunted across one normal brush and this third brush. It is still in parallel with the generator output, but now the amount of current received by the field will be varied in accordance with generator speed and the third brush position, as follows:

An armature coil is also an electromagnet. The current produced in the coil establishes a magnetic field of its own. The flux lines of this field are at right angles (90°) to the flux lines of the field coil which the armature coil is cutting. At slow speed, with the resulting low-current production, this armature field is slight. It has little effect on the magnetic field of the field coil. But as the speed—and the strength of the armature field—increases, the armature field's flux lines, pushing at right angles into the field coil's flux lines, will distort the latter. As the field flux lines are pushed to one side, the generator becomes slightly out of time. The armature coil is now cutting the field flux lines a little later (farther around the circle of its rotation) than formerly—a little later than the point at which the brushes are set to take off the maximum current. Therefore, a little less than maximum current is taken off. The generator is tending to limit its own output.

The third brush takes advantage of this tendency. It is so positioned (and the commutator is so wired to the armature coils) that it will take off a maximum voltage for supplying the field coils with current just when the generator is running at an optimum selected speed. This position usually is halfway between the other two brushes, and the optimum speed usually is the slowest speed at which the generator is expected to charge the battery. Hence, the field coils receive maximum possible excitation at the slow generator speed, and the output voltage is as high as possible for this speed. At higher speeds, the same limiting factor already described will operate and, though this factor is not sufficient to prevent voltage increase as the speed increases, it does slow down the voltage increases. The third brush doesn't change this, but it does provide a selected starting point (generator speed) at which voltage is brought up to as high a value as possible.

Most third brushes are held in a *movable brush holder* which can be rotated a limited amount around the armature. Thus, the starting position of the third brush can be varied. By varing its position, the generator speed at which a charging voltage is obtained can be varied. This won't make the generator produce a greater or lesser maximum voltage; it will only make it produce a greater or lesser voltage at a given slow speed. For purposes of adjustment, the third brush generally is on top where removal of the band that covers the brushes will expose it. It is thinner than the other two brushes.

Maintenance

Service maintenance, outside of a specially equipped shop, is very limited. Brushes can be replaced if worn out; the third brush can be adjusted for position; the commutator can be cleaned (brightened) if not too badly dirtied or worn; and, of course, the armature shaft bearings should be lubricated as required. To service brushes or commutator remove the generator *cover band* (which usually snaps on around the brush end of the case).

For brightening the commutator, use only fine sandpaper (*not emery,* which contains metal particles that might wedge between the bars to short them). With the generator rotating, reach in and lightly sand the whole commutator surface until it is bright. Brush pigtail leads can be disconnected, and new brushes can be slipped into the holders. Be sure to position each correctly and to position its spring, also, to hold it down against the commutator. New brushes should be run in until they seat on the commutator. Sparking at the brushes will be excessive until they do seat.

If a commutator is badly grooved or worn out of round, the brushes will arc excessively until the commutator is blackened over and current ceases. If not already too small around, the commutator can be rerounded and smoothed by grinding in a special jig that will simultaneously rotate it. Afterwards, the separators have to be *undercut* (tops lowered a hairbreadth below bar tops) with an undercutting tool.

Bearings are serviced like any others; also, the generator inside must be kept clean. Excessive oiling will wet the parts, make dirt adhere to them, and might result in shorting out the wiring. If coil wires or the soldered commutator leads do become shorted and burned out, only a shop specially equipped to do rewinding can make the necessary repairs.

AN ALTERNATOR-BATTERY SYSTEM

When used with the type of battery system previously described, an alternator simply replaces the generator to produce ac rather than dc. Since dc must be used for charging of the battery, some type of *rectifier* must be included in the system. This is a device which will allow current to flow through in one direction but not in the other and which, therefore, converts the alternator ac into a (one-direction only) dc. The rectifier may be a separate unit or may be combined with the regulator (a *rectifier-regulator*). Rectified (direct) current may then be used for operation of the entire system (all components operated on dc); or, the dc may be used only for battery charging and operation of an electrical starter, while unrectified ac is used throughout the rest of the system.

The only functional difference between an alternator and a generator is in the method by which the generated current is collected to flow through the external circuit. Instead of a segmented commutator with two or three brushes, an alternator has a fully round (unsegmented) *collector ring* with a single brush. The armature is called a *rotor* and the field coil assembly is called a *stator*. Automotive alternators have built-in *diodes* (rectifier units) to rectify the external-circuit current as it leaves the alternator.

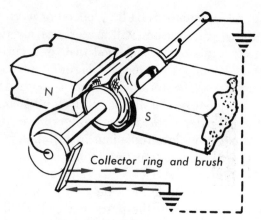

Typifying ac generation.

THE CUTOUT

Regardless of how a generator is constructed, there are times when it is not charging the battery—when it is stopped, or revolving too slowly. If the connection between it and the battery were permanent, at such times as these the battery voltage would exceed generator voltage. In such case, the battery would discharge through the generator, which might operate it as a motor or might not; but, in either case, this would run down the battery. To prevent this a *cutout* is used, usually installed on top of the generator. This is a small relay (coil-operated switch) with the coil constructed somewhat like an induction coil. It is a fine-wire coil with a heavy-wire coil on top of it, around the same core.

The fine-wire (*voltage*) coil is connected in series with the generator field coils. When the current through the field coils—and, therefore, the generator output voltage—builds up to sufficient value, the electromagnetic effect of this voltage coil will become sufficient to pull the *contact arm* down against its spring and close the *relay contacts*. Generated current will now flow through these contacts to the battery; but it also flows through the heavy-wire (*current*) *coil*. Its effect is to further increase the electromagnetic strength, to ensure that the contacts remain closed.

Now, if the generator voltage drops below the battery voltage, the battery will send a reverse current through the cutout current coil to the generator. Reversing the current reverses the polarity established by this coil. Its polarity is now in opposition to the polarity of the voltage coil. This opposition cancels out all the electromagnetic effect, and the contact arm spring pushes the arm up to open the contacts and stop current between the battery and generator.

Maintenance is limited to servicing the contact points if these become pitted or dirty. They can, within reason, be resurfaced and leveled by squeezing them gently together with a fine-tooth file (or fine emery paper) in between, and filing them until they are properly restored. Setting of the contacts after filing (or if arm or spring has become damaged) is very critical as the resistance to movement offered by the arm can affect the amount of generator voltage required to close the contacts. Each manufacturer gives instructions which must be specifically followed. If the coil is damaged, install a new cutout.

A VOLTAGE REGULATOR

When used, a voltage regulator serves the purpose of a cutout. It also functions to keep

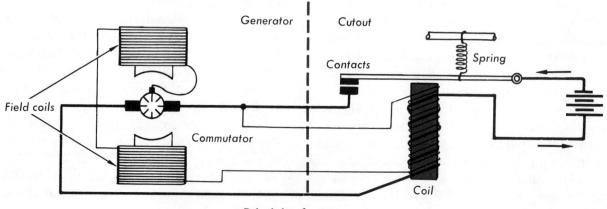

Principle of a cutout.

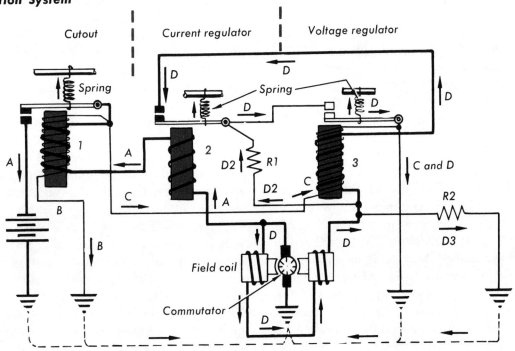

A type of voltage regulator.

the generator voltage output from exceeding a predetermined maximum value, regardless of the speed of generator operation. (A regulator is not required if the generator will operate at one constant running speed.) Its purpose is to keep the generator from developing enough voltage to possibly damage itself and other parts of the ignition or accessories systems. Too high a voltage causes excessive arcing across breaker points. Quite a few makes are manufactured, each operating in a slightly different manner. Some have only two coils, but the majority have three. One of the latter types is illustrated and explained as follows:

Coils 1, 2, and 3 are called the *cutout, current regulator,* and *voltage regulator,* respectively. The cutout and voltage regulator coils are each wound with both a primary (heavy line in the illustration) and a secondary (light line in illustration), while the current regulator has a primary only. The cutout contacts close when this relay's coil is energized; but energizing of either the current regulator or of the voltage regulator coils opens the respective relay's contacts.

At a start, No. 1 contacts are open; No. 2 and No. 3 contacts are closed. Generated current ("A" arrows) flows through No. 2 to energize this coil, but is insufficient to open No. 2 contacts. It flows on through No. 1 primary, then (since No. 1 contacts are open) back through No. 1 secondary and to ground ("B" arrows). This flow energizes No. 1 coil to close these contacts. At this time the current flows through the contacts and into the battery to charge it, instead of back through the high resistance of No. 1 secondary (as in the beginning).

At this stage the field coil current ("D" arrows) flows through No. 3 primary, around through the still closed No. 2 contacts, through the closed No. 3 contacts, and on to ground. It is insufficient to energize No. 3 coil to the degree required to open No. 3 contacts.

Now, if generator voltage drops below a predetermined level, the opposing battery current will tend to flow backward through the closed No. 1 contacts, and backward through No. 1 primary, through the circuit ("A" arrows) to the generator, and to ground. At the instant this backward battery voltage equalizes generator voltage, there will be no current in No. 1 primary or secondary. The coils will be de-energized, and the No. 1 contacts will be spring opened to prevent current flow. This is the cutout action.

If the generator output exceeds a predetermined amount, the increased current through No. 2 will energize the core to a sufficient degree to open the contacts of this relay. With the contacts open, the field coil current ("D" arrows) can no longer flow through them. Instead it must flow ("D2" arrows) through the resistance R1, then through the still closed No. 3 contacts to ground. Adding this resistance to the circuit reduces the current flow through the field coils. It also reduces the generator output, which, in turn, reduces the energy of No. 2 coil and allows No. 2 contacts to again close (thus starting the cycle all over again). As a result, as long as the generator output tends to exceed the predetermined amount, this cycle is repetitive. And the contacts vibrate (open and close) in a manner that will just allow the generator output to be maintained at a predetermined value. This is the current-regulation action.

The small fraction of generated current which flows ("C" arrows) through No. 3 secondary to ground aids the primary of this relay. As soon as the generator voltage exceeds a predetermined amount, the current in this secondary, together with the field coil current ("D" arrows) in the primary, will magnetize the core sufficiently to open the No. 3 relay contacts. When this happens, the field coil current ("D2" and "D" arrows) can no longer flow through No. 3 contacts to ground. Instead, it must flow ("D3" arrow) through the larger resistance R2 to ground, thus still further reducing this current to reduce generator output. Again, the current reduction results in a reduction of the relay-coil magnetism (No. 3 coil, this time). A cycling is created which will cause the No. 3 relay contacts to vibrate, thus maintaining generator output voltage at a predetermined level. This is the voltage regulation action.

Maintenance of regulators is the same as for a cutout, with the exception that a regulator is even more delicately adjusted and requires exact knowledge of the manufacturer's requirements if it is to be readjusted. Most regulators have the wiring diagram inside the cover, together with pertinent information. Aside from cleaning contact points, however, it usually is advisable to discard a malfunctioning regulator rather than attempting to repair it.

A LEAD-ACID STORAGE BATTERY

Construction Features

The commercial lead-acid battery is composed of either three or six cells housed in a single case. As each cell normally provides approximately 2.2 volts, these are rated, respectively, as 6- and 12-volt batteries. The voltage is due entirely to the materials used in the cells. These are lead peroxide for the positive electrode (*anode*) and spongy lead for the negative electrode (*cathode*). However, each cell is composed of a number of positive electrodes and a number of negative electrodes, each in the form of a *plate* (a thin, rectangular piece of the metal). All the positive plates are connected together and all the negative plates are connected together, but the plates are alternated (a positive, then a negative, etc.) and are separated by thin wood or composition *separators*. The number of plates, and their square-inch areas are the factors which determine the ampere-hour rating of a battery and its maximum discharge current. Plate thickness determines the life of a battery.

In addition to the plates and separators—and the case in which these are housed—a battery must have an *electrolyte*. The latter is the chem-

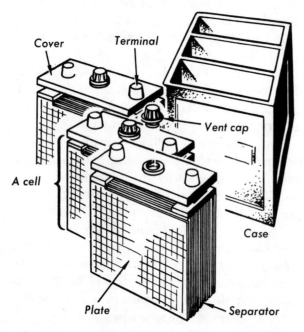

Components of a 6-volt lead-acid storage battery.

ical agent responsible for the electrical action. It consists of a *dilute solution of sulfuric acid.* Distilled water is used to dilute the acid because it is free of all minerals. Tap water from a city settling basin may be used if it is known to be free of minerals; but well (especially artesian well) water, which contains much iron, must *never* be used, as the mineral content would cause internal battery shorts. As the water dilution is subject to evaporation—especially in warm weather or if the battery is much used— new dilution must be added as required to keep the plates and separators submerged. Allowing the battery to run dry will damage the plates and may also buckle the separators (if these are wood). New acid is never required, unless some has spilled out.

When the components are assembled they do not at once generate electricity. *Charging* is required. A battery is charged by running a suitable electric current through it, backward from the negative to the positive. Charging causes the electrolyte to chemically decompose the negative plates and to carry the chemical products to (and deposit them on) the positive plates. This action statically charges the two sets of plates. When, thereafter, the battery terminals are connected into a circuit, the process is reversed, and the chemical action restores the chemical products from the positive to the negative plates, simultaneously causing a continuous discharge of electrons from the positive to the negative, then around the external circuit, from the negative to the positive again. This will continue until the full charge is dissipated and the battery is again run down (*discharged*).

Maintenance

A battery should *never* be left long in an uncharged condition. The sulfuric acid (which is then all in the plates) will quickly cause *sulfation*, which destroys the plates. Condition (amount) of charge is determined by testing the *specific gravity* of the electrolyte, which increases with the degree of charge. A *battery hydrometer*, especially calibrated for this purpose, is used. When the battery is charged, the electrolyte will have a specific gravity of 1.275 to 1.300 (called "1275" to "1300") as against

1.000 ("1000") for plain water (which the dilution approaches as the battery is discharged).

NOTE: So-called dry or precharged storage batteries are available. The only different feature of such a battery is the manner of preparing it for shipment from the manufacturer. It is shipped with the acid in a separate container, to be added when the battery is readied for service. Dilution must also be added, and the battery must then be charged. However, a shorter charging time is achieved.

Direct current must be used for recharging. Also, the *vent caps* (which are removed for refilling each cell) must be off while charging so that gases formed by the process can be vented. And each cell must be filled with dilution (called *battery water*) to cover the tops of the plates (but don't fill the vent tops, or some of the dilution will boil over and lose acid).

Charging can be done by either the *constant-current* or the *taper-charge* methods. The first is accomplished with charging apparatus that will supply 5 to 10 amperes (per each three cells if more than one battery at a time or larger batteries are being charged) under suitable dc voltage continuously until the specific gravity of the electrolyte is raised to the correct reading. The second (called *quick charging*) requires apparatus which starts charging at a very high amperage (100 or more amperes), and then tapers off as the full-charged condition is approached. In any event, total amperage up to the ampere-hour-rating of the battery must be "pumped through" the battery to chemically condition it for the output capacity we refer to as its fully charged condition.

NOTE: An engine-driven generator recharges a battery at up to a 35-ampere (or greater) rate depending on speed of operation and voltage regulator control. This high rate is required in many applications (such as automotive use) since the battery may be discharged at as high a rate as 400 amperes for periods of several minutes when starting a cold engine.

A third charging method (*trickle charging*) can be accomplished with a small ac-to-dc rectifier (available under various trade names) which supplies a relatively small amperage suitable for long-period recharging of a battery that is not fully run down.

Accumulated dirt (if moist) and grease can conduct electricity. They will create a short between battery terminals if allowed to accumulate on the battery top. On the other hand, *battery corrosion* (an acidy substance) will collect on the terminals and block current, especially if the connections are not good and tight. Then, too, the dilution, which is water, can be frozen. The closer its specific gravity is to the specific gravity of water (battery in a discharged condition) the more nearly will its freezing temperature approach that of water; but if a battery is kept fully charged, or nearly so, the dilution will not likely freeze at temperatures above 0°F (at sub-zero temperatures precautions must be taken). Freezing of a discharged battery is a hazard as this will crystallize the plates and/or crack the case. Crystallized plates are useless. A cracked or damaged case destroys the battery by allowing the dilution to leak out. Any portion of the plates allowed to become dry while they are charged is subjected to chemical decomposition. Drying will quickly destroy charged plates.

Battery voltage is not constant. It is only "at par" when the battery is properly charged. When half-discharged a 6-volt battery may deliver as little as 4.5 volts. It can also be overcharged to deliver as much as 7.5 volts for a few minutes. Excessive overcharging will, however, burn it up by causing internal shorts through the separators. This is the principal reason for providing a voltage regulator in applications in which the danger exists.

If kept properly cleaned and charged, a new battery will render good service for some time (different guarantees are made for the different makes). Eventually, however, the battery will wear out. Some of the chemical products become dissipated with each charging and discharging. In time, the negative plates become too eaten away (thinned and/or full of holes) to give further service. The separators, too, lose their insulating qualities. One or another of the cells gives out, and the whole battery must be discarded (rebuilding isn't worth the cost). Discharging a battery too rapidly can hasten this

deterioration due to the excessive heat generated. Enclosing a battery which is subjected to heavy discharges (so that there is no air circulation to dissipate the heat of discharge) can also hasten the deterioration. Discharging produces internal heat in proportion to the rate of discharge. Maximum permissible heat is approximately 115°F.

AN AMMETER

Usual practice is to incorporate, in any system of this type, a special-design ammeter which will register both the charging current of the generator and the discharging battery current to all system parts except a starter. The starter draws such a heavy current that registering of this also on the one ammeter would be impracticable. Various systems have the ammeter connected in many different ways.

AN IGNITION SWITCH

A switch of some type is always employed in the circuit (called the *primary circuit*) which goes from the battery through the breaker points to the primary winding of the ignition coil and back to the battery (generally through ground). As the battery is a constant electrical source, if this switch were not used to open the circuit during periods of engine idleness and the breaker points should stop in a closed position, current would continue to flow until the battery ran down. This would damage the battery and, possibly, the breaker points and/or coil.

Such a switch also serves two other useful purposes. First, by opening the primary circuit and thus cutting off the ignition, it affords a quick, easy method of stopping the engine. And the engine cannot be started until this switch is turned on again. Therefore, by making the switch key-operated (usually done), it is able to serve the second purpose of protecting the owner against unwarranted use of his engine.

AN IGNITION COIL
The Principle Involved

When current flows through a coil a magnetic field is created surrounding the coil. As the cur-

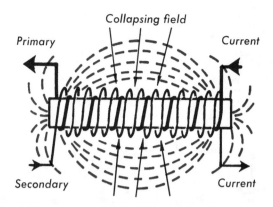

Induction-coil principle.

rent increases, this field expands; a decrease of current allows the field to contract. Consequently, if current is turned on and off this field correspondingly expands from zero to maximum size and then collapses back to zero. A moving (expanding or contracting) field generates an emf. If a second coil in a separate circuit is positioned within this moving field, the generated emf (*induced voltage*) will produce a current through this second coil and its circuit. This phenomenon is called *current induction,* and a device containing two such coils (a *primary* and a *secondary coil*) is called an *induction coil.*

If a secondary coil contains more loops than the primary, the induced secondary voltage exceeds the primary voltage. However, if secondary voltage is greater than that of the primary, then the secondary amperage is correspondingly smaller. The secondary output (volts × amps) always approximately equals primary input.

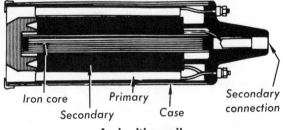

Iron core   Primary   Secondary
Secondary   Case   connection

An ignition coil.

Construction Features

An ignition coil is an induction coil having the primary and secondary wound around the same core and encased in an insulated metal or bakelite shell. The core usually is a bundle of soft iron wires. The two coils are wound on separate insulating cylinders so that they telescope together around the core. The secondary (usually the inner coil) consists of 10,000 to 20,000 turns of fine wire; the primary has 150 to 250 turns of heavy wire. This great difference in numbers of turns allows the coil to boost a 6-volt battery current through the primary to a very high secondary voltage (in some cases, up to 25,000 volts). Hence the coil produces an output current of negligible amperage but with sufficient voltage to jump a spark plug gap (8000 to 12,000 volts generally are required, depending on the type of plug) and create the ignition spark.

Two screw-post terminals—one marked BAT, and the other DIST—serve for connecting the coil into its primary circuit. A single snap-in fitting is provided for the secondary wire that carries current to the distributor and hence to the plug(s). The other secondary coil end is grounded, either through the coil case, which must then be firmly grounded, or through a separate terminal so marked. Coils are classed as having positive or negative polarity, depending on which secondary coil side is grounded. Some coils contain built-in ignition switches in the primary circuit.

Maintenance

The very high secondary voltage makes leakage loss inevitable if the coil case becomes cracked or if the area around the secondary snap-in connection becomes moist or dirt- or grease-coated. For best operation an ignition coil must be leakproof. One that is externally damaged or internally shorted is not worth repairing.

A DISTRIBUTOR WITH CONDENSER

Function and Operation

The following discussion refers only to mechanical distributors. *Electronic ignition distributors* are discussed later in the chapter.

As previously explained, for an induction coil to induce a current in its secondary winding, the magnetic field of its primary winding must be in motion, and the secondary winding must be simultaneously a part of a closed electrical

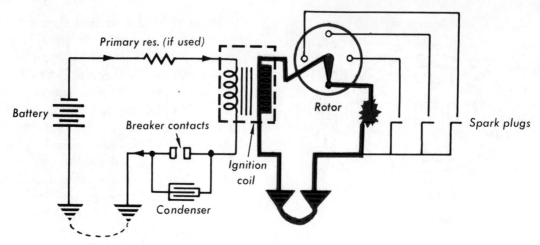

Basic ignition circuit for a four-cylinder engine.

circuit. In an ignition system, the magnetic field of the coil primary is initiated by first sending battery current through the primary winding to build up the field, and then by opening the primary circuit to allow the field to collapse. During the instant of its collapse, the secondary winding is connected into a closed circuit with the spark plug to be fired. Therefore, the high-tension current induced in this circuit flows around the circuit by jumping the spark plug gap and thus firing the plug.

It is the job of the distributor (*timer-distributor*) to make these events occur in proper relation to the spark requirement of each cylinder in an engine. For this purpose the distributor contains *breaker contacts* in the primary circuit, and a *rotor* to distribute secondary current to the proper spark plug circuit. The condenser serves two purposes—it assists the collapse of the primary field, and it protects the primary circuit opening mechanism by preventing the battery current from arcing across the gap as the contacts open. Arcing would quickly burn up the contact points.

Basic primary and secondary ignition circuits are illustrated. At the beginning of a firing cycle, the breaker contacts close, and battery current flows through the coil primary to build up its field. At the right instant, these contacts open. Ordinarily, current does not cease at the instant two such contacts start to move apart; it tends to bridge (jump) the widening gap and to continue flowing until the resistance created by the widening gap is too much for the current volt-

age. (Air does have slight conductance qualities.) The amount of this resistance increases, beginning with the split-second the contacts start to open.

During the time the contacts are open, the condenser is statically charged with a high potential due to the back emf developed across the coil.

Hence, in the split second that the gap resistance starts to create a voltage drop in the battery current, the potential (voltage) stored in the condenser will be greater than the battery voltage across the contacts. This being the case, the condenser instantly begins discharging backward through the circuit, in opposition to the battery current. Thus it effectually heads off and instantly stops the battery current, so that the contacts will make a clean break, without any arcing of current across them.

The current created by the condenser discharge builds up rapidly to maximum value as the battery current ceases and no longer opposes it. This current surges back through the primary coil, through the battery and ground,

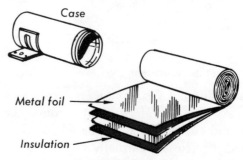

A typical condenser.

and into the opposite side of the condenser, to recharge the condenser in the opposite direction. Thus recharged at the opposite side by its own discharge, the condenser again discharges, this time in the other direction through the primary circuit, back through ground, the battery, and the primary coil into the first side of the condenser again. This action would continue indefinitely—with the current "bouncing" back and forth through the primary circuit—except that the current is opposed by the various resistances in the circuit. With each "trip" the voltage drops, until the voltage becomes zero and the current ceases. Hence, we say that the condenser discharge constitutes an *oscillatory* current of rapidly *diminishing value.*

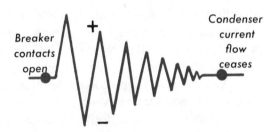

Typifying one cycle of diminishing value oscillatory condenser current flow.

NOTE: Some systems include a *fixed resistance* of certain value to limit battery current in the primary circuit and to more quickly diminish the condenser current. Others do not.

With the first surge of condenser current, this current travels backward through the primary coil to reverse its polarity and hasten the collapse of the field. The following surge is again reversed, to oppose the effects of the first surge, etc. In this manner, the condenser operates to hasten, then to resist, the collapse of the magnetic field. It thus adds considerably to the movement of the field. Since the rapidity and duration of field collapse determine the voltage and duration of the current induced in the secondary circuit, the condenser action increases the strength and the duration of the secondary current.

Just as the condenser oscillatory cycle begins, the rotor of the distributor closes contact in the circuit to one of the spark plugs. This contact

remains closed long enough for the secondary current to jump the spark plug gap. The contact breaks and, at the same time, the breaker contacts close (thus making a definite end to any lingering condenser discharge, and again sending battery current through the primary coil). The cycle repeats itself, with the rotor contacting a second spark plug electrode, and so on, continuously at the rate of 6000 cycles per minute for a four-cylinder, four-stroke-cycle engine, running for instance, at 3000 rpm.

Construction Features

A typical four-cylinder distributor is illustrated. The *rotor* and the *cap*, together, constitute the spark distribution mechanism. There is a central *tower* in the cap which holds the *lead* (wire) from the coil secondary winding. An *electrode* (terminal) at the base of this tower is in constant contact with a spring-type *conductor* on the rotor. In revolving, the rotor sweeps past and makes electrical contact with the electrodes at the bases of the towers around the cap circumference. These towers hold the leads to the various spark plugs—there is one for each plug.

The *cam*, on the same shaft with the rotor, has as many *lobes* (corners) as there are cylinders. A spring-loaded *breaker arm* is pivot mounted on the *breaker plate* so as to ride against this cam, which rocks it outward each time a lobe comes around to the arm. There is a *movable contact point* on the arm end and a *stationary point* on an adjustable mount adjacent to the first point. These two points, made of very hard metal, constitute the *breaker contacts,* which are separated to form a gap and break the primary circuit each time the arm is rocked outward. The primary coil connection to the movable point is made at the external *primary terminal,* which is wired to the arm (or current is carried to the arm through the spring); and the stationary point is grounded to complete the primary circuit. The *condenser,* also shown mounted on the breaker plate, is likewise grounded at one end and connected to the terminal at the other end.

NOTE: Breaker arms and other parts are also made in many other designs. The condenser may be mounted on the outside or at any location in the system.

The cap, rotor and cam all have fixed positions, so that each can be mounted in only one way. *Snap fasteners* hold the cap to the body. The rotor and cam are on the *shaft,* which usually is held in precision bearings of the plain type. There may be a *coupling,* which joins two separate parts to form the one shaft through use

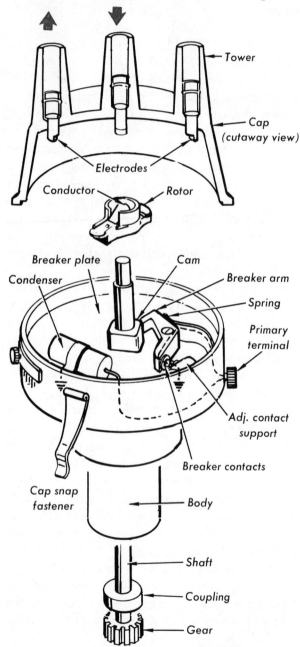

A basic four-cylinder engine distributor.

of a soft iron *shear pin,* which will break and prevent further damage if anything goes wrong inside the distributor. A small *gear* on the shaft end provides the means of driving it.

The distributor for any four-stroke-cycle engine has to rotate at *one half* crankshaft speed (since each cylinder fires only on every other revolution). The coupling between the distributor shaft gear and a driving gear on the crankshaft provides the correct gear ratio. Due to the 2-to-1 gear ratio, a 1° distributor shaft movement corresponds with a 2° crankshaft movement, etc. When used with a two-stroke-cycle engine, a distributor shaft revolves at the same speed as the crankshaft (a 1-to-1 gear ratio).

Adjustment Features

A distributor must fire the spark in each cylinder at precisely the correct instant. This may be just when the piston reaches TDC of its compression stroke or—as discussed in Chapter 4 under "Combustion Lag"—sooner or later, depending on several factors. With some engines the distributor is set in one optimum unvariable position specified by the manufacturer; with other engines this first setting is established, but is then varied (either by manual or automatic means) to advance or retard the spark as required by engine speed and/or load. In either case, we shall refer to the first setting as the *initial timing adjustment* and to any subsequent settings as the *spark advance adjustment.* Then, in addition to these, there is also a *contact gap adjustment.*

All adjustments are begun with the initial timing adjustment. Since the cap, rotor, and cam are all in fixed positions, all that is needed to make this adjustment is to install the distributor

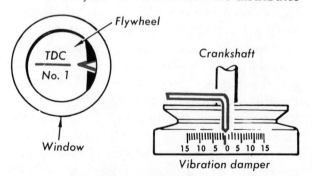

Typical timing marks.

with its shaft gear correctly meshed with the gear that drives it, so that the rotor will be contacting the spark plug electrode for a piston known to be in its firing (TDC) position. To accomplish this a *timing mark* generally is provided to indicate when a selected reference piston is at its TDC position. The mark is a scribed line on the flywheel (or other engine rotating part) which can be aligned with an arrow placed on an adjacent housing. Installation is made by aligning the mark and arrow and then seating the distributor in place and meshing its gear so that the rotor will be pointing to the desired electrode.

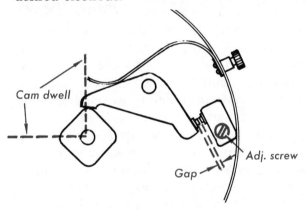

Contact-gap adjustment.

With the distributor installed, the contact gap is next adjusted. The maximum-width gap obtained with the contacts full open will, of course, determine two things: first, how long the contacts remain closed out of each cycle; second, the exact instant of contact opening. We call the duration of time that the contacts remain closed the *cam dwell*. This is important because it determines how long battery current will flow through the primary coil to build up its magnetic field. It determines the degree of *coil saturation* and therefore the strength of the resulting spark.

The contact gap is extremely critical. It is adjusted, usually, with a *feeler gauge* to exact specifications. Afterwards it should be checked on a *distributor test stand* capable of measuring coil saturation for comparison with manufacturer's specifications. After the gap is adjusted, the instant of opening will be determined by the relationship between the breaker arm and cam (the instant at which a cam lobe starts to

rock the arm). Final setting of this is established during the spark advance adjustment.

Spark advance adjustments can be: (1) a nonvariable preoperating adjustment which establishes a single advanced setting for all engine operating conditions; (2) a variable range of spark advancement controlled by the operator through a *spark lever;* (3) automatic spark advancement obtained by built-in controls. The first is used only when an engine is designed to run at one (only) operating speed under known conditions. Most small engines have the first type, though many outboard motors have the second arrangement (manual).

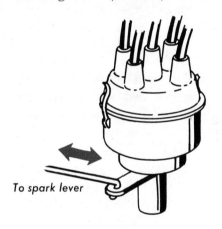

Manual spark-advance control.

With the first type, any precise advance setting called for by the manufacturer is established by using the timing mark already described. In this case, however, instead of a single TDC line, the mark will include additional lines at each side of this TDC line (which may or may not be numbered). Each additional line forward (in direction of rotation) from the TDC line will indicate that the reference piston is then so many degrees ahead of TDC. Lines behind the TDC line indicate that the piston is so many degrees past TDC. (The manufacturer will specify the degrees indicated by each line if these aren't numbered.) Setting is made by moving the correct line into alignment with the arrow and then by adjusting the breaker cam and cam relationship so that the contact gap is just starting to open. To accomplish the latter, either the distributor body or the breaker plate will be designed so that it can be rotated the required degrees with respect to the shaft, and be locked in this position.

When a manual control is provided, the general practice is to have the whole distributor body rotatable about 40° around the shaft. The spark lever is linked to the distributor to effect this rotation at the will of the operator. Movement usually is between the stemlike lower part of the distributor body and a holding bracket. It is such that both advanced and retarded spark settings can be obtained by lever operation.

Automatic Controls

Three different types of automatic spark advancement controls are in use: a *centrifugal advance control*, a *vacuum brake control*, and a *vacuum advance control*. The centrifugal control may operate to rotate the body, the cam, or the breaker plate with respect to the shaft; both the others are described later. In any case, bearings are provided to make the rotation smooth and accurate. Their proper operation is very critical.

A centrifugal control is simply a two-flyweight governor with springs, comes like ones previously described, mounted on the shaft below the breaker plate inside the distributor body. The weights are linked to the movable body, cam, or breaker plate so that they will rotate it and advance or retard it in relation to the shaft, relative to the speed of engine operation.

Often used with the centrifugal control is the vacuum brake control, which functions to regulate operation of the centrifugal control in accordance with the load condition of the engine. It consists of a small spring-loaded plunger connected to a diaphragm mounted in a housing attached to the side of the distributor body. The diaphragm is actuated by suction through a tube connected with the carburetor intake manifold. Whenever the engine is under load, the resulting drop in manifold pressure will cause atmospheric pressure on the diaphragm to move it outward (against the spring) so that a friction pad attached to the plunger will rub against a disc attached to the centrifugal advance mechanism. The resulting braking action slows the spinning weights and retards their advancement of the spark.

The vacuum advance control takes the place of the two just described and accomplishes both purposes simultaneously. Two springs are used to hold the movable breaker plate in retarded position. Their action is opposed by movement of a diaphragm, which acts to advance the breaker plate in relation to both speed and load. The suction that operates this diaphragm comes from a tube connected with a passageway in the carburetor that opens into the intake manifold at one end and into the venturi at the other end. As a result, the diaphragm is sensitive to both speed and load variations and establishes a compromise breaker plate advancement.

With any of the preceding controls, it is necessary to adjust the breaker arm and cam relationship to a specified starting value. Timing marks like those described for a nonvariable-type setting may be provided. As a rule, however, this starting setting is accomplished with instruments and must be done in strict accordance with manufacturer's instructions. After the starting setting is made, the *advance curve* (rate of advancement to be obtained by the automatic control) must also be adjusted. This, too, must be done per manufacturer's instructions.

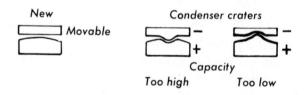

Breaker contacts.

Maintenance

Considerable maintenance may be required in connection with any distributor. First, there is lubrication; then, adjustment; and, finally, repair or replacement of damaged parts. Many close inspections and checks are required to track down possible troubles. Indeed, with possible exception of the carburetor, the distributor is probably the most difficult part of an engine to service properly. Proper service is impossible without a good ignition test stand having provisions for all the various tests required. We cannot, here, go into all of these tests; we can only touch on the highlights.

The shaft bearings require lubrication, but the wrong type of lubricant, or too much of it, will cause a malfunction by getting all through the distributor to short out the circuits (especially the high-tension secondary circuit). Parts must be grease-free and clean. Bearings and other shaft parts are replaceable. If there is any wobble at all in the shaft, parts must be replaced as required.

Any cracks in body, cap, or rotor, or any moisture or dirt may lead to loss of current and a poor-quality spark. All parts must be sound and clean. Then, too, all wire insulations must be good, all connections must be tight, and all contacts must be sound and clean. This refers both to the rotor and electrode contacts and to the breaker contacts. The latter, especially, must be in good condition. If pitted or dirty they may be cleaned and readjusted, though generally it is best to replace them.

NOTE: The movable contact is flat while the stationary one is convex. They meet at just one point, in the exact center. If reground, they must be fitted in this manner. They are, however, always available in matched sets for replacement. They must be carefully installed to make contact squarely as described.

Electrodes are not replaceable. If burned or otherwise damaged, the cap must be replaced. The rotor, also, must be replaced if burned or damaged.

The condenser must be electrically sound with no internal leakage or shorts which are very likely to develop in time. It also must be the *exact required capacity*. If its capacity is too high, a crater will form in the positive breaker contact; if too low, a crater will form in the negative contact. The latter development will also indicate condenser malfunction due to leakage.

NOTE: If battery positive terminal is grounded, the grounded contact point is the positive one; and vice-versa.

Breaker-arm spring tension must be correct, and the arm must seat squarely against the cam. The contact gap has to be correct to obtain desired cam dwell and coil saturation. All other adjustments must be exactly as specified by the manufacturer. If automatic controls are used, each has to be adjusted precisely in accordance with manufacturer's instructions. A centrifugal control usually is adjusted by bending or repositioning the spring anchors; vacuum controls are adjusted by altering spring tension with an adjusting nut or by resetting spring anchor posts. All such springs must be in perfect condition, and all control parts usually are replaceable.

SPARK PLUGS

Construction Features

Many different types of spark plugs have been manufactured, each with its special advantages such as longer life, easier cleaning, etc. There have been separable types, which come apart for cleaning, and nonseparable types, and those with porcelain insulators and some using other materials, etc. On the whole, however, the only characteristic of a plug that needs to concern us is its ability to handle the high-tension current without leakage loss, so that it will deliver a sure, full-volume spark. Engine manufacturers make exhaustive tests and always recommend exactly the right make and style plug (or plugs, if there is a selection) for their engines.

NOTE: A special type of modified plug has been developed for two-stroke-cycle engines (accompanying illustration), and such a plug usually is recommended by the engine manufacturer. In any event, the performance of practically any two-

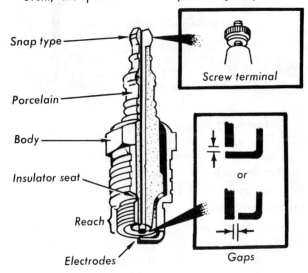

A nonseparable conventional-type spark plug.

stroke-cycle engine will be improved by use of such a plug. Even small four-stroke-cycle engines can use this modified plug to advantage.

About the only choice left to the mechanic is among *hot, normal, and cold plug styles*. All make and specification plugs are available in a range of styles. The difference between the styles is in the height of the *insulator seat* (inside the plug body) above the electrodes. Below its seat the insulator tapers down to the ungrounded (center) electrode, leaving an air gap between the insulator and the body. When a spark jumps across the electrodes it generates considerable heat. Any of this heat lodged in the center electrode has to be carried off through the insulator into the body, and then out through the metal of the cylinder head. It can't jump the gap between the insulator and body. Therefore, the longer this portion of insulator is (the higher the insulator seat), the hotter the plug will be during operation.

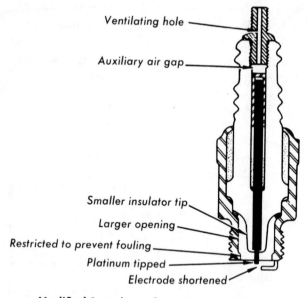

Modified 2-stroke-cycle engine spark plug.

A manufacturer usually recommends the type (hotter or colder) of plug best suited to his engine under normal conditions. If an engine is put to abnormal use, however, a different style may be required. Should a plug run too hot, excessive deterioration will result (its life will be shortened). Should it run too cold, thick carbon formation and misfiring may result. En-

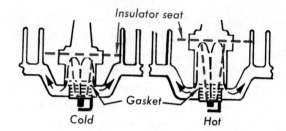

Plug selection.

gines run at above normal speed, under excessive load conditions, or where cooling circulation is below normal, may require cooler-than-average plugs. The reverse is also true.

Most modern plugs are *nonseparable* and have *snap-type terminals*. The *reach* length varies according to cylinder-head design requirements (an improper reach will place the electrodes too high or too low in the cylinder). Electrode *gap* types vary between the two types illustrated. All plugs require a *gasket* (under the body shoulder) to make a leakproof and good heat-conducting installation. It is very important that a plug be installed to proper tightness (preferably with a torque wrench) to provide these requirements. Since the gasket flattens out under the proper pressure, a new gasket should preferably be used each time a plug is reinstalled.

Maintenance

If a plug is fouled with carbon deposits, it can be cleaned in a good sandblast cleaner. Scraping and/or sanding electrodes to brightness is, at best, only a stopgap method. It does not effectively remove carbon and residue deposited up inside where they will affect the heat conductance and running temperature of the plug. After cleaning, the grounded electrode (outside one) must be adjusted (bent) as required to reestablish the specified gap. When checking the gap, be sure to check the actual air gap, not just the vertical distance between electrode surfaces.

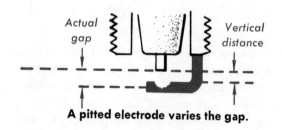

A pitted electrode varies the gap.

If an electrode is pitted, the gap will differ from this vertical distance, as measured by a feeler gauge.

There is no certain method of determining whether the usefulness of a spark plug has ended or not. All plugs deteriorate with use and become electrically unsound due to breakdown of the insulator. Any slight crack or even crazing of the insulator is cause for discarding a plug. If one appears to be good, however, the only way to determine its usefulness is to clean, adjust, and then test it—either in a test stand or in the engine. A plug must produce a reliable, full-bodied spark. If there is any doubt, it is best to replace it.

WIRING HARNESS AND WIRES

In any electrical system, the proper current depends on good, sufficient-size properly insulated *conductors* (the wires which carry current from place to place). In an ignition system, the high voltage carried by the secondary circuit wires—coil to distributor and distributor to plugs—makes it extremely imperative that these wires be of proper size and well insulated. Also, any wire in or about an engine is subjected to grease, dirt, destructive heat, etc., which can break down its insulation. It is therefore important that all wires be kept clean, out of contact with hot surfaces, and in sound condition.

Equally important are the wire connections. Any loose or dirty connections will interrupt current. Wire end pieces (clips, fasteners, etc.) must be firmly fastened to their wires and must fit snugly on or in their terminals. For the reason that engine vibration will break off wiring stiffened with solder, the latter is sparingly used in an ignition system. The secondary coil terminal is often formed as a loop and the spark plug (flexible) lead wire is simply wrapped through it. Modern coils have a plug-in connection. Screw-type terminals must be tight. And special attention is always required to properly tighten the clamp-type fasteners used to connect the wires to a battery.

When wiring harnesses are provided, deterioration of one wire requires replacement of the entire harness. Other wires are replaced individually. If you are ever in doubt, substitute a larger, rather than a smaller, size wire. Check the manufacturer's coding for proper connections.

MAGNETO SYSTEMS

With the magneto systems generally in use for small engines there is no battery to be charged. This means that there is no need for a cutout or a voltage regulator. The magneto serves to generate the current needed for ignition and either incorporates the features of an induction coil, or a separate coil (like the induction coil previously discussed) is used.

NOTE: Magnetos for systems containing batteries are discussed under "Magneto-Dynamos."

A grounded, single-wire circuitry is always used. Hence, we can characterize the simplest magneto system as consisting of just a magneto together with a spark plug for each cylinder (or a plug or plugs for each rotor housing), and a single-wire lead to each plug. An *impulse coupling* (a mechanical device explained at the end of this chapter) is used with some engines.

There are several basic types of magnetos: a *flywheel magneto*, a *rotor (magnematic) magneto*, a *rotating-coil magneto*, and a *magneto-dynamo*. These are discussed separately.

A FLYWHEEL MAGNETO

Construction Features

The parts of a typical single-cylinder, two-stroke-cycle engine magneto are illustrated. Current for the spark is generated by rotation of *permanent (alnico) magnets* past a coil. The magnets are cast into the flywheel and the coil is mounted on a *coil core (or armature)* fastened to a stationary *stator plate*. The coil is wound like an induction coil, with a primary (about 175 turns of heavy wire) and a secondary (about 10,000 turns of fine wire). Current for the primary is generated current, created by

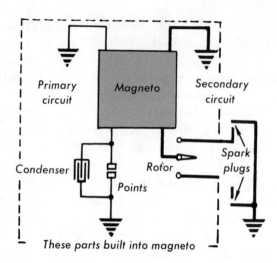

A 2-cylinder magneto ignition system.

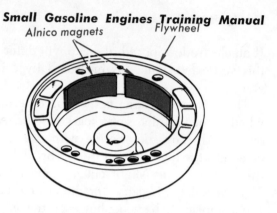

Flywheel with permanent magnets.

the motion of the magnets past the coil. The primary circuit is broken once each revolution by opening of the *points* as the *breaker arm* rides under spring tension against the *cam* mounted on the engine crankshaft. Breaking of this circuit collapses the primary magnetic field and induces a very high-voltage current in the secondary. This current is fed through the wire lead to the spark plug.

The *condenser* is shunted across the points to serve its customary purpose. Lubrication of the cam is obtained by means of the *lubricating pad*, which wipes the cam with special lubricant (Lubri-Plate, usually). When the flywheel is assembled on the crankshaft end, a special *dust shield* is used to prevent seepage of dust into the in-

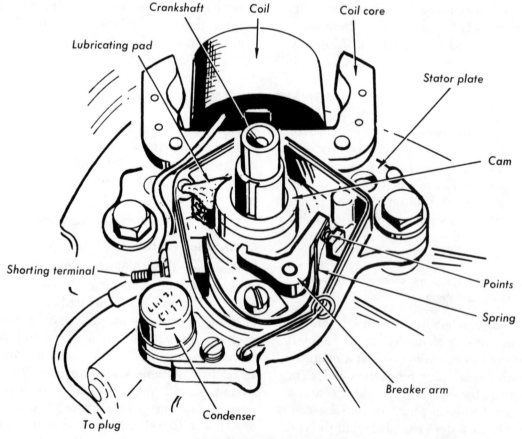

A typical flywheel magneto.

terior. One side each of the primary, secondary, condenser, points, and (of course) the plug is grounded. The *shorting terminal* is used to attach a wire to the engine control panel whereby the other side of the primary can also be grounded—to stop magneto operation and kill the engine.

Adjustment of the stationary point is provided, but the particular magneto shown has no spark advancement provision. (If required, spark advancement would be provided by mounting the breaker arm on an adjustable breaker plate which could be positioned in any one of several fixed positions.) The timing is built into each magneto to suit the requirements of its engine. No timing is required other than to assemble the parts correctly and to adjust the points according to specifications.

Operation

Three stages of magneto operation are illustrated. In No. 1, the flywheel has rotated to position the two permanent magnets (S and N), with poles as indicated, over the left and center legs of the coil core. This action establishes a magnetic field through the coil core, the permanent magnets, and the flywheel rim (as indicated by the arrows). No appreciable current is induced in either winding of the coil (on center leg) at this time. (Only the primary winding is shown in this view.) The breaker points are closed.

> NOTE: Each permanent magnet must, of course, have two poles. For simplicity, we speak only of the exposed magnet pole; the other pole is at the imbedded side.

When the flywheel rotates into No. 2 position, however, the approach of the "S" permanent magnet to the center leg of the coil core tends to reverse the coil core polarity and make this center leg into an "N" pole. This reversal tendency starts to collapse the original magnetic field, but the instant it starts to collapse, a current is generated in the primary coil winding through the still closed breaker points. This current makes an electromagnet of the center leg, a magnet having its "S" pole at the top, where it was during stage No. 1. Thus, the primary cur-

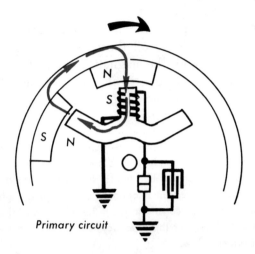

Primary circuit

Magnetic field established.

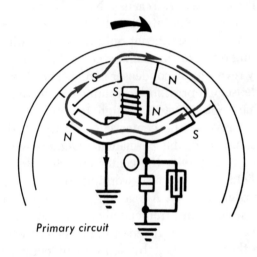

Primary circuit

Primary current "holds" field.

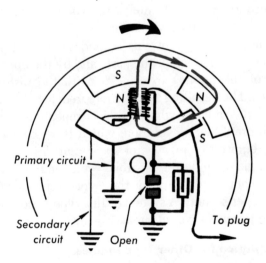

Primary circuit

Secondary circuit   Open   To plug

Points open, field reverses and induces secondary current.

Three stages of magneto operation.

rent tends to oppose the magnetic field reversal. This current *chokes* off the opposing magnetic current and holds the magnetic field as it was in stage No. 1. The electrical effect is much the same as would be the mechanical effect of compressing a stout spring.

> NOTE: There is a similar tendency to produce a current in the secondary winding, which is around the same center leg. But due to the spark plug gap the secondary circuit is open to all except a high enough voltage current to jump the gap——and the force (voltage) of this tendency is too weak, so no current can flow.

The choking effect continues (and builds up) until the flywheel reaches No. 3 position, at which time the points open to stop primary current. At this instant, not only does the built-up choking effect of the primary cease, but it is suddenly reversed by condenser action. The charged condenser, as previously explained, immediately discharges backward through the primary circuit to reverse the center leg polarity. And at the same time, the permanent magnets—which are now fully over the center and right legs of the field core—are also working to reverse the center leg polarity. Altogether, these forces create a sudden and tremendous movement of the magnetic field as it reverses—a movement which is further prolonged by subsequent condenser oscillation, as previously explained. This movement induces an extremely high voltage current in the coil secondary winding.

The occurrence of this secondary current is timed (by location of the parts and by the opening of the points) to coincide with the piston cycle. The resulting spark is created just when needed. And the whole cycle is repeated with each crankshaft revolution, which is, of course, the required recurrence of spark-plug firing for a single cylinder, two-stroke-cycle engine. At 3600-rpm engine speed the cycle repeats 60 times per second. A system of this type is said to have *normally closed* breaker points.

Adaptation for Other Type Engines

Two-Cylinder, Two-Stroke-Cycle Engine— When used with such an engine, the magneto is required to produce two sparks per revolu-

tion. The customary method is to provide two separate groups, 180° apart, each containing a coil with core, a set of points with breaker arm, and a condenser. Each coil feeds one of the plugs, and the flywheel magnets, in revolving, will cycle each in turn to produce the two separate sparks per revolution.

One-Cylinder, Four-Stroke-Cycle Engine— Such an engine requires one spark every other crankshaft revolution. However, during alternate revolutions the piston is in the same position on its exhaust stroke that it is in (the following revolution) on its compression stroke at the time firing takes place. Therefore, there is no harm done if an extra spark occurs at this time. Consequently, many manufacturers use the previously described magneto as it is, allowing the extra spark to occur during exhaust. This usage does, however, require the magneto and plug to do unnecessary double work.

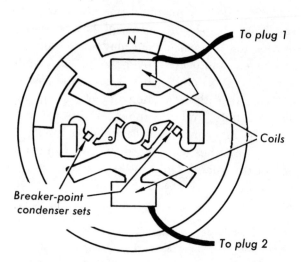

Two-cylinder, 2-stroke magneto.

Multicylinder, Four-Stroke-Cycle Engine— The magneto described can be used for a two-cylinder, four-stroke-cycle engine as it is, if a *distributor-rotor* with two *electrodes* is used to separate the two sparks produced during two crankshaft revolutions, and to direct each to one (only) cylinder. When there are four cylinders, a dual-coil magneto as described for a two-cylinder, two-stroke-cycle engine is used, together with two such distributor-rotors. Such a rotor is mounted on or driven in time with the engine camshaft. One half of its revolution, therefore,

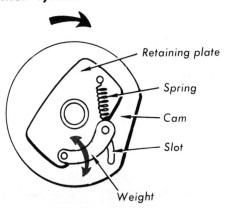

Principle of a variable cam.

corresponds with one crankshaft revolution and one of the magneto cycles; the remaining half of its revolution corresponds with the second crankshaft revolution and the following magneto cycle. During each half revolution it contacts the electrode leading to one (only) plug.

Other Variations

There is also a *normally open* breaker point flywheel magneto in which the breaker points close to permit primary current during a relatively short interval. The magnetic circuits and sequence are different from those already described, but the end result is the same.

Some breaker arms ride directly on the cams, as previously shown; some ride against *cam fol-*

lowers which, in turn, ride on the cam. Also, various types of stationary or movable point adjustments are provided.

Coils are wound, insulated, and even positioned in various manners. With some, the lead wires to the plug are separable; with others, the lead wires are an integral part of the coil assembly.

In some cases hardened-steel permanent magnets, instead of the alnico magnets, are used. If these lose their magnetism (as all may, in time), they can be easily remagnetized (called recharging), whereas alnico magnets can be recharged only by special factory equipment. Most magnets now are cast into the flywheel but some are separable.

Although the magneto described has no provision for spark advancement during engine operation, such provision can be made by using a variable cam. One such cam is illustrated. Centrifugal force of rotation causes the *weight* to swing outward against the *spring*, thus pulling the *cam* around to rotate its lobe with respect to the shaft. The weight and spring are mounted on a *retaining plate* secured to the shaft above the cam (which is loose on the shaft) to hold it in place. The cam is both rotated and positioned by a *pin* in the weight which engages in a *slot* in the cam.

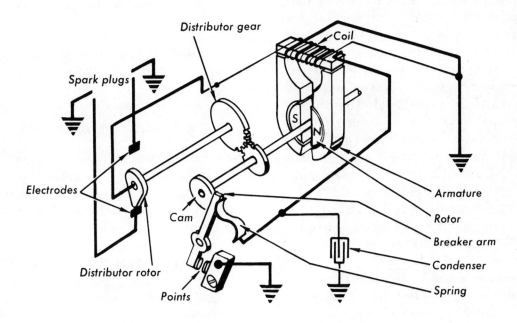

Components of a rotor magneto system.

ROTOR (AND MAGNEMATIC) MAGNETOS

Much the same principle of operation is involved in this type magneto as in the preceding one. With this type the permanent magnets are located centrally on a shaft (the crankshaft or an auxiliary shaft) and rotate within the arms (*laminations*) of an armature. They are called the *rotor*. A *coil*, an induction type the same as used in a flywheel magneto, wound on a *core* is fastened across the two armature laminations so as to create one large, horseshoe-type electromagnet. The *cam* is on the same shaft with the rotor and operates a spring-loaded *breaker arm* to open the *breaker points*, much as described before.

The particular magneto illustrated has a *distributor-rotor* with two *electrodes* to adapt it for use with a two-cylinder, four-stroke-cycle engine. This distributor-rotor is mounted on the camshaft or is operated in time with the camshaft on an auxiliary shaft driven from the crankshaft through a *distributor gear*. When used with a single-cylinder, two-stroke-cycle engine, no distributor-rotor is required, and adaptations for other engine types follow the same principles outlined for the flywheel magneto.

This type of magneto has been adapted for use with single-cylinder, four-stroke-cycle engines without the rotor-distributor, by separating the breaker arm, cam, and condenser (which involve the majority of service problems) from the balance of the magneto. These parts are housed in a receptacle on the engine exterior, while the rotor and armature remain associated with the crankshaft, as required. The breaker arm is actuated by a plunger (cam follower) which rides on a cam secured to the camshaft or camshaft gear. Such an arrangement is particularly well adapted for use with a variable cam, similar in principle to the one previously described. With this arrangement a spark advancement—from engine starting to engine full speed —of up to 25° is achieved. The device is called a *magnematic magneto*.

There are many other possible variations. The breaker arm can be actuated in a number of ways, and a distributor-rotor, if used, can also be placed in a variety of locations. One variation

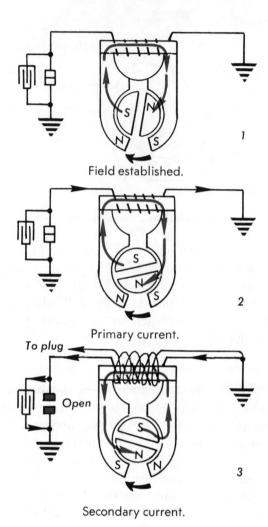

Field established.

Primary current.

To plug

Open

Secondary current.

Stages of rotor magneto operation.

that is much used is a *carbon-brush* type of distributor-rotor. This is much like a segmented commutator, with a brush or brushes, as required.

ROTATING-COIL MAGNETOS

Now that the development of alnico has produced such powerful and reliable permanent magnets, this rotating-coil type of magneto is practically obsolete. It differs from the other two types principally in the fact that its rotating member is the coil and its stationary member is the permanent magnet. The coil is of the usual induction type, and a collapsing primary magnetic field is used to induce a high-voltage secondary winding current, much the same as with the types previously described. As they operate in the rotating primary-coil circuit, the breaker points usually are mounted on the rotating shaft (the cam being stationary), and current is carried out of the rotating secondary through a *collector ring and brush,* which is simply a one-segment—full round—commutator with a brush. The *ground switch* shown would be used to open the primary circuit and stop the ignition. As many variations and adaptations to various engine types are possible with this magneto as with the other two.

MAGNETO-DYNAMOS (AC MAGNETOS)

Used with some engines (especially motorcycles, motorbikes, etc.) for which a small surplus current is required for lights, a horn, etc., magnetos of this type are actually combinations of the magneto principle with the principle of an ac alternator. These are variously called ac magnetos, ac generators, and dynamos. Properly, they should be called *magneto-dynamos.*

This type is a magneto because it contains the customary *induction-type coil.* However, this coil has only a two-leg core. In order to obtain the required reversal of polarity in the core, a number of permanent magnets having alternate "N" and "S" poles are distributed evenly around the flywheel—there may be six, as shown, or any even number from four up. Reversal is obtained

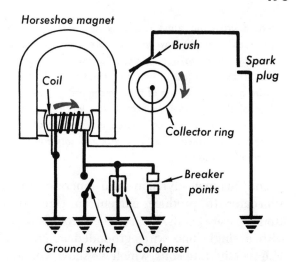

A rotating-coil magneto.

when the first two of any three magnets move on around to allow the next two to be opposite the core ends. During this interval of movement a field is established through the core by the first two magnets, then the next two magnets tend to reverse this field and so generate primary current. Finally, the contacts open, allowing the field to collapse and reverse, and generate the high-voltage secondary spark current. All other events are exactly as described for other flywheel magnetos.

NOTE: There is one other difference to be mentioned. Each three magnets in passing by the core

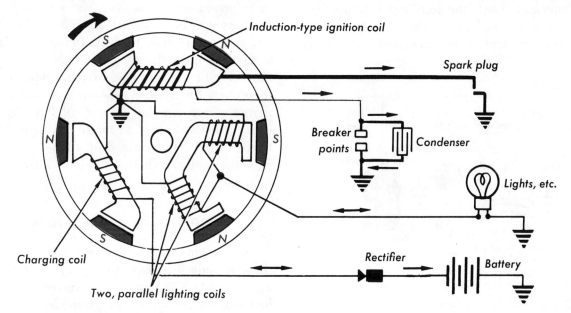

Typifying a 6-pole dynamo designed for ignition, lighting, and battery charging.

would have the same tendency, just described, to produce a spark. However, the breaker points are *normally* open to stop primary current and prevent this. They close just in time to permit a single cycle (with one, only, group of three magnets) per revolution of the flywheel. In fact, if this were operated only as a magneto, only three magnets would be required, in which case the breaker points could be normally closed.

The magneto-dynamo also operates as an alternator to produce current in three other (and separate) coils. Each of these coils is wound with a single fine wire, grounded at one end. The two *lighting coils*, which we show wound on separate legs of the same core, but which could just as well each have a separate core, are connected in parallel in order to increase current output (just as batteries are connected in parallel). Assuming it is the same size as each lighting coil, the one *charging coil* therefore produces but half as much current in its separate external circuit. We show it wrapped at the center of a horseshoe core, but it could consist of a pair of coils (like the lighting coils) connected in series, to have the effect of one double-size coil producing twice the amperage. Coil sizes and arrangements can be varied to develop the voltage or the amperage desired.

Now, when any three poles pass the two ends of a core (either the lighting-coil core or the charging-coil core), the resulting buildup, collapse, and reversal of the core's polarity will generate a current in the associated coil or coils

—provided the external circuit is closed. As the core polarity is reversed continuously in regular sequence, the resulting current is ac. No collector ring is needed to direct this current into the external circuit, as the coils are all stationary. Whenever the external circuits are opened (by switches), current will cease in the opened circuit.

> NOTE: In the particular application illustrated, the ac of the lighting coils is used to operate lights and a horn which operate equally well on ac or dc. The charging coil current, however, is used to recharge the battery, which requires dc. This current, therefore, is passed through a rectifier to convert it to dc. No cutout is needed since the battery can't discharge back through the rectifier. No voltage regulator is needed because the back emf produced within the dynamo has its own limiting effect.

Another system, which serves a battery, starter and lights, as well as the ignition, employs a 12-volt *flywheel alternator* which has only two ac coils (no ignition coil). One coil furnishes current, through a rectifier, for charging the battery whenever the engine is running. The battery then furnishes all power needed for starting, ignition, lights, etc. However, since use of lights might quickly discharge the battery, the second coil is also brought into use for additional charging of the battery whenever the lights are turned on. Current from this coil passes through the contacts of a relay (electrically operated switch) and a second rectifier to the battery. The relay is a separate unit; the second rectifier usually is housed with the first one, together with a fuse to protect the system. The relay contacts are normally open so that no current flows in this circuit; however, the master switch is designed to direct battery current through the relay and close the contacts whenever the lights are turned on.

AN ENERGY-TRANSFER MAGNETO SYSTEM

This system uses a flywheel coil similar to those already discussed, to generate a low-voltage current. This current is converted into a high-voltage current for plug firing by a separate ignition coil, like the coil used in a battery sys-

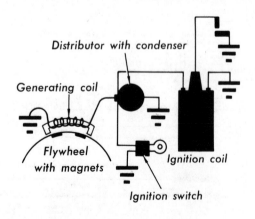

An ET magneto system.

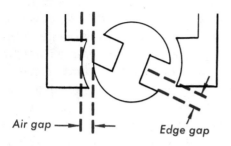

Air gap — Edge gap

Rotor adjustments.

tem. A separate distributor with condenser, also similar to ones used in battery systems, is also used. A typical system is illustrated.

In this system the generating coil is mounted at the perimeter of the flywheel. Operations of the distributor and ignition coil are exactly the same as already explained in the battery-system section. The ignition switch, when off, grounds the primary circuit to stop the engine. If there are two or more spark plugs a separate generating coil is used for each.

MAGNETO ADJUSTMENTS AND MAINTENANCE

It is necessary when assembling some magnetos (especially rotor types) to adjust *edge gap* and armature air gap.

With a rotor magneto, *edge gap* is the distance between one rotor magnet edge and the designated edge of one lamination when the rotor shaft is in a specified position of revolution. With a flywheel type it is the distance between a specified coil-core leg and the edge of a designated flywheel magnet (shown by a mark on the flywheel) when the flywheel is in a specified position of revolution. If construction is such that variations are possible, edge gap adjustment must be accomplished as instructed, for it determines the timing of magnetic field buildup and primary current generation.

Air gap is the clearance between adjoining faces of the stationary and the rotating members. If construction permits sideways movement of either member, air gap must be adjusted as specified. This ensures proper mechanical clearance and keeps magnetic field lines properly equalized throughout.

In addition, the breaker-point gap must be adjusted (as the distributor contact gap is ad-

justed), and the position of the breaker arm with respect to the cam must be adjusted. Point gap determines the time interval of secondary current generation and affects the point opening time, whereas breaker-arm position more directly determines the point opening time. Point opening time, called the *static spark-advance setting*, regulates the instant of spark plug firing. Many different ways are provided for checking and making these adjustments. Both are extremely important and must be accomplished precisely as the manufacturer instructs.

NOTE: In most flywheel magnetos, the plate on which the breaker arm is located can be repositioned. It is to be positioned in accordance with aligning marks or reference points that are specified. With rotor magnetos it generally is necessary to match up gears. Quite often, special locking keys to hold a shaft or part in a reference position are required.

Whenever a variable cam arrangement is provided to obtain automatic spark advance, the spring(s) and the weight and cam movements will have to be checked and adjusted per instructions.

All parts of a magneto usually are replaceable. As with distributors and generators, parts must be electrically (as well as mechanically) sound, clean, and dry. Connections must be tight or properly soldered. Points, replaceable in matched sets, must be smooth and bright—two flat points usually are used. Springs must have proper tension. Where dust covers and/or seals are provided, they must be correctly installed.

Permanent magnets will lose their charges if abusively handled, or if stored too close togther or too close to any electromagnet or current-carrying conductor. The usual rule is to separate stacked flywheels, for instance, by at least one inch. Recharging is possible, as previously mentioned.

AN IMPULSE COUPLING

The spark strength of a magneto is necessarily relative to its speed of rotation. It is relatively quite weak at engine cranking speed. Further, if a magneto is statically timed to produce

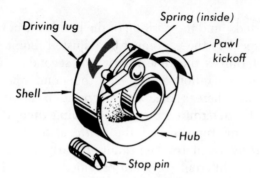

A single impulse—spark impulse coupling.

its spark some 20° to 35° ahead of TDC, which is the desirable advance at full-power running speed for most engines, then this static setting is much too advanced for the slow cranking speed (can result in engine backfiring). It is to overcome these two magneto disadvantages and to provide easier starting, that impulse couplings are used. Not all engines are so equipped, but many are. The real need for an impulse coupling depends on engine design, which determines the seriousness of the magneto disadvantages.

The coupling (usually used with rotor magnetos) is installed between the engine and the rotating magneto part, so that the driving force for magneto rotation must pass through the coupling. The coupling consists of a *shell* and a separate *hub* coupled together by a powerful *spring*, but otherwise free to rotate separately. If the hub (at the magneto side) is held stationary, the shell (at engine side) can be rotated until the spring becomes wound up. Then, if the hub is released, not only will the shell rotation be transmitted through to it, but, in addition, the unwinding spring will kick the hub

around at still greater speed. In this fashion the hub rotation can first be retarded, then be speeded up. Its movement, transmitted to the magneto, will result in first retarding the spark, then speeding up the magneto to create a strong spark.

A *pawl*, fastened at the hub side, together with a *stop pin* are used to hold the hub stationary until the spring is wound, and then to release it. The accompanying illustrations show the sequence of events during one revolution of a cranking operation.

The illustrated coupling is for a one-spark-per-revolution magneto. If two sparks per revolution are generated, two pawls, 180° apart, are used. In this case, the shell makes only a 180° revolution to rewind the spring each time. An even greater number, in a larger-diameter coupling, can be used if required. Note that the pawl has a long (heavier) end. Once the engine is started, the centrifugal force of its running speed draws this longer end outward against the inner edge of the shell to withdraw the short end from contact with the stop pin. Thereafter, the coupling serves only as a solid joint to transmit the drive uninterruptedly through the shaft.

The amount by which the shell winds the spring prior to release of the hub determines how much the magneto spark is retarded from its static spark advancement setting. This is measured as an angle between the hub keyway and two points on the shell and is called the *lag angle* of the coupling. The delayed spark is called the *impulse spark*. A coupling having the proper lag angle must be used. Also, a coupling

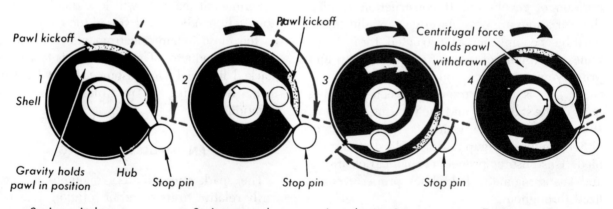

Operation of one-pawl impulse coupling.

must be properly positioned with reference to engine and magneto operation, so that the kick-off of the hub will coincide with firing of the spark plug. Usual practice is to reference the occurrence of the impulse spark and of the regular (advance) spark to the position of the *drive* lugs on the back of the shell. There is no standardization, however. You must position any coupling on its shaft and retime the shaft with the crankshaft, if necessary, according to specific manufacturer's instructions.

SOLID-STATE IGNITION UNITS

Variously referred to as *solid-state, electronic or breakerless* ignition units, these parts eliminate the need for mechanically operated breaker points. Their use in an ignition system therefore eliminates the deterioration and readjustment problems associated with breaker points; moreover, an extremely high (up to 400 volts) energy ignition spark having a longer duration when needed is produced by these units. Such a spark is especially suitable for any high-speed engine. It also reduces the incidents of engine misfiring and thus tends to make the exhaust freer from pollutants.

TERMINOLOGY

New electrical terms are needed for explanation of these units. These are defined as follows:

Capacitor. Another term for condenser, previously described.

Diode. A device which will allow electric current to flow in one direction, but will block a reverse-direction flow. Also see "Rectifier."

Gate-Controlled Switch. Also called *silicon switch, SCR* and (more properly) *gate-controlled silicon rectifier.* A device which permanently blocks electric-current flow in one direction and also blocks it in the other direction unless a second, small *trigger current* opens a *gate*, thus permitting current to flow through the gate in one direction only.

Heat Sink. A mechanical component on or in which heat-generating electrical units are mounted, and which will absorb and dissipate the heat so as to protect the electrical units from damage.

Pulse Current. The discharge from a capacitor.

Pulse Transformer. Used in place of, but cannot be interchanged with, a conventional ignition coil. Also contains a primary and a secondary coil and functions to transform a relatively high-voltage primary pulse current into a higher-voltage secondary current for firing of a spark plug.

Rectifier. Any of several types of units, or an assembly of such units, which will permit an electric current to flow through in one direction only. A diode is a rectifier unit; there are also *dry-metal* types using silicon or some other metal having the proper electrical characteristics. One rectifier alone in an ac circuit will convert half of the ac (*half-wave rectification*) into a dc of

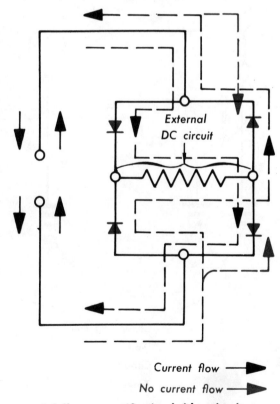

External
DC circuit

Current flow ➤
No current flow ➤

A full-wave rectification bridge circuit.

half the original amperage value. Four rectifiers in a bridge circuit will convert all of the ac (*full-wave rectification*) into a dc of equivalent effective value.

SCR. See "Gate-Controlled Switch."

Semiconductor. Any material which permits partial or controlled flow of an electric current.

Solid State. The branch of electronic technology that deals with the use of semiconductors.

Thermistor. A semiconductor device that will decrease in resistance as its temperature rises and which can therefore be used in a control circuit to automatically compensate for temperature increase.

Thyristor. An electrical "safety valve" which does not pass current in either direction and is used in a circuit to protect other components against voltage surges.

Trigger Current. A timed, small-value current used to open the gate of an SCR, and to thus initiate the sparking of a plug.

Zener Diode. A type of diode which will permit current to flow freely in one direction and will also permit a reverse current flow when the voltage reaches a predetermined level.

UNITS AND OPERATION OF A CAPACITOR-DISCHARGE SYSTEM

This type of magneto system differs from the conventional systems previously described, not only because it uses solid-state units but, also because the secondary (plug firing) current is induced by the rapid buildup (rather than by the collapse) of the primary current. To accomplish this, a relatively high-voltage stored within

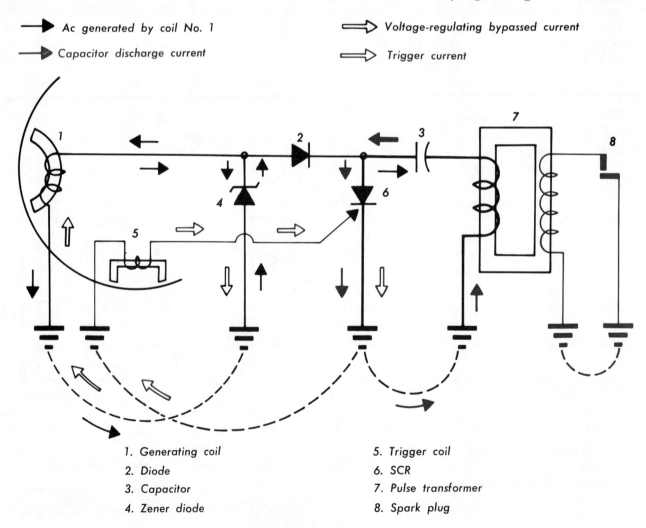

→ Ac generated by coil No. 1

→ Capacitor discharge current

⇒ Voltage-regulating bypassed current

⇒ Trigger current

1. Generating coil
2. Diode
3. Capacitor
4. Zener diode
5. Trigger coil
6. SCR
7. Pulse transformer
8. Spark plug

A typical CD circuit.

a capacitor is triggered to discharge through the primary coil of a pulse transformer, thus suddenly building the transformer primary field to a high value and inducing an exceedingly high-voltage secondary current for firing of the plug.

This system employs an induction-type ignition coil excited by permanent flywheel magnets (refer to "Magneto-Dynamos," preceding) to generate an ac for the ignition system. A small, second *trigger coil* is also positioned to be excited by the flywheel magnets. Other components generally consist of a diode, a capacitor, a zener diode, an SCR and a pulse transformer. The transformer may look like a conventional ignition coil or have a different appearance. The other components usually are incorporated in a heat-sink (or control) assembly. These units may be mounted inside the flywheel.

The accompanying illustration diagrams a typical single-cylinder system. Half of the ac generated by the generating coil (1) flows through ground and the zener diode (2) to charge the capacitor (3); the other half simply flows through ground and the zener diode (4) back through the coil. Because the zener diode will pass current in both directions if voltage reaches a predetermined value, the circuit through this diode serves to limit the voltage applied to the capacitor; excess voltage will "close" the zener and bypass part of the charging current through ground and back to the coil.

After the capacitor has been charged, the flywheel will revolve its magnets away from the generating coil around to the trigger coil (5). The resulting excitation of this coil generates a small trigger current of sufficient value to open the gate of SCR (6), thus allowing the capacitor to discharge through ground and the primary winding of the pulse coil (7). This discharge creates a rapid voltage rise in the primary, which induces a secondary-coil current of very high voltage for firing of the spark plug (8). If the engine has two or more spark plugs, separate trigger coils—one for each plug—are used.

A FLYWHEEL ALTERNATOR SYSTEM

This system differs from the preceding in that a *trigger module* is used in place of the trigger

coil. The module is mounted outside of, but adjacent to, the flywheel in such a position that it can be electrically triggered by a *flywheel trigger projection* attached to the flywheel. Housed in the module are three diodes, a resistor, a sensing coil with magnet, and the SCR. These perform the same functions described for the charging and triggering circuits in the preceding. There are two terminals: "I" for connection to the ignition switch and "A" for connection to the alternator. Reversal of connections will permanently damage the module.

A pulse transformer which also houses the capacitor is used with this system. This is mounted away from the flywheel, wherever convenient, and is connected to the ignition switch and the spark plug(s).

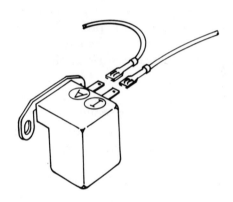

A typical trigger module.

When two or more plugs are to be fired, there is a flywheel projection for each plug. An air gap of exacting specification must be established between the module and each projection. This is obtained by adjusting the module position.

ELECTRONIC IGNITION FOR ALTERNATOR-OR GENERATOR-BATTERY SYSTEMS

Conversion of a conventional battery system from contact ignition (with breaker points) to electronic ignition is obtained by alteration of the distributor and the addition of an *electronic control unit* in the primary circuit of the ignition coil. The secondary circuit is not altered.

An electronic distributor has a *reluctor* instead of a rotor, and a *pickup coil with permanent*

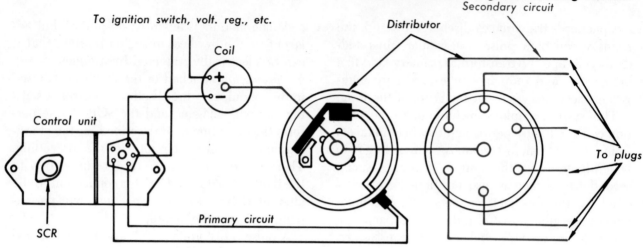

A typical electronic-ignition battery system.

magnet in place of breaker points. There is no condenser. The control unit contains an SCR (called a *switching transistor*) and the other elements (similar to those previously described) necessary to operation of the circuit.

There are as many *teeth* in the reluctor as there are plugs to be fired. As each tooth of the rotating reluctor passes by the magnet, a voltage pulse (pulse current) is produced in the pickup coil. This pulse is transmitted to the SCR which then functions to interrupt current flow in the primary circuit. The interruption induces the high-voltage secondary current which fires the plug associated with the particular reluctor tooth. Units are so designed that no adjustments are required.

ELECTRONIC EQUIPMENT ADJUSTMENTS AND MAINTENANCE

Due to the many differences in electronic-equipment circuitry and parts construction it is impractical to outline general service procedures. Generally, the only adjustment, other than already mentioned is the proper matching of timing marks during reassembly—and the only maintenance possible is the replacement of entire faulty components as specified by the manufacturer. Service checks require use of an *automotive-type timing light* or an *electronic ignition tester*, in accordance with instructions provided for each system.

PRACTICAL CHECKING GUIDE

A small-engine ignition system can fail in three ways: (1) wrong spark timing; (2) too lean a spark; (3) no spark. The first is a mechanical problem to be corrected per manufacturer's specifications, but the other two are electrical failures which may be due to a broken wire or connection, a poor contact, a shorted wire, or a malfunctioning electrical component. Chapter 11 discusses ignition troubleshooting without special tools. The following are some practical checks with meters.

NOTE: The meters recommended are a dc volt-

meter, a dc ammeter, and an ohmmeter—each of proper low range. A continuity checker, a flashlight with leads arranged to light the bulb when wire ends are touched together, can be substituted for the ohmmeter where only either 0 or ∞ should be indicated. "Ground" is the engine chassis.

Battery—Voltmeter should read rated voltage within ±5%.

Magneto—Use ohmmeter. Ground to ungrounded side of breaker points or condenser should read less than infinity (an infinite reading indicates an open in primary). Ground to terminal for wire to plug (or distributor) also should read less than infinity (an infinite reading indicates an open in secondary). In either case, a zero reading on the ohmmeter would indicate a short.

Condenser—Use ohmmeter. Grounded terminal to chassis should read zero (otherwise grounding is poor). Ungrounded terminal to chassis or grounded terminal should read infinite (or capacitor is defective).

Coil—Use ohmmeter. BAT to DIST should read less than infinity (infinity would indicate a primary open; zero, a short). Secondary connection to ground should also read less than infinity (infinity would indicate an open in secondary or a poor grounding; zero, a short).

Voltage Regulator—An ammeter in line to battery should indicate current *toward* battery with engine running.

Breaker Points, Etc.—Any switch or points, when closed, should give zero ohmmeter reading across terminals (otherwise, a poor contact is indicated).

Wiring—Any wire should give a zero ohmmeter reading, end to end (or a break is indicated). Unless it is grounded, an ohmmeter reading from wire to chassis should be infinite (or grounding is indicated).

Ordinary meters cannot detect a high-tension short. Such a short (causing a lean spark) might become visible as a blue-white halo of light if the engine is run in the dark. Watch spark plug(s), coil, distributor, and condenser.

CHAPTER 9

The Exhaust, Lubricating, and Cooling Systems

EXHAUST SYSTEMS

There are but two principal parts in any engine exhaust system, the *exhaust manifold* and the *muffler*. An exhaust manifold serves the primary purpose of collecting the exhaust gases issuing from the exhaust valve(s) or port(s) of the cylinder(s), and of conveying these to the muffler. A muffler serves the sole purpose of deadening the loud noises caused by, and carried with, the still expanding gases as these issue violently from the cylinder(s).

The important thing about both an exhaust manifold and a muffler is that the exhaust gases *must not be restricted* by an alteration of the manufacturer's design. Undue restriction can cause direct power loss, by forcing the engine to pump the exhaust gas past the restriction. A restriction will also cause indirect, accumulative power loss and possible engine damage, by building up the internal engine temperature to a dangerous degree.

EXHAUST MANIFOLDS

In addition to the primary purpose of collecting the exhaust gases, an exhaust manifold may also serve as a source of ready heat. This heat may be used for warming the mixture in the intake manifold, as previously mentioned, or for warming other parts of an engine, its accessories,

or its connected drive train or load. Exhaust gases quickly attain very high temperatures, and while the heat thus represented is excess heat insofar as the combustion and power of the engine are concerned, it can be put to use as it is being carried away from the cylinder(s).

Multicylinder engines usually do make some use of this heat. The chief use is for warming the intake mixture. As already noted in connection with the intake manifold, the more complicated systems used with large (principally automotive) engines are arranged to direct a measured or variable amount of the exhaust gases around the intake manifold. When the

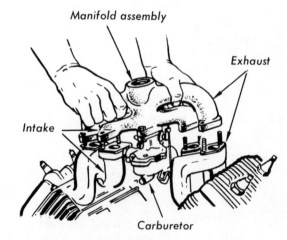

A typical multicylinder manifold.

amount is variable, a bimetallic thermostat (like the other described in Chapter 7 under "Automatic-Type Choke Systems") controls a shutter arranged to direct more or less of the exhaust around the intake manifold as the temperature of the intake requires.

Multicylinder engines of the small class generally use the exhaust to heat the intake simply by association and arrangement of the parts. A typical example is illustrated. Here the physical attachment between the integrally cast intake and exhaust manifold, plus the encirclement of the intake by the exhaust pipe, accomplishes the heating.

With single-cylinder engines there is no need to collect exhaust gases. An exhaust manifold serves little purpose. In fact, in most such engines this manifold is nothing more than a pipe nipple or elbow attached to the cylinder to support the muffler. Or it may be an integral portion of the cylinder block casting (accompanying illustration).

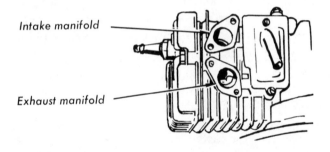

Intake manifold

Exhaust manifold

Typical single-cylinder engine manifolds.

MUFFLERS

The noise created by an open exhaust is due partly to internal engine noise traveling with and through the exhaust gases, partly to continued expansion of the gases during exhaust, and partly to the thunder clap effect of the gases striking the atmosphere. Mufflers are scientifically designed: (1) to absorb the internal engine and expanding gas sounds, and (2) to accumulate the gases just long enough to allow them to issue forth into the atmosphere in a steady, less violent stream. And both purposes have to be accomplished with a minimum of back pressure (restriction that takes a buildup of force to over-

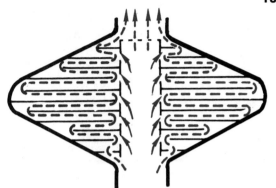

One design of baffle-type muffler.

come), so as not to consume more of the engine power than absolutely necessary.

NOTE: Any muffler does consume some engine power. However, a muffler that is the right size and in good condition will not consume more than the manufacturer allowed for in his design of the engine. Mufflers are sized according to their capacities for passing a certain volume of gases at a certain pressure (as measured in the exhaust manifold). Up to 20% of an engine's power may be allowed for creating the necessary muffler operating pressure; but up to 50% or more may be consumed by a wrong-size muffler or one in a poor condition.

With small engines, mufflers are generally much oversized to consume a very minimum percent of engine power. And those used with two-stroke-cycle engines have to be approximately twice the size of those for comparable four-stroke-cycle engines due to the greater volume of noise to be muffled in a two-stroke-cycle engine.

Mufflers are of two general types. The earliest mufflers used a number of internal *baffle plates* so arranged that the gases flowed through and among these plates, which absorbed much of the noise while also smoothing out the puffs to a steadier slower velocity stream. These are still used on some engines. The newer design mufflers, more generally used with small engines, use an internal labyrinth of tubes with bypass holes. While the main flow passes through the series of tubes or other devious passageway arrangement which absorbs engine noises, the excess exhaust puffed out at the peak of an exhaust stroke is forced through the bypass holes to short-cut the main flow—and thus fill in the gaps between puffs, and smooth out the flow

off

off

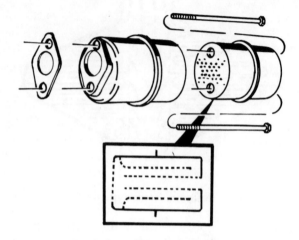

Typical small-engine muffler.

issuing from the muffler. With both types of mufflers, much of the cushioning effect results from the friction created by the muffler, together with the fact that the muffler affords space for the gases to complete their expansion and start to cool. A very long exhaust pipe could accomplish this same purpose.

NOTE: Engine mufflers are designed also with consideration for the use to which the engine will be put, and the nuisance factor of its noise. On this basis, most lawn-mower engines are very carefully

muffled; farm pump engines for use at remote installation points may not be so well muffled—and outboard motors do not use mufflers at all, since the exhaust is expelled under water, where its noise is lost—and, incidentally, where from its rocket effect it adds something to the craft's propulsion.

EXHAUST SYSTEM MAINTENANCE

There are two things to consider: (1) keeping the exhaust system fully open from end to end, and (2) keeping it in repair so as to avoid undue noise.

Several things can close an exhaust system: carbon accumulation in the manifold or muffler; other foreign-matter accumulations; and dents, warpage, or other damage to parts. Carbon or mud, etc, can seriously block the system. These must be soaked and blown out whenever visibly present.

Exhaust parts are subjected to extremes of heat and acid conditions. They will wear out from corrosion and/or abuse (especially a muffler). New parts should be installed whenever such deterioration allows gases to escape or in any way blocks the passage of the gases.

LUBRICATION SYSTEMS—IN GENERAL

A WORD ABOUT LUBRICATION

A Lubricant Is a Specialized Product

In general, whenever we think of lubricants in connection with an engine we immediately picture a can of motor oil and, possibly, also a can of grease (for the transmission gears, etc.). Actually, the list of specialized engine lubricants—ranging from kerosene on through to the heaviest, toughest grease—includes several hundred varieties. Almost any liquid has some lubricating qualities (even water); so also do a good many of the "soft" solids, such as tallow. Even ice is slippery. In a broad sense, anything that will reduce surface friction and permit two contacting surfaces to slide together with noticeable freedom is a lubricant.

Obviously, a lubricant has to have special qualities which adapt it to each lubricating job.

Take ice, for instance. At below freezing temperature it is an excellent lubricant for skates, sleds, and similar runners. If thick enough, it holds and cushions the runners. Some of the ice molecules coat the runners to fill all cracks and other imperfections and make the runners' con-

Ice is a lubricant.

tacting surfaces glass smooth. Then the heat generated by the friction of movement thaws just enough of the contacting molecules to allow them to slide freely. Thaw ice a bit more, however, and it becomes all water—which will no longer support and lubricate a skate or a sled, but will support and lubricate the sailing of a ship. Again, subject water to the high temperatures usual inside an engine and it immediately vaporizes into steam (which would support and lubricate the flight of an airplane, but is useless for sailing a ship).

The preceding remarks indicate the first and principal difference between lubricants and the types of jobs they will do. We can begin a long list of special requirements by saying that a lubricant must function at the required temperatures. This single requirement rules out water, kerosene, and similar highly volatile lubricants for use in an engine. An engine lubricant must, first of all, withstand the very high temperatures found in an engine, and retain its lubricating qualities. But this is only the beginning of the list. It is as endless as are the demands created by our progress in developing smaller, lighter-weight, larger-horsepower engines with their many refinements of construction and operation. Today's engine lubricant must, indeed, be highly specialized, and this is why there are so many kinds of lubricants for so many different specific applications.

WHAT MOTOR OILS AND GREASES ARE

Motor oils and greases are, primarily, petroleum products. At the start of Chapter 7 we mentioned how gasoline is refined by fractional distillation, and the fact that lubricating oils, petroleum jelly, and paraffin, are some of the other products refined from the same crude oil at later stages of the distillation process. These, then, are the heavier-end hydrocarbons separated from petroleum at extremely high temperatures. The lubricating oils consist of molecules still light enough to flow at room temperatures, while the greases consist of molecules too heavy to flow except at higher temperatures.

We say that these are primarily petroleum products to indicate that they are not wholly so. After distillation, the oils and greases are subjected to other processing to alter the characteristics of their molecules. They are graded and separated into groups of certain molecular weights and qualities. *Additives* (other substances having desirable qualities) are blended with the various groups. And, in the end, the industry produces the very long list of specialized lubricants that we mentioned earlier. Making proper allowances for the rash advertising claims of competing refineries and processors, each and every distinct type of lubricant so produced *does* differ from every other one in some respect. It does have certain distinctive qualities which adapt it to doing a specific job better than another type of lubricant.

The General Classifications and Their Importance

In order to have a basis for the comparison of lubricants produced by various competitors, the motor industry—through the Society of Automotive Engineers (S.A.E.)—has established certain basic requirements and ratings. First, and most important of these, in the *viscosity* rating of a motor oil. Correctly speaking, viscosity is the resistance of an oil to flow at a given temperature. All oils flow more readily at higher temperatures or will cease to flow at all at a

"Lighter"   "Heavier"

Types of oil.

low enough temperature. That is, the viscosity becomes greater as the temperature is lower. Therefore, standard viscosities are measured at the arbitrary temperatures of 100°F, 130°F, and 210°F—and the oils are then given an *S.A.E. rating* (grade) of viscosity.

The preceding grading also takes into consideration an oil's *pour point*, that is, the temperature at which it will perceptibly flow from a receptacle without assistance. It also takes into account other factors of a more scientific nature. On the whole, however, and certainly insofar as the average oil consumer is concerned, we can state that the S.A.E. grading of an oil tells us how light- or heavy-bodied the oil is, how quickly it will start to flow in winter or summer, and how thin or thick a film it will create at the high running temperatures inside an engine. These gradings are known as S.A.E. 10, S.A.E. 20, etc. The smaller numbers indicate lighter oils and the higher numbers therefore indicate the heavier oils.

In addition to S.A.E. grading, oils are also rated among lubrication engineers according to viscosity index, flash and fire points, carbon residue, sulfur content, neutralization number, saponification number, and, sometimes, even color. Viscosity index relates to how a temperature change affects viscosity; the flash point is the temperature at which the oil will give off ignitable vapors, and the fire point is the temperature at which these vapors will support a flame, etc. All these factors have meaning to any technician interested in selecting a particular oil for a particular job, but the average consumer without a laboratory at his disposal must rely on the recommendations of his engine manufacturer as to the type and grade of oil to use.

Greases, too, are rated by S.A.E. grade and by types. Again, the grades indicate weights, light or heavy, and the types usually refer to the additives which qualify a grease for a particular kind of use. For instance, axle grease is the oldest, commonest type of general-use grease; water-pump grease has enough waxy additive to resist saponification (lathering, like soap) in the presence of water; transmission oils, high-pressure gear greases, etc., all have their special additives and qualities.

We have mentioned all the foregoing factors involved, only so that you will appreciate the fact that oil and grease selection is a highly technical and extremely important matter. When a manufacturer specifies a particular grade of oil or grease, or a particular type, he is not merely making a suggestion in favor of some brand or of some whimsical opinion of his own. If the specification happens to point to just one marketed lubricant, you can be sure that this is the only lubricant that the manufacturer could find that has all the qualities he deems desirable. However, lubrication recommendations generally are more inclusive and simply state the grade of oil or grade and type of grease, leaving the brand selection up to the individual. In such cases, as much as is specified (the grade, type, etc.) is important and must be adhered to if the engine is to perform correctly.

THE FUNCTIONS OF A MOTOR OIL

In the foregoing we have stressed the fact that there are differences in grading and typing of motor oils and greases. *Also*, that the important differences may, at times, even apply to the qualities of a particular marketer's product. Fortunately, with respect to motor oils, competition has spurred all reputable refiners to keep pace. On the whole, one brand of engine oil is approximately as good for our engines as the next brand. Motor oil recommendations by engine manufacturers are therefore limited to the mere specification of S.A.E. grade. But this one specification is extremely important.

It is a popular *mis*conception that a motor oil serves but one purpose—to lubricate. True, this is its principal purpose and its end purpose; but to accomplish this major purpose it has also to fulfill three other requirements. In all, then, an oil has to perform *four functions* in an engine. And because of this, the grade of oil used, as well as all the other qualities which we will take for granted in any standard brand oil, is important.

Lubrication

As we have said, almost any liquid will lubricate to some extent. Lubrication is a quality of

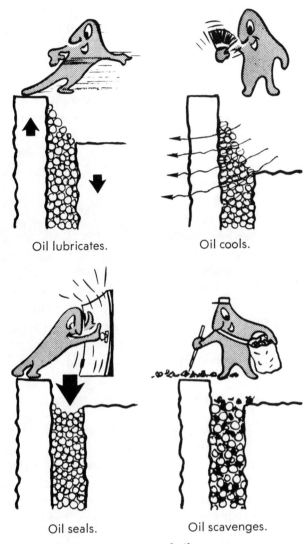

Oil lubricates.   Oil cools.

Oil seals.   Oil scavenges.

Functions of oil.

lubricate surfaces which it is thin enough to seep between, to fill all the infinitesimally small spaces between the molecules of each surface and make the surfaces glass smooth. And it must not only be thin enough to thus plate and smooth each surface; it must also be so thin that when this is done there will still be a sufficient number of the oil molecules in between the plated surfaces for the liquid flowing action to take place between these molecules.

In short, to serve the true purpose of lubrication, the oil must be as light a weight (small S.A.E. number) as will still have the qualities of adhesion, surface tension, lack of volatility, etc., required to keep its body and stay with the superheated surfaces to be lubricated.

Cooling

Because oil does lubricate the piston movement by forming a thin film on the cylinder walls, much of the excess heat of combustion must pass through this film on its way out. Moreover, heat is generated by friction in all other moving parts, and these are also subjected (to more or less a degree) to the excess heat of combustion. Consequently, one of the prime functions of the oil in an engine is to conduct heat, to help the heat get out of the engine.

A thicker liquid will absorb more heat than a thinner one, but it will also hold this heat more readily. And the quality we want here is not the absorption of heat, but the transfer of heat. So, again, a thinner oil will serve the function of cooling better than a thicker one.

Sealing

During compression and combustion, molecules of compressed or burning and expanding mixture push downward on the piston, ring tops, and on the space between the rings and cylinder wall that is occupied by the oil film. The film, then, has to be tough enough to withstand the bombardment of these molecules attempting to dislodge it. And, here, for this purpose, a more viscose, heavier oil is better as it will, obviously, offer more resistance than a thinner one. If this toughness were the sole function of a motor oil, we would have to favor the higher S.A.E. number grades.

a liquid primarily because liquid molecules are in a state (Chapter 2) which permits them to flow past each other. On the whole, the warmer and/or thinner a liquid is, the less resistance it offers to this flow. But the quality of flowing is not the sole requirement of a good lubricant. Its molecules must also have considerable *adhesion* so that they will cling to the surfaces to be lubricated; the liquid must have enough *surface tension* to form a protective film; it must not be too volatile, etc.

In a motor oil all these requirements amount to just this: Without sacrifice of the necessary adhesive and surface tension qualities, etc., the oil must nevertheless be as thin as possible to perform the best job of lubrication. It can only

In an engine, however, the other oil functions are far more important. Then, too, if the piston(s), rings, and cylinder walls are in good condition, the "line that the oil must hold" is a very, very thin one. Use of a lighter oil may permit some *blow-by* (escape of combustion molecules through the oil film into the crankcase), but this will be at a minimum. And a lighter oil will re-form its line more rapidly after each incident than a heavier one could. So, again, we must conclude that a lighter oil is better. If an engine is in such bad shape that only a heavy oil will help to seal in compression and combustion, the real remedy is to repair it.

NOTE: Use of heavier oil to stop compression and combustion blow-by may give a temporary appearance of solving the problem. But the sacrifice of the other three functions of the oil, plus the fact that carbon deposits will accumulate rapidly from the burning of the oil, will soon overcome all benefits. There is no surer way to make an engine deteriorate rapidly than to use too heavy an oil.

Scavenging

Scavenging is one of the most important functions of an oil. Every internal-combustion engine "breathes in" and creates dirt. Despite the air cleaner, etc., microscopic particles of grit and dust from the air do get inside. When you consider that the smallest visible particle is about one ten-thousandth inch in diameter, while the protective oil film may be as little as four one-millionths inch thick, you will realize how even invisible hard particles of sufficient size to protrude through the oil film can pass the air cleaner, etc., and get inside to do damage.

In addition to these particles, there are water and carbon to contend with. It is estimated that one gallon of water is produced in an engine for every gallon of gasoline consumed. True, most of this blows out the exhaust as vapor, but some remains inside to dilute the oil. And carbon also will accumulate. This carbon is from incomplete combustion and from the burning of the oil itself. At times, fortunately, the carbon is soft and fluffy and blows out the exhaust; but quite often it is hard and gritty and becomes deposited in the oil and on the metal parts.

All these contaminants plus metal particles worn off the engine parts, plus other foreign substances that do get inside, add up to making an internal-combustion engine a fairly filthy machine. Its own filth would soon destroy it by abrasive action if it were not for the oil. A good oil scavenges all this filth and deposits it as sludge in the bottom of the oil pan or on the oil filter screen. Once again, a thin oil has much better washing action than a thick one.

Summing Up

All things considered, then, when an engine manufacturer specifies a grade S.A.E. 10, 20, or 30 oil there is good reason for using this light an oil in preference to a heavier grade (even though the heavier grade will last longer and save a few pennies on the oil). Then, too, you can readily understand just how short the life of an engine will be without proper lubrication —and how extensive the damage will be.

SHOULD AN ENGINE CONSUME OIL?

Yes, an engine will very definitely consume oil. At the high temperatures encountered, no oil ever produced will fail to partly vaporize, partly flash and burn, when fully exposed to the engine combustion. The mere force of combustion will blow away some oil molecules.

In a good engine, the oil film deposited on the cylinder walls need be but microscopically thick. Hence, when the piston makes a downstroke and leaves some film on the walls above to be exposed to combustion, very little oil is lost. Only when sloppy rings and scored or out-of-round cylinder walls permit a thick film to be deposited, will the oil consumption become excessive. There is no rule of thumb. We can only repeat that a good engine will necessarily consume some small percentage of its oil, while a poor engine will consume proportionally more.

HOW LONG SHOULD OIL BE GOOD FOR?

Here is another moot question about which much has been said on all sides. It is pointed out by one group that oil never loses its lubricating quality. You cannot wear it out, so to speak. On the other hand, you can (and an engine certainly does) contaminate it with acids, water, dust, metal particles, carbon, etc., thereby robbing the oil of its qualification as a good lubricant to perform its four functions.

Let us beg off from a direct answer to this question and give the only honest answer we know: "Oil in an engine will last only so long as it remains sufficiently free of contaminants to be unretarded in performing all four of its functions." If the particular engine is exceptionally well equipped to cleanse its oil, the oil mileage may be considerable; but if the engine is poorly equipped, the mileage will be proportionally reduced. And the state of engine repair

(how filthy it actually is in producing contaminants) will have a definite bearing on the oil mileage.

Actually, oil mileage should have little importance to any owner of a small engine. Even at worst, the small amount of oil used per year won't allow much saving from any reasonable endeavor to stretch the mileage. This makes it poor economy not to change oil or not to use as much as specified, when you contrast the pennies of saving from being oil stingy, with the dollars of waste from possible engine damage. And failure to have any fixed plan at all for changing and/or adding oil is such sheer wasteful carelessness that it is hardly worth discussing.

LUBRICATION SYSTEMS—TWO-STROKE-CYCLE ENGINES

GENERAL TYPES

There is *no* lubrication system, as such, in a two-stroke-cycle engine. Lubrication of all engine moving parts is accomplished by adding motor oil to the gasoline. This mixture, on entering the crankcase, lubricates the crankshaft, connecting rod, their associated bearings, and the lower part of the cylinder walls. When subsequently entering the combustion chamber, the mixture lubricates the upper cylinder walls. These are all of the internal engine parts that require lubrication.

To accomplish lubrication, the oil molecules become plastered onto the metal parts as the mixture carries them through the engine. This is possible since they are heavier and have more inertia than the gasoline molecules, and since they have more adhesion. In the crankcase, the oil will build up to a more or less stable thickness film on all parts, and remain so unless a subsequent deluge of gasoline without oil content should wash the parts clean. The same is true of the combustion chamber, except that combustion will vaporize and consume much of the oil, so that any washing action would have less oil to wash away.

When proper oil in proper ratio to the gasoline is used, the carbon formed by the burned

oil in the combustion chamber will be negligible in amount (compared with the washing action of subsequent refills of the chamber by fresh mixture). Also, it will be a soft, sooty carbon that is easily blown out with the exhaust, rather than a hard, gritty type that would accumulate on the spark plug, rings, and walls. In short, while the burned oil must form carbon, the results will not be cumulative and detrimental to any marked degree if proper oil and quantity is used.

Usual specifications call for between 1/3 pint and 3/4 pint of oil per gallon of gasoline, depending on the particular engine. They also usually state the grade (S.A.E. number) of oil to use, or specify some special (low-carbon) preparation like "Outboard Motor Oil." The grade isn't as important as with a four-stroke-cycle engine (following), since the oil is diluted by the gasoline. But the best practice, nevertheless, is to follow exact specifications. Also, the *quantity* of oil per gallon of gasoline is very important. Too little will underlubricate the engine, while too much will simply multiply the carbon problems and might soon result in wetting (with oil) and fouling of the spark plug. A sign of too little oil is an overheated, brassy-sounding engine; of too much, is a smoky, dampened-sounding engine without its usual power. *Always measure the exact ratio of oil specified.* Don't rely on size judgment, color of mixture, or any other such haphazard methods.

AUTOMATIC-MIX TYPES

With a few makes of two-stroke-cycle engines, particularly those used in motorcycles, the proper mixing of oil with the gasoline is not left up to the operator. It is accomplished automatically, by the engine itself, and is done in direct, variable relation to the speed of engine operation on the principle that at higher speeds an increased ratio of oil is needed.

With such engines, the fuel tank has two sections—one for gasoline and the other for oil, and the two are kept separate. Oil is fed from its tank section into the mixture in the crankcase, by a *variable-speed pump* which squirts the oil onto the connecting-rod bearing assembly from which it drips to become mixed with the fuel mixture. The rate of oil flow (pump speed) is regulated by linkage connected with the throttle. Therefore, at higher speeds, the flow rate is increased.

The pump is located in the crankcase and is driven by a worm-drive gear arrangement from the crankshaft (accompanying illustration). It is a simple piston-type pump having a spring that returns the piston to starting position after each downstroke, and ball-check valves to con-

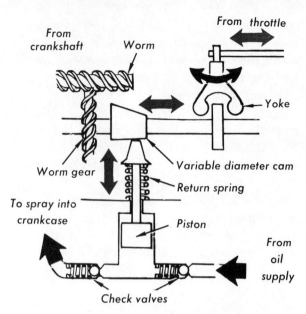

Principle of variable-stroke oil-feed pump.

trol oil flow. The piston shaft rides on a variable-diameter *cam* which will push it down (at each revolution) an amount depending on which area of the cam is in contact with the shaft end. This is determined by a yoke controlled from the throttle control, the movement of which will reposition the cam to expose various parts of its surface to contact with the piston shaft.

NOTE: The explanation and illustration are only intended to convey the principles of operation. Different designers arrive at the same result in a variety of ways.

LUBRICATION SYSTEMS—FOUR-STROKE-CYCLE ENGINES

GENERAL REQUIREMENTS AND TYPES

Oil Fills and Changes

Every four-stroke-cycle engine requires that motor oil be added, separately from the gasoline, into the crankcase—to maintain an established level of oil within the crankcase. Generally, a *dipstick* (measuring rod) or an *oil-level gage* of some type is provided for measuring or registering the oil level. Oil is poured into the crankcase through some form of *filler tube* with a removable cap or plug. Oil of exactly the specified grade (and type, if indicated) and the indi-

cated level *must* be adhered to (within reasonable limits) for proper engine performance.

NOTE: The oil-level gage may consist of any one of several types. It may be a marked line on the filler tube or the top of this tube, or it may be a marked vertical glass tube installed so that it will fill with oil simultaneously and to the same extent as the crankcase. Generally, in single-cylinder engines, the filler tube is simply an integrally cast, tubular projection from the crankcase base having a threaded plug. The gage is the top level of this projection (accompanying illustration). For larger engine arrangements refer to the illustration, "A Typical Force-Feed System."

(Fill to level of filler tube top)

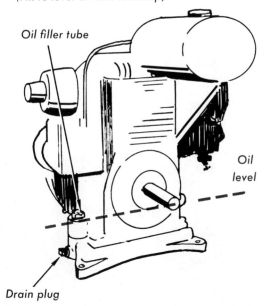

Oil filler tube

Oil level

Drain plug

A typical filler-tube and oil-level arrangement.

Each engine manufacturer will always specify the quantity of oil required for each new filling and will make a recommendation as to how often the oil should be changed. With a new engine, the first oil change is nearly always required after a relatively short period of service because the wearing-in of new engine parts usually contaminates the first oil quickly with metal and carbon particles. The time between subsequent changes will be somewhat longer. A *drain plug*, located at the lowest point in the oil system, is always provided.

Whether specifically recommended by the manufacturer or not, it is a good, economical practice to flush the crankcase and oil system after draining out used oil, whenever the appearance of the used oil indicates an excess accumulation of sludge. (And it is usually best to flush after the first oil fill is drained, regardless of the oil's appearance.) Sludge, an accumulation of all the solid contaminants, settles to the bottom of the system. If there is enough of it to form a detectable settling at the bottom of the pan into which you have drained the oil, or to slow the draining oil with visible clots, you can be certain that enough will have remained inside to immediately contaminate the fresh oil and do possible damage. *Alawys examine* the drained oil. It is an excellent indicator of the engine's condition.

For flushing, use a commercial flushing oil or kerosene. Refill to the oil level. Preferably (and especially if the engine has an oil pump) operate the engine a short time—long enough to circulate the flush through all oil-carrying parts. Then drain, and refill with the required lubricating oil. And while you are about it, don't overlook servicing the oil strainer, if such a strainer is provided (as explained later).

IMPORTANT: As mentioned before, all engines consume oil at some rate, depending on the engine design and condition. Consquently, it is absolutely essential that the oil level be checked periodically, and new oil added as required.

If an engine crankcase holds a sizable quantity of oil (so that economy becomes a factor), and if the oil change periods are relatively far apart, adding oil is both economical and practicable. (However, always look at the oil on the dipstick, and don't add fresh oil to visibly dirty oil; change instead.) With a very small engine, on the other hand—one that only holds a total of a pint or so—there is no economy in adding oil. It is much better simply to drain and refill. If such practice should call for too frequent oil changes, the engine definitely needs an overhaul.

The Effects of Oil Grade on the Engine

It is the practice of engine manufacturers to specify one grade of oil (say S.A.E. 30) for summer, another (say S.A.E. 20) for winter, a third (say S.A.E. 10) for subzero weather, plus possibly, the addition of kerosene to thin the oil in accordance with the temperature drop below some subzero degree listed. There is a reason for these variations.

If an oil is too heavy for the weather (outside temperature), its naturally increased viscosity due to cold will hamper its flow and might even stop it entirely. At starting of the cold engine it will retard movement of the engine parts as molasses would and make starting difficult, if not impossible. After starting, there will be a long interval during which the engine must run without benefit of lubrication—until the oil warms up sufficiently to again flow as intended. There is no need to say what damage this can do to the engine.

Too thick.   Too thin.

Flow of oil.

When too thin an oil for the weather is used, the engine will start and operate normally, at first. As soon, however, as the engine's running heat accumulates and is added to the weather heat, the oil will thin out excessively like water. At this stage its film tension will be much less than required, and its adhesive qualities will be reduced. As it won't cling to the metal parts properly, it won't do a thorough job of lubrication. Moreover, lack of film tension will increase blow-by and accelerate the accumulation of contaminants. In a short time the oil will become uselessly filthy, and the damage being done to the engine will mount accordingly.

NOTE: If an engine is idle all fall, summer oil may still be in it when it is put back into service in winter. In such case, it is just common sense to change to the required grade of oil before using the engine, even if the old oil does have a lot of mileage left in it. And the same applies when going from colder to warmer weather.

Detergent and Nondetergent Oils

Many of today's automotive oils have detergent additives which help the oil to accomplish its scavenging function. You can use one of these detergent oils or a nondetergent type, as you prefer, in any four-stroke-cycle engine. However, whichever type is used during engine break-in should always be used thereafter. It is not advisable to switch back and forth between the two types.

Never use a detergent oil in a two-stroke-cycle engine. The cleaning additive will not function to cleanse the engine since the oil is mixed with the gasoline. It will foul the spark plug instead,

since it won't burn and be blown out with the exhaust.

There Are Five General Types of Systems

Small engines utilize five different types of oil systems, each based on a slightly different principle of oil circulation to the engine parts. We will discuss each of these systems separately, from the standpoint of oil circulation, and then we will discuss the component parts.

A SPLASH SYSTEM

This system is the simplest. It is much used in vertical-type single-cylinder engines. The engine crankcase is filled to an established oil level. Movement of the connecting-rod end, dipping into this reservoir of oil with each crankshaft revolution, is relied on to splash oil to all parts requiring lubrication.

To make the splashing effective, there is an *oil dipper* at the connecting-rod end. This may be an extension of one of the connecting-rod cap screws, a projection from the cap, or a separate attached piece.

A CONSTANT-LEVEL SPLASH SYSTEM

This system is the same as the preceding splash system, with this exception: A simple, plunger-type *pump* operated by the engine is arranged to maintain a constant level of oil in the *oil trough* from which the oil dipper splashes the oil. The oil trough is located at the required level and position within the crankcase for the oil dipper to pass through it with each revolu-

Oil reservoir   Oil dipper

Typical splash system.

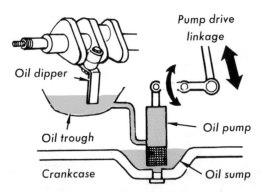

A typical constant-level splash system.

tion. Its location may be above or to one side of the bottom center of the crankcase. Hence, this type system can be used for horizontal-type engines as well as for vertical types.

The pump draws its oil from an *oil sump* (a low area), usually located at the bottom center of the crankcase. It is operated by mechanical linkage, either from the crankshaft or the camshaft.

AN EJECTION-PUMP SYSTEM

This system (also) utilizes the splash principle. Instead of an oil dipper action, however, it substitutes the action of an *oil pump*. The oil pump sprays oil onto the moving engine parts, from where it is splashed to other engine parts. In some cases, the spray is directed at the re-

volving connecting-rod end; in other cases it is directed up to the camshaft so that it will drip back onto the connecting rod and crankshaft. In all cases, however, the object is to create a moving mist of oil that will cover all internal engine parts.

An oil dipper may be used with this type of system. In such a case the ejection spray is directed at the dipper, which serves to splash it around. Another variation of this system employs an oil-slinger cup, attached to the crankshaft, to force feed oil into the crankpin, which is hollow and open at both ends. The oil is then thrown out of the crankpin after lubricating the connecting rod bearing, to spray other engine parts. An oil pump operated from the camshaft ejects oil into the slinger cup.

Most often, simple plunger-type oil pumps are used. Some installations do, however, use gear-type pumps.

A DRY-SUMP SYSTEM

This type of system is employed in some foreign-make engines. No oil is stored in the crankcase; instead, it is stored within a separate *oil tank* (or box), which may be positioned in or outside of the crankcase at a convenient location. Oil is delivered to the engine parts in the crankcase by an oil pump, and the condensed oil that settles to the bottom of the crankcase is returned to the tank, also by an oil pump. The two pumps may be separate units or, as in the

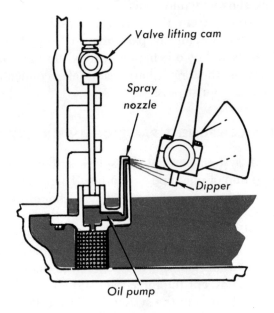

One type of ejection-pump system.

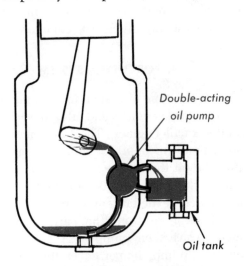

Principle of a dry-sump system.

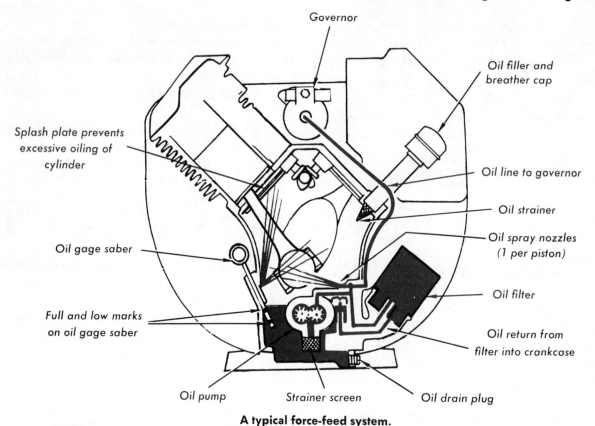

Governor

Oil filler and
breather cap

Splash plate prevents
excessive oiling of
cylinder

Oil line to governor

Oil strainer

Oil spray nozzles
(1 per piston)

Oil gage saber

Oil filter

Full and low marks
on oil gage saber

Oil return from
filter into crankcase

Oil pump   Strainer screen   Oil drain plug

A typical force-feed system.

accompanying illustration, they may be one, *double-acting piston-type pump.*

With this type of system, the oil pumped into the crankcase usually is delivered in the form of a spray, as in the "Ejection Pump System." Or it might be force-fed to various parts (refer to a "Force-Feed System," following). In either case, very little liquid oil is allowed to gather at the crankcase bottom, since the pump returns it to the tank as rapidly as it accumulates.

A FORCE-FEED SYSTEM

This is the type of system used in automotive and other large engines and, also, in some multi-cylinder small engines. The basis of any such system is an *oil pump* of sufficient capacity to pump the oil through separate *oil lines* to some or all of the engine parts requiring lubrication. Oil is stored at the bottom of the crankcase, and drips back here from the various areas into which it has been pumped. A gear-type or a vane-type pump, immersed in the oil, is used. An *oil-filter* (discussed later) usually is used.

The accompanying illustration shows a typical system as used with one make of small engine. This system force-feeds (by oil line) only the governor; the connecting-rod bearings and crankshaft bearings are spray-fed by the nozzles shown, while other parts are mist-fed by the splash from the revolving crankshaft, etc. In other systems, all parts may be force-fed by oil lines and by oil channels drilled through such parts as the crankshaft, camshaft, connecting rods, and wrist pins.

THE COMPONENT PARTS OF AN OIL SYSTEM

All systems, except the splash system, use an oil pump of one type or another. There are three types of pumps in general use—the *piston* or *plunger type,* the *gear type,* and the *vane* or *impeller type.* All are quite simple in construction, but the piston type uses the simplest form of drive mechanism. It is, therefore, most used with all except the force-feed system.

In addition to the system itself and the pump (if used), other engine components which must be considered in connection with a lubrication system are: (1) *oil strainers*, (2) *oil filters*, (3) *oil breathers*, and (4) *oil seals*. All systems require strainers, breathers and oil seals; but use of an oil filter is generally limited to the multicylinder engines employing some form of force-feed system.

OIL PUMPS AND LINKAGES

The Piston Type

This is much like other piston-type pumps already described. Usually, the piston *returning spring* is located inside the cylinder (accompanying illustration), though it may be outside, around the piston shaft. Or, there may be no return spring, and two-way action of the operating linkage will be depended on to return the piston for each of its suction strokes. Spring-loaded ball-check valves are used, as shown, to govern the direction of flow.

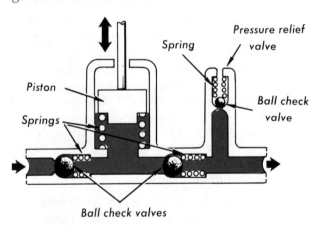

Principle of a piston-type oil pump.

If used with a force-feed type of system, the pump will be equipped with a poppet-type *pressure relief valve* of the type shown. Otherwise, this valve will be omitted.

Linkages to such a pump vary a great deal. The simplest type is a direct extension of the piston shaft to ride on one of the cams used to operate the valves (shown in a previous illustration). Cam-operated rocker arms of various designs are also used, with the cam located on either the crankshaft or the camshaft. Several different arrangements are shown.

The double-acting piston-type pump mentioned in connection with a dry-sump system is usually gear driven, from the crankshaft or camshaft. There are several designs of such a pump. One typical design is illustrated.

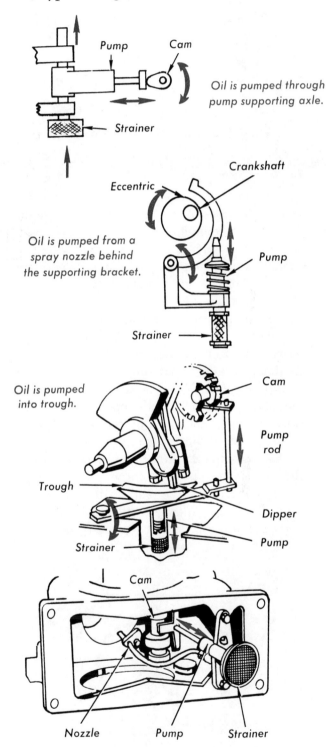

Typical piston-drive linkages.

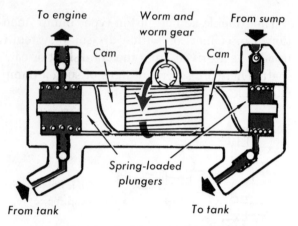

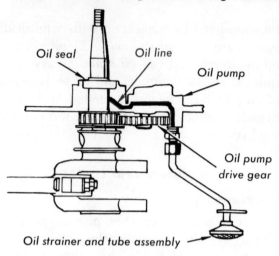

Principle of a double-acting piston pump.

Typical gear-type pump installation.

The Gear Type

A pump of this type requires no valves, nor springs. It consists entirely of two equal-size gears meshed together and enclosed in a housing contoured closely to the two gears so that oil cannot escape from between the gear teeth and adjacent housing sides. An oil inlet is provided at one open area adjacent to the gears where they mesh, and an oil outlet is similarly located at the other open area. When revolved, the two gears will turn in opposite directions, away from the inlet opening. They thus suck oil into their housing through this opening, carry it around in the tooth spaces to the outlet opening, and, then, since the oil cannot pass by the meshed teeth, they squeeze it out by the outlet opening.

One gear, the *drive gear*, is driven externally through its shaft by some form of gear arrangement connected to the crankshaft or camshaft. The other is an *idler gear,* driven by the first. Such a pump can deliver oil at very high pressures, with a continuous flow. If the flow should

be impeded, the pressure could build up to a damaging degree. Therefore, most such pumps are fitted with a pressure relief valve of much the same design as shown with the piston-type pump.

> NOTE: Large engines are generally equipped with an oil-pressure gage which, by registering the pressure of the oil circulated through the force-feed system, will indicate at a glance whether or not the oil is circulating freely. Any blockage or break in the closed lines would vary the normal operating pressure. As small engines operate only at low oil pressures (5 to 15 psi, as a rule) and with only limited force feeding, gages are not used.

The Vane (Impeller) Type

Like the gear type, this pump also needs no valves other than a pressure relief valve, and only one spring. The circular *impeller* is enclosed, like the gears, in a housing; but it is set to one side of the slightly greater-diameter

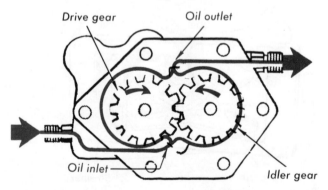

A typical gear-type oil pump.

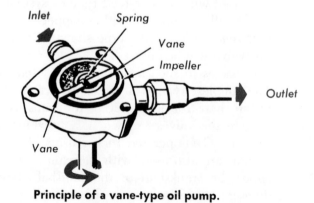

Principle of a vane-type oil pump.

housing opening. The two *vanes* are free to slide back and forth through the impeller. Their spring holds them out to contact the housing opening sides. Thus, when rotated, the vanes will slide to follow the circle of the housing opening, and this action, combined with the off-center position of the impeller, results in the formation of two pockets that start at "no" size, grow to maximum size, then return to "no" size. The growth of each pocket coincides with its passing the inlet opening, so that it sucks in oil, and its diminishing coincides with its passing by the outlet opening, so that it then squeezes the oil back out.

Pumps of this type also produce considerable pressure. They may be driven by any of the arrangements used for gear-type pumps. There are also several variations of the design shown and explained here, but the principle of each variation is the same.

Oil-Pump Maintenance

While repair parts for oil pumps are usually offered, it is generally best to replace an entire pump assembly if it is malfunctioning. Pumps, especially the piston types, are relatively small and inexpensive assemblies. Moreover, any defect in one part will likely result in damage to other parts. Most engine manufacturers advise means of quickly checking pump operation to determine its condition and methods of cleaning. And a pump should be cleaned, checked, and, if need be, replaced, every time that engine service permits. Also check the pump operating linkages and oil lines to make certain these are functioning properly.

OIL STRAINERS

Any fine-meshed screen located where the engine oil must pass through it to strain out solid particles is a strainer. We have shown a number of typical strainer installations in previous illustrations. Almost all fill pipes are so equipped to strain fresh oil being poured into the engine. Practically all oil pumps are also equipped with strainers, usually at the inlet side of the pump or at the end of the inlet tube if a tube is required. This strainer eliminates solid particles

Typical oil strainer.

from the oil being pumped to engine parts, and keeps these particles settled down in the sump where they can be removed with the next oil change.

A strainer or the entire pump-strainer assembly is removable for cleaning. Naturally, the wire mesh must be kept reasonably free of sludge and dirt for the oil to flow freely. A pump strainer should be checked (for good condition) and cleaned every time that engine service permits.

OIL FILTERS

An oil filter is actually just another strainer. It is an improved, more thorough-acting, easily renewable type strainer that is offered as an accessory on some of the larger small engines which hold enough oil to make adding oil between changes worthwhile. Those which hold very little oil are seldom so equipped. Whenever

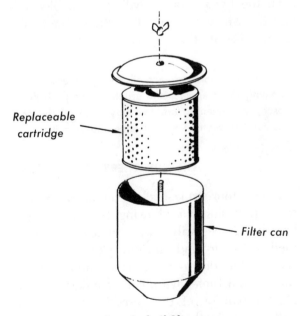

Replaceable cartridge

Filter can

A typical oil filter.

furnished, the oil filter is used in addition to strainers. It does not take the place of the usual strainers.

Oil filters are similar in design to air cleaners. They are usually shaped like a can, through which the oil (from a force-feed pump) is forced to circulate following a routing that passes it through the filter element. This element (the filter *cartridge*) is replaceable. Cartridges are made of various materials, all designed to strain the oil finely to remove all dust and other solid impurities. Note, however, that no filter or strainer can ever remove any lighter ends of gasoline, water, etc., which might accumulate to dilute the oil or alter its characteristics.

It is intended that a filter cartridge will become dirty, and eventually clogged. When this happens, the filter will do much more harm than good, by impeding the oil flow. And, of course, the filter cartridge should be changed for a new one often enough (per manufacturer's recommendations) to avoid this. To allow for human neglect, some systems using filters have built-in bypass tubes and valves which will permit the oil to continue flowing even after the cartridge is clogged (generally referred to as bypass-type filters). Obviously, however, a bypass is not intended to permit ruthless neglect of the filter. Should the filter be allowed to remain clogged too long, chances are that even the bypass valves will become blocked—and the whole system will partially (or totally) fail to lubricate the engine.

NOTE: If a nondetergent oil has been used consistently in an engine, and a detergent type is then used, the cleansing additive may wash all the accumulated dirt into the oil filter and clog it.

OIL BREATHERS

Every four-stroke-cycle engine *must* have some provision for allowing the crankcase to breathe. Necessarily, the crankcase is assembled to be air- and oil-tight. Consequently, on each piston downstroke, some compression of the air and fumes within the crankcase results. Heat expansion of the gases in the crankcase will cause further compression. In addition,

blow-by (even in a snug engine) will force some gases into the crankcase to further increase the pressure inside. Then, too, this blow-by contains harmful acids together with explosive gases which could accumulate and explode in the crankcase, unless dissipated.

The breather, then, serves the principal purpose of relieving pressure within the crankcase. It also serves the secondary function of eliminating the lighter harmful ends of blow-by, which will rise and be the first portions of the internal atmosphere to be expelled. Some engines have an *open-type breather*, while others have a *closed type*.

NOTE: The elimination of crankcase compression is much more important in a single-cylinder engine than in a multicylinder one. If there are two or more pistons, while any one or more are downstroking, one or more others will be proportionally moving in an upstroke to offset the piston movement contribution to crankcase compression. But in a single-cylinder engine there is no such offset. Each downstroke creates a positive amount of compression.

If not relieved, the crankcase compression in a single-cylinder engine would, from the start, seriously strain all seals, gaskets, etc., in the crankcase assembly. This could result in immediate oil leakage. And the longer the engine would run the worse this would become, due to blow-by pressure buildup.

Open-Type Breathers

A breather of this type is simply an opening from the crankcase to the atmosphere located

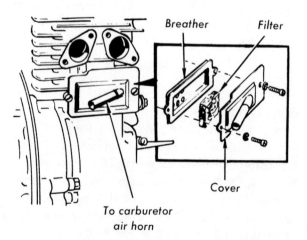

A typical single-cylinder engine open breather.

where oil cannot splash out. It is also designed to prevent dust, etc., from the atmosphere from entering the crankcase. For this purpose, a fine-wire screen or air-filtering element of some type usually is provided.

When practicable, this type of breather is usually located at the top end of the oil filler tube (see previous illustration, "A Typical Force-Feed System"). It is embodied in the design of the *filler tube cap,* which has screened openings to allow air to pass in and out of the tube.

Another much-used location, especially in single-cylinder engines, is the *valve chamber* (area at side of cylinder block which is open to the crankcase to permit lubrication of the valve cam followers, etc.). A typical valve-chamber breather assembly is illustrated. The *breather* is mounted inside the chamber opening and a steel-wool *filter* is provided. A *cover* encloses these. The cover opening in the case illustrated is a tube connection, and a tube is used to connect this with the carburetor air horn so that only filtered air can enter. Other arrangements may omit the air horn connection, or place a similar breather assembly at an entirely different location.

Closed-Type Breathers

A closed type is one which allows air to pass out of the crankcase, but does *not* allow air to pass in. It is used exclusively on single-cylinder engines. The reason for this construction is to maintain a slight vacuum in the crankcase whenever the engine is operating, thus making certain that there will be no pressure whatsoever to cause oil leakage. The slight vacuum is created because air that is expelled during the piston downstroke or expelled because of heat expansion will not be replaced through the breather.

This type also may be located in the valve chamber cover or elsewhere. It may consist of a *flapper (reed-type) valve* assembly or of a *ball-check valve* assembly—both identical to ones already described elsewhere in this text. Then, again, it may be a special type such as the following.

In the dry-sump system previously described, the breather is a *rotary valve* (illustrated) driven off of the crankshaft. There is a passage (slot) through the valve (which is merely a shaft) that mates with a similar slotted passage through the valve bearing when the valve is rotated to the correct position. When the slots are mated the piston will be making a downstroke. Thus, excess oil mist and vapors in the crankcase are expelled through the passageway

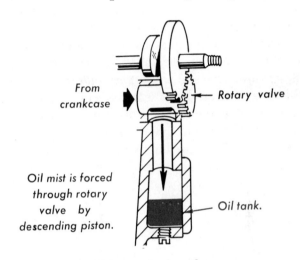

Rotary-valve breather.

provided to reenter the separate oil tank (which is open to the atmosphere). Correct timing of the valve is, of course, essential.

OIL SEALS

Where the crankshaft projects through the crankcase an oil seal is required to prevent oil leakage. This seal is important in that it must function properly to prevent loss of oil. It is, in effect, a special type washer placed around the crankshaft and protected by a retainer which locks it into the crankcase hub. Subjected to the wearing action of crankshaft rotation, it is expendable. Oil seals do require occasional replacement.

The seal itself may be made of leather, rubber, or synthetic rubber. The metal retainer is usually a tight (force) fit in the crankcase hub and requires careful attention regarding its mounting position, per any specifications furnished.

HEAT—ITS RELATION TO AN ENGINE

MEASURING HEAT

Thermometers

A thermometer registers temperature—that is, the average degree of heat in any substance we stick the thermometer into. We have *Fahrenheit* (F) scale thermometers on which the freezing temperature of water is +32, and its boiling point is +212. The scale of this thermometer is divided into degrees (°) each of a "size" such that there are 180 of them between these two points. And we have *centigrade* (C) scale thermometers on which water's freezing point is 0, its boiling point is +100—and the scale of which is divided into 100 degrees between these two points. These are the two scales most widely used; other scales are used in some scientific work.

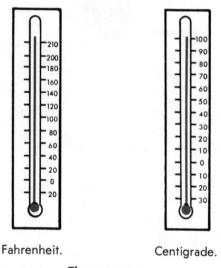

Fahrenheit.   Centigrade.

Thermometers.

The Calorie

However, a thermometer does not tell you how much heat there is in an object. If you have a boiler full of water at 100°C and dip out a cupful, the cupful will also be 100°C, but the cupful will cool very rapidly in comparison to the slower cooling of the boiler full. There is a great deal more total heat in the boiler than in the cup.

To define the quantity of heat contained in something scientists use the term *calorie*. The calorie is defined as the amount of heat required to increase the temperature of one gram of water by 1°C. Therefore, our cupful of water, if the weight of water in it is 100 grams, contains 100 calories of heat, while the boiler may contain 100,000 or more calories of heat (depending on the weight of its water, measured in grams).

NOTE: The calorie has become very familiar to us in connection with food. When spelled with a capital "C" (*Calorie*—called the large Calorie) it is used by biologists and dieticians to rate the amount of heat energy (fat-building capacity) stored in food. One (large) Calorie equals 1000 (small) calories.

Specific Heat

Now, it takes a relatively long time to heat water. Iron, aluminum, and other substances heat much faster. That is, it takes less time and less heat, for instance, to increase the temperature of one gram of iron by 1°C. So, once again, we use the term "specific" (meaning comparative) to indicate just how quickly (or slowly) other substances heat in comparison with water. We say that water has a *specific heat* of 1. Then, if some other substances heats twice as fast as water (takes half as much heat to increase 1°C in temperature), its specific heat is 0.5, etc.

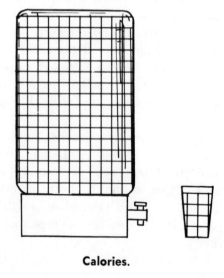

Calories.

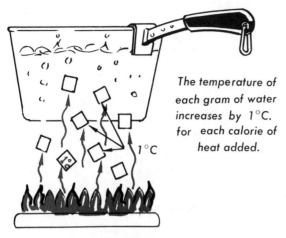

The temperature of each gram of water increases by 1°C. for each calorie of heat added.

Specific heat.

The specific heats of all known substances have been carefully measured. For instance, the specific heat of iron is 0.113. Therefore, if we want to know the number of calories of heat in a block of iron we multiply its temperature (in degrees centigrade) by its weight (in grams) by its specific heat (0.113). Or, if we want to know how many calories of heat it will take to heat the block of iron from 100°C to 150°C (a total of 50°C) we multiply its weight (in grams) by its specific heat (0.113) by 50.

BTU

In this country the calorie and also the centigrade thermometer are used principally for scientific purposes. For commercial purposes, instead of the calorie we use the *British thermal unit (Btu)* which equals 2.52 calories—and we use the Fahrenheit thermometer. (One Btu is defined as the amount of heat required to raise 1 lb of water 1°F.) Furnaces, boilers, etc., are rated in Btu per hour. For instance, if a furnace is rated at 25,000 Btu/hr, this means that

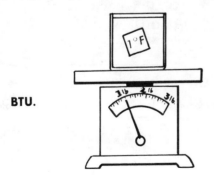

BTU.

the furnace will produce enough heat in one hour to raise the temperature of 1 lb of water from 0°F to 25,000°F.

HEAT IS AN EXPANDER

NOTE: We have previously learned (Chapter 2) how heat causes the expansion of all substances, and that cold (the lack of heat) causes their contraction.

Expansion of Solids

We have discovered that different solids (aluminum, iron, etc.) expand at different rates when heated; yet each solid will expand uni-

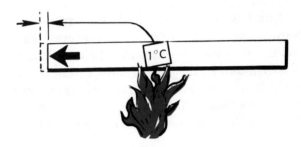

Coefficient of linear expansion.

formerly. That is, for every 1°C increase in temperature a given solid will expand a given amount. We call this the *coefficient of expansion* of the material. For instance, an aluminum bar will expand in length by 0.000023 ft for each 1°C rise in its temperature. Therefore the *coefficient of linear* (straight-line) *expansion* of aluminum is 0.000023. For iron it is 0.000011.

All the materials that we use (to build bridges, engines, etc.) have been carefully measured in laboratories, and their coefficients of expansion are known. By using the known coefficient of expansion of the material we can predict exactly how much (for instance) an aluminum

Coefficient of superficial expansion.

piston will grow in length or diameter when heated from (let's say) 0° to 2000°C.

With solids, we are interested not only in the growth along a straight line, but also the growth of a surface (in two directions) and the growth in volume (in three directions). Consequently, in addition to the coefficient of linear expansion we also have a *coefficient of superficial expansion* (two directions) and a *coefficient of cubical expansion* (three directions). Since the molecules are pushing out equally in all directions, the superficial expansion is approximately double linear expansion; and cubical expansion is approximately triple the linear expansion.

Expansion of Liquids

With few exceptions, liquids also expand uniformly (at much faster rates, however, than solids). We have also measured their coefficients of expansion. In the case of liquids, though, we are only interested in their coefficients of cubical expansion—that is, in the increase in volume; so these are the only coefficients that are used .

Coefficient of cubical expansion.

Water is the best known exception to the rule of uniform expansion. It does not expand uniformly; in fact, as everyone knows, water reverses the rule and expands (instead of continuing to contract) just before it turns to ice. (It expands while cooling from 4°C to 0°C.) A few other liquids have similar peculiarities.

NOTE: In the case of water, this peculiarity is very fortunate. Because it expands to freeze, ice becomes lighter (less dense) than the 4°C water around it and rises to the top. Otherwise, our rivers, lakes, even the oceans, would in time freeze solid from the bottom up—and the summer sun could do no more than melt a few feet at the top.

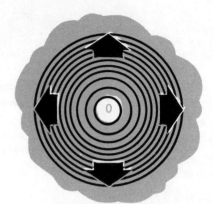

Coefficient of gas expansion.

Expansion of Gases

All gases have *exactly the same* coefficient of expansion, which is uniform *at all temperatures*. This fact was discovered by a Frenchman, Charles, and is known as the Charles' law. Again, we are interested only in the coefficient of cubical expansion. He proved that this, for all gases, is 0.003665, that is, 1/273. In short, starting with a certain volume at 0°C, a gas expands 1/273 of its volume for every 1°C increase in temperature, or contracts 1/273 of the volume it had at 0°C for every 1°C decrease in temperature. Theoretically, at —273°C. (absolute zero) all gas would disappear, but, of course, this is only imaginary as each gas will liquefy first and then contract more slowly.

HEAT IS AN EVAPORATOR

As we have learned, evaporation of some solids and all liquids is the result of the activity of the molecules. A certain number of the sur-

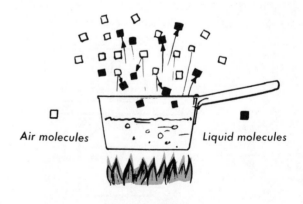

Air molecules   Liquid molecules

Heat evaporates.

face molecules, in bumping against their neighbors, rebound and fly off into the surrounding air. If the air is stagnant they will do this until there are so many of them in the air (in between the molecules of the air) that each new one that bounces off the solid or liquid will collide with one already in the air and bounce right back into the solid or liquid. When this state exists, we say that the air is *saturated.*

If we add heat, two things occur. First, the molecules become more active; they can bounce higher and with more force. Second, the air expands and makes more room in between its molecules for molecules of the solid or liquid to occupy. Therefore, the rate of evaporation increases and, at the same time, the saturation point of the air is raised. The solid or liquid will evaporate more quickly, and more of it will then evaporate into a certain volume of stagnant air.

HEAT CONTRIBUTES TO CHEMICAL CHANGES IN MATTER

We have learned that matter retains all its same characteristics so long as only its molecular activity is affected, but that it changes form (becomes some other form of matter) if a chemical change in its molecules occurs. We have learned that there are several ways in which molecules can be altered. One of these ways is by heat.

HEAT BUILDS UP IN AN ENGINE

Heat Absorption

Heat (Chapter 1) is an active form of energy. As such, it can move about from substance to substance, or can accumulate within a substance. When it accumulates, we say that the substance has *absorbed heat* (become warmer). Some substances absorb heat more readily than others, and most metals have the quality of absorbing heat rather rapidly. Other substances, like air, absorb heat slowly. When two substances having different absorption qualities are in contact, any heat energy present will be absorbed more rapidly by the substance having superior absorption quality.

Different substances absorb heat at different rates.

NOTE: This is why an open gas flame can be directed to heat the pan on the stove, intsead of the air around it. It is also the reason that rocks, pavement, etc., become so much hotter in the sun than the air around them.

As the temperature is increased, a substance may change state (solid to liquid to gas), but there is no known limit to the amount of heat which can be absorbed. (Gases in our sun contain millions of degrees of heat.) Yet, fortunately, all things which are in contact tend to equalize their content of heat. If a very hot iron is left standing in air, in time the excess heat from the iron will move out into the air until both are at the same temperature. Or, if the hot iron were plunged into water—which absorbs heat much faster than air—it would be quenched (cooled) quite rapidly by having its excess heat move out into the water. Thus nature prevents, on the whole, the undue accumulation of heat in any substance. It keeps heat energy ever on the move from object to object.

The Sources of Heat in an Engine

Combustion is in itself a source of active heat. The chemical change which constitutes combustion (burning) is a process that releases the latent heat energy stored in molecules and converts this to an active form of energy. In short, combustion creates active energy in the form of heat. The energy begins to move. Now bear in mind that this active, on-the-move energy has to show itself in some form or other. Bear in mind, also, that all force—including pressure—is, in reality, simply a form of energy showing itself; and that energy can neither be created from nothing nor destroyed. While latent, it may remain hidden; but in its active form it has to show itself in some way.

When the chemical change of combustion unleashes latent heat and "brings it to life," the activated energy shows itself mostly as heat energy. This heat energy is absorbed by the newly transformed gases and activates their molecules in a manner to show itself as kinetic energy—the energy which makes the gases expand. When, and to the same degree as, this expansion overtakes piston movement, this kinetic energy shows itself as pressure (in the gases). The resulting buildup of pressure is partly converted back into kinetic energy and transferred to the piston in the form of a terrific push (momentum); but most of the pressure energy is excess since the piston can only move so fast and so far, which limits the amount by which the compressed gases can go on expanding. This excess kinetic (or pressure) energy then changes back to heat energy, the only form in which it can continue being active.

Consequently, as a result of combustion and the very considerable excess pressure developed by it, we are left with a great excess of heat energy. This heat energy is part and parcel of the combustion gases, but the surrounding metal is cooler and is also capable of absorbing heat very rapidly. It immediately absorbs a large percentage of the heat from the gases—before they are expelled by the exhaust stroke—and becomes very hot.

Moreover, there are still two other sources from which the engine metals absorb heat. First, there is the ambient (surrounding) air temperature to consider. Whatever the temperature of the air might be, the engine metals will have, prior to engine operation, absorbed enough heat from the air to be at the same temperature. Then, too, while the engine is running, the metals will further absorb heat released by any friction that is occurring. Friction converts the kinetic energy of the two rubbing surfaces back into heat energy, by activating the molecules of the two rubbing substances.

All in all, the metals of an engine are subjected to a very considerable amount of heat energy. In fact, it has been estimated that the most efficient internal-combustion engine made converts only 25% of its latent fuel energy into useful mechanical power; all of the remaining 75% is transformed into useless heat energy which is mostly absorbed by the engine metals. When you consider that we are speaking in terms of thousands of degrees per combustion incident, and that heat absorption is cumulative, you can well understand that the engine metals would very quickly absorb enough heat to change their state to liquid or gas, if the heat were not dissipated quite rapidly.

THE UNDESIRABLE EFFECTS OF HEAT IN AN ENGINE

We already know that an excessive accumulation of heat can liquefy the engine metals. But short of this, assuming that no such excessive accumulation will occur, what other undesirable effects does heat have? It has several. Some are related to the fact that it causes expansion and evaporation; others to the fact that it contributes to chemical changes.

Consider the first, which is expansion. The coefficients of cubical expansion of different metals are different. Various metals are used in

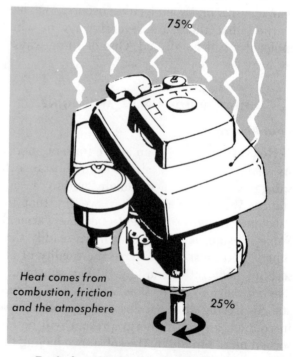

75%

Heat comes from combustion, friction and the atmosphere

25%

Typical engine-heat sources and losses.

Conduction.　　　　　　Convection.　　　　　　Radiation.

Heat transfer.

an engine. Hence, each will expand differently. If enough heat is absorbed by them, this variation in expansion will greatly alter the fits and clearances planned by the engine designer. Parts that are supposed to run with certain fits or tolerances will become too tight or too loose. Moreover, even metals having practically identical coefficients of cubical expansion will, in expanding outward in all directions, close the planned gaps between parts. In short, excess heat can expand the metals to close all the running spaces so that moving parts no longer have clearances in which to move. The moving parts will then *freeze* (stick). The protective oil will

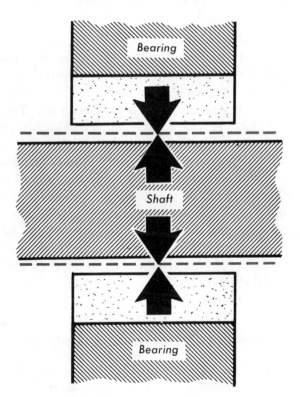

Heat can "freeze" moving parts.

have been squeezed out so that the freezing is accompanied by much scraping, scoring, and other damage.

Now consider the second effect of heat, which is evaporation. Excess heat will vaporize the protective oil films and leave moving parts exposed to friction and all its damaging results.

The third effect is chemical change. Excess heat will accelerate whatever damaging chemical action is occurring as the result of accumulated acids, etc., from the burned fuel and oil.

HEAT TRAVELS IN THREE WAYS

Heat will always leave a hotter object to go to a cooler one, as already said, until the temperatures of the two—in fact, of everything around—are the same. Everything in the universe would be one average temperature if it weren't for two factors: (1) it takes time for heat to travel; (2) nature effects chemical changes which store up heat as latent energy within the molecules of matter. This latent heat doesn't travel until another chemical change converts it back into active energy and starts it traveling again.

To leave one object and go to another, heat has three methods of travel—*conduction, convection, and radiation.*

Conduction means contact heating, such as occurs if you should burn your hand by touching a hot iron. Convection means the mass movement of heated particles of air or a liquid, such as would occur if you immerse a hot iron in water while holding your hand in the water to feel the increase of heat. Radiation is the jumping of heat from one object to another, such as occurs if you warm yourself by standing near a

Conduction heating.

hot stove; it occurs daily as the sun warms our earth by radiating heat to it.

Let us "picture" a group of men, a pile of bags at a warehouse door, and a freight car on a siding a few hundred feet from the bags. The problem is to get the bags loaded into the freight car. This can be done in three ways that we shall call "conduction," "convection," and "radiation."

If the men form a line and pass the bags from hand to hand, this is conduction. In the case of heat, it is the molecules of matter which "pass" the heat from one to another. Naturally, then, the more active and/or close the molecules are, the oftener they bump one another, the faster they will pass (conduct) the heat. The molecules in solids are close together, so they conduct heat fairly fast—the speed depending on just how active the individual molecules of a certain solid are. But (for instance) when heat reaches the end of the line of the molecules in a solid, the adjacent molecules of air are much farther apart. From this point on, the heat isn't conducted so fast; it is bottled up in the solid, just as the bags would be bottle necked if the last few men in the line should have to walk with each bag a dozen or so feet to hand it to the next man.

Convection (current heating).

If each man picks up one bag and walks (or runs) with it to the freight car—this is convection. Since the molecules of a solid can't walk or run (can only shake and bump), a solid cannot convect heat. Convection can only be accomplished by currents of flowing molecules in a liquid or gas. Wind and ocean currents, for instance, convect heat. And the heat, itself, may cause a current. As we have already learned, heat expands matter, makes it less dense and lighter. Therefore, when liquid or gas molecules become heated they rise above cooler, heavy molecules around them—and we have a convection current.

Radiation (radiant) heating.

Should each man pick up one bag and throw it from the pile into the freight car, this is radiation. Our sun throws heat at us (at least, if there are any "messengers" to carry the heat across the empty space, we don't know of them). Theorists have supposed the existence of such messengers, have called them "ether;" but this is still just theory. All we are certain of is that objects which are hot do seemingly throw heat to other surrounding objects, even through a vacuum. We also know that the color, surface polish, and other qualities of an object determine how much radiated heat it will catch and how much it will bounce off. Black, for instance, catches (absorbs) more radiated heat than white; a mirror absorbs less than frosted glass, etc.

THE PROBLEM OF ENGINE COOLING

From the foregoing it should be obvious that every internal-combustion engine *must* have a planned cooling system to dissipate its excess heat. The size of a cooling system (its capacity to dissipate the heat) will depend on several factors:

First, there is the engine type. A two-cylinder engine has twice the number of heat-producing combustions per revolution as a single-cylinder engine; four cylinders result in four times as many; etc. The speed of operation of any engine will determine the number of combustions per minute, and the horsepower produced is, of course, directly related to the size of each combustion. All these engine-type factors, therefore, must be considered. A high-speed, multicylinder big-horsepower engine certainly requires considerably more cooling capacity than a small single-cylinder one.

Second, there are the engine metals to consider. If the metals used have the characteristic of passing heat slowly, it will be much more of a problem to cool them than it would be if they passed heat rapidly. Iron, for instance, passes heat slowly in comparison with aluminum, which is one of our best metals from the standpoint of conducting heat rapidly.

Third, a manufacturer must consider the probable use of his engine. The climate and surroundings in which it will be operated may make a considerable difference in the ambient temperature. And whether it is to make short runs or to run continuously may make a difference. While it is true that these conditions generally lose much of their importance once an engine has leveled off at its running temperature (which is so high that average weather changes, for example, seem negligible in comparison), there are extremes under which they can become quite important. Moving an engine from the North Pole to the equator, for instance, will make a great difference in its operation if the cooling system isn't adjusted.

In addition, the design of an engine and of its lubricating system will also have considerable effect on the capacity of the cooling system required.

AIR-COOLED ENGINES

Practically all single-cylinder engines and even most of the small multicylinder engines are air cooled. Air cooling is the most economical from a production standpoint. It is also quite adequate—despite the poor heat-conduction quality of air—for a properly designed small engine. Due to variations of engine size, there are two general types of air-cooling systems—the *open-draft and exposure type,* and the *enclosed forced-draft type.*

AN OPEN-DRAFT AND EXPOSURE TYPE

Used largely with the single-cylinder engines which generate relatively little heat, this system depends primarily on exposure of sufficient engine metal surface to the atmosphere and, in most cases, to a breeze of some type. Even though the air that is in contact with one small area of the engine metal may not conduct much of the heat away from the metal, by greatly increasing the area of metal exposed to the air the total amount of conduction can be raised to the required amount. Any breeze will contribute convection cooling. Also if the engine is mounted onto another massive metal object, some of its heat will be conducted away through this object, and some more will radiate out to the object.

Engines of this type have large *cooling fins* cast integrally with the cylinder head and/or block. These fins greatly increase the metal area exposed to the atmosphere. Moreover, the metal having this increased area is the metal which

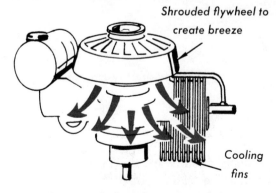

Shrouded flywheel to create breeze

Cooling fins

An open-draft and exposure system.

surrounds (and has to absorb the heat from) the combustion area. These engines, also, are usually mounted on massive objects like lawn mowers or moving objects which create a breeze to aid the cooling. Then, too, fast-cooling aluminum usually is used in their construction, and their design will be such that a fairly wide range of engine temperature variation will have no ill effect.

NOTE: Fins are sometimes erroneously called "radiant" fins. But the amount of heat that any fins will radiate is negligible. Their only useful function is to conduct heat out to the air for conduction or convection cooling.

As previously said, any breeze present will add convection cooling, and a breeze is, in most cases, provided. It may arise due to engine use, as in a fast-moving vehicle, or simply due to the rising of heated air to create an air current around the engine. In many cases it is created by the engine flywheel. When this is done, some

(Note that all parts are enclosed)

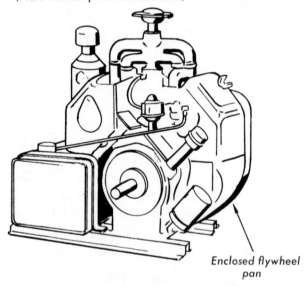

Enclosed flywheel pan

A forced-draft system.

form of *shrouding* (cover) is usually provided to direct the breeze as desired, over the fins.

In any engine of this type, the number and sizes of the fins are very important elements of the cooling system. If shrouds and/or a flywheel breeze are used, these also are important to the cooling. And if the manufacturer has depended on the engine mounting, this too can be quite important. Detract from any one of these elements (by breaking off fins, removing shrouds, blocking off the breeze, and/or putting the engine to a different use) and the cooling system may fail to cool adequately.

A FORCED-DRAFT TYPE

The only difference between this type and the preceding type is that the engine is designed to create a positive, carefully channeled forced draft, to add convection cooling. A *fan*, driven from the crankshaft, may be used; or the *flywheel* may be designed to serve this purpose. Such an engine is always shrouded even around the cylinders and there may also be air baffles to direct (channel) the draft to various engine areas. Obviously, both the breeze-creating device and the shrouds are very important parts of the cooling system.

SYSTEM MAINTENANCE

Aside from the obvious necessity of keeping any functional parts (fins, shrouds, fans, etc.) in good repair, the only other maintenance requirement of an air-cooled engine is *to keep it clean.* This is, however, extremely important. Dirt, grease, rags, and all the other filth and litter which can accumulate around an engine are very poor conductors of heat. Their presence will seriously reduce the conduction cooling and may entirely block any convection cooling relied upon.

WATER-COOLED ENGINES

Outboard motors are the principal small engines that are water cooled. This circumstance arises from the fact that water is always handy

for their use. Water, which conducts heat much more rapidly than air, is an excellent cooling medium both for conduction and convection cooling. It does, however, generally present the disadvantage of requiring a separate water-circulating and water-cooling system, together with all the problems of system stoppages, leaks, freezing, etc., which go along with the handling of water.

All large engines are presently water-cooled. The system generally used requires separate water-circulating channels through the cylinder block and head, a water pump to ensure forced circulation, a radiator to both store the excess of water needed and to air-cool the water, and a fan to create a convection cooling breeze through the radiator. A bimetallic thermostat valve may be included to regulate the amount of water flow. If a ready source of ample fresh water is available, the radiator and fan may be omitted. Whenever the water will be in contact with ferrous metals, rust inhibitors are required;

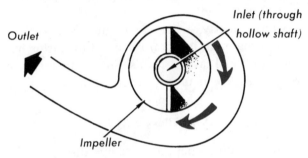

A centrifugal water pump.

and the system must always be protected against damage that could result from freezing of the water.

NOTE: Thermosyphon systems have been used. Instead of by pump, circulation is created by the rising of warmer water and the settling of cooler water (in short, by gravity).

Any land-based small engine that is water cooled must have the same type of system as a large water-cooled engine. Motorboat engines, however, have at hand a ready supply of fresh cool water and do not require radiators or fans. Their water-cooling systems consist only of

water channels through the engine, and a water pump.

The *water channels (or jackets)* are simply enclosed passageways and areas designed to hold as much water as practicable (and necessary) in direct contact with the heated metal surfaces of the cylinder and head.

When and as required, hoses or tubes are used to connect the two ends of the water jacket system with the cool water source, in a manner to result in partial thermosyphon circulation when this is practicable. The *pump* is located wherever convenient in one or another of the two hose lines.

The pump is operated by mechanical linkage from the crankshaft or camshaft, if there is one. A pump may be of the piston type or of the vane (impeller) type—both described previously under "Lubrication"—or may be a centrifugal (rotating-vane) type. The latter has blades which whirl the water outward by centrifugal force into an outlet shaped to receive it. This water is picked up by the blades from an inlet at their center, usually through the blade shaft or hub. Even ordinary propeller pumps have been used.

NOTE: One type of outboard motor uses a combined air- and water-cooling system. Air-cooling is of the open-draft type. To aid this, water from the motor propeller is raised up into the engine leg around the exhaust pipe to quench it, and reduce the overall amount of heat to be dissipated by the air-cooling. No pump or water jackets are used.

Water pumps are exactly like oil pumps, insofar as service is concerned, except that a water pump requires lubrication. The type of lubrication depends on the type of pump and its manner of installation. If it must be separately greased or oiled (is not lubricated by the engine lubrication system), the manufacturer will specify the requirements. Service for other portions of a water system consist solely of keeping the system clean, leak-proof, and in good operating condition.

CHAPTER 10

Starting and Power Transmission

STARTING AND STARTERS

THE NEED FOR A STARTER

Summing up the information contained in preceding chapters, we find that any engine which is in good operating condition will start and run, if these three requirements are furnished: (1) fuel-air mix, (2) ignition, and (3) momentum. It takes all three to produce the conditions under which an engine will develop mechanical power—the power to keep itself running, as well as the excess power with which it will pull a load. The fuel-air mix supplies a source of energy, and ignition unleashes this energy at the timed intervals required. But it is the momentum of the engine itself which causes it to inhale the proper fuel-air mixture, compress this mixture as necessary, time the ignition, and finally, to exhale the used charges to make way for new ones.

In a live (running) engine, momentum is part and parcel of the running operation; but a dead engine must be given sufficient momentum to get it going. An external force must be applied to make it inhale, compress, and exhale charges until at least one such charge is fired and the engine comes to life to supply its own momentum for continued operation. Supplying this external force is the task of a *starter*.

When a battery ignition system is used, the ignition spark is always ready and waiting for its timed occurrence; but when magneto ignition is used, momentum is also required to generate electrical current.

Consequently, when an engine is equipped with a magneto ignition system, the starter has the additional task of furnishing the momentum required for generating the ignition current.

STARTING SETUPS

In preceding chapters we have discussed the facts that certain starting requirements are necessary in connection with both fuel and the ignition system. Let's briefly review these.

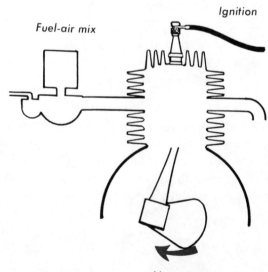

Fuel-air mix   Ignition

Momentum

Three requirements for starting and running.

Insofar as the fuel system is concerned: When an engine is operating at idling or running speed and at an operating temperature, the carburetor will function automatically to supply the proper fuel-air mixture; but at cold starting temperature and at the very slow rpm practicable with a starter, a special enriched mixture attained by choking is needed. Small-engine carburetors are not, as a rule, equipped with automatic chokes. The operator must accomplish the choking.

Insofar as the ignition system is concerned: A magneto system is generally left on since there is no current anyway, without momentum; but a battery system (or a battery starting system) is always turned off to stop an engine and must be turned on by the operator for starting. With either system a retarded spark is preferable for starting, while an advanced spark is necessary for running.

To prepare for starting, then, the carburetor must be choked, the ignition must be on, and the spark timing must be retarded as required. After starting, the choke and spark timing must be readjusted.

With modern automotive engines, choking and spark timing are automatic, leaving only the turning on of the battery ignition system for the operator to do. And even this has been combined with starter operation, so that one key and practically a single operation controls the entire starting and readjusting procedures. Even large industrial engines and a few of the better-equipped multicylinder small engines are similarly simplified. But the great majority of small engines require a sequence of two or more operator-controlled steps for starting, and for the subsequent readjustments to running conditions.

In some cases, priming and then choke control —two separate steps—are required for fuel system preparation and readjustment. In other cases, choke control alone is required. And there may be no steps to take in connection with the ignition system—or there may be just one step (turning it on), or two steps (turning it on and adjusting the spark timing). With some engines all steps require the manipulation of separate controls. There may be a manual priming pump (bulb), a choke control, an ignition switch, and a spark control. Or there may be a single lever

An "All Engines" Check List

Item	To Start	To Run	To Stop
FUEL	On	—	Off
PRIMER	Operate	—	—
IGNITION	On	—	Off
CHOKE	Closed	Open	—
SPARK	Retard	Advance	—

control marked OFF, START, IDLE, and RUN (or some other arrangement and/or selection of similar positions), which combines all the necessary steps.

NOTE: In addition to the preceding controls, most small engines also have a gasoline cock (valve), which it is advisable to turn off when the engine isn't operating to make certain that gasoline won't leak from the tank. Turning this on is, then, an additional starting step.

No two makes of engines have exactly the same kinds and arrangements of controls. Each manufacturer will specify his engine controls and the sequence of their use. It will be necessary for you to look to these instructions for proper starting and operating procedures.

REQUIREMENTS OF A STARTER

Regardless of type, any starter must meet *two* requirements: it must be capable of producing the engine momentum required for starting, and it must disengage from the engine after starting. Every starter operates by rotating the engine crankshaft either directly, or indirectly through reduction gears.

With a large four-stroke-cycle engine, it takes considerable starter torque (turning force) to overcome the friction of tight bearings and cold oil, plus the opposing forces created by the high compression pressure developed in the cylinders —not to say anything of the dead weight of the many parts to be set in motion and the valve springs, etc., to be compressed. On the other hand, in a single-cylinder, two-stroke-cycle engine not all these factors are present. And those which are are somewhat less. Whatever those factors do total up to, starter torque must be sufficient to rotate the crankshaft at suffcient velocity for proper carburetion and ignition.

Starter operation need not be long sustained, nor even constant. Any engine in good condition should start after cycling two to (at most) twelve times (if not overchoked). And, as long as the successive cycles are not so long delayed as to permit compression to be entirely dissipated (by slow leakage out of the combustion chamber), starter operation can be intermittent.

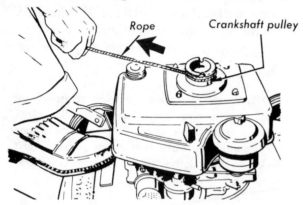

A windup rope starter.

Once an engine starts it is necessary that the starter disengage, primarily to prevent damage to the starter. When manual starting is used, this doesn't apply since there is no starter (as such) to be rotated; but disengagement is necessary to protect the operator from possible injury. In the case of geared-down starter units, if these were not disengaged after starting, the very high rpm at which the engine would rotate them would soon wear them out.

TYPES OF STARTERS

There are many different kinds of starting devices or starter units used with small engines. All of these, however, can be grouped into five general classifications:

1. *Manual*—Rope types
2. *Manual*—Crank and kick types
3. *Mechanical*—Windup types
4. *Electric*—Friction-clutch types
5. *Electric*—Centrifugal and Bendix-drive types

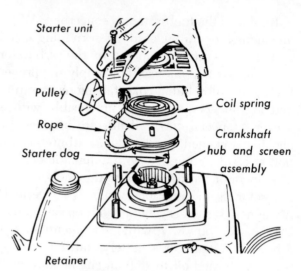

A direct rewind starter.

MANUAL ROPE-TYPE STARTERS

Used principally with lawn-mower engines and for outboard motors, rope-type starters are divided into *three* separate groups.

A *windup rope starter* consists simply of a fixed position pulley on the crankshaft, together with a rope which can be wound around this pulley and then quickly unwound with a straight pull to spin the crankshaft. The pulley is notched, to hold the rope end in a manner that will ensure disengagement of the rope at the end of a pull. This pulley may be located on any exposed portion of the crankshaft, but is usually at the end opposite the drive end. With such a starter the operator must rewind the rope for each new try. He also must pull the rope with a brisk, steady pull to impart the required momentum to the crankshaft.

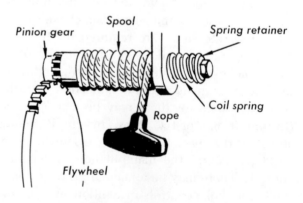

A geared rewind starter.

A *direct rewind starter* is exactly like the preceding, except that the rope end does not become disengaged from its pulley at the end of a pull, and the rope is rewound on its pulley by action of a coil spring within the starter unit. Generally, the starter is a removable unit mounted over a windup rope pulley which could be used after removal of the starter, in an emergency. However, this is not always the case. In either case, the rope pulley is designed to rotate the crankcase pulley (hub and screen assembly, in the accompanying illustration) through some type of spring-loaded starter dog or ratchet which will disengage whenever the rope is not being used for starting. Each time the rope is pulled, its initial movement—working on an internal friction brake—revolves the retainer just enough to thrust the dog outward to make engagement.

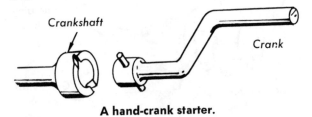

A hand-crank starter.

A *geared rewind starter* is essentially the same as the last one mentioned except that the starter unit drives the crankshaft through reduction gears (which give more leverage to the operator). In the type shown the gearing is from a pinion gear in the starter unit to a ring gear encircling the flywheel (magneto housing). Also, a threaded spool holds the rope, and the rewind spring is inside the spool. Provision is made to disengage the pinion from the ring gear.

MANUAL CRANK AND KICK TYPES

Crank starters are seldom used except as alternate provisions with electric-starter-equipped multicylinder engines (for use in case of electric starter failure). Though more difficult and unpleasant to use than a rope type, when used by an able operator the direct mechanical leverage offered by a crank does provide the extra-positive sustained momentum sometimes needed to start a multicylinder engine.

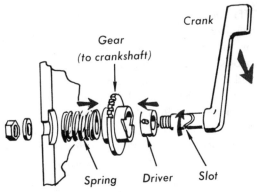

One type of kick starter.

Kick starters are leg-operated cranks, used principally with motorcycle and similarly mounted engines. Such starters usually work through gears to the crankshaft. There are many arrangements for disengaging the crank, two of which are illustrated.

In the first arrangement a slot in the crankshaft moves a driver into engagement with the gear during initial movement of the crank. A spring thrusts the driver back to disengage it whenever the torque between the crank and gear is relieved by engine rotation of the gear, or by stoppage of crank movement. A second coil spring (not shown) returns the crank to the starting position.

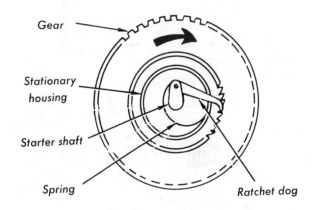

Another type of kick starter.

In the second arrangement initial rotation of the crankshaft moves a ratchet dog outward through its slot in the stationary housing to engage the ratchet teeth inside the gear hub. A coil spring withdraws the dog to disengage it afterwards. Again, a spring (not shown) returns the starter to starting position.

NOTE: The first is a fully disengaging type, since engine rotation of the gear, as well as cessation of crank movement, will effect disengagement. The second is not fully disengaging. It will remain engaged as long as pressure is applied to the crank, even though the engine has started. Pressure is, however, relieved from the crank at the end of its downstroke, unless the operator should attempt to swing it back up through the other side of its circle. A stop is usually provided to prevent this.

MECHANICAL WINDUP STARTERS

This unit takes the place of a rewind starter, and offers the advantage of a steadier spring action to replace the rope action. Operation is similar to that of a rewind starter. However, instead of a rope and a light rewind spring there is a heavier windup spring. As the operator winds this up with the handle provided, a spring-loaded ratchet-type catch holds the spring in tension. When ready, the catch is released by button or lever movement and the windup spring rotates the starter pulley, which is constructed like the rope pulley of the rewind starter.

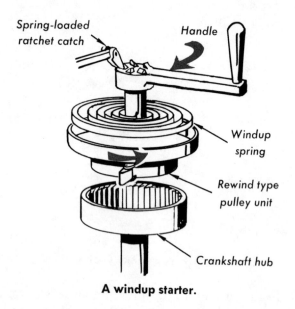

A windup starter.

ELECTRICAL FRICTION-CLUTCH STARTERS

Starters of this type have been offered as accessory units for operator mounting to replace rope-type units. They are also factory installed on some models of lawn mowers. The electric starter motor may be a 12-volt dc type. In this case a small auxiliary battery, for recharging by some outside source, is furnished. Or the motor may be a 110-volt ac type for plugging in to a household outlet. In either case, very little current is consumed if properly used. Instructions packaged with the unit warn against making more than half a dozen brief starting attempts to guard against burnout from overheating. Obviously, the lawn-mower engine must be in good enough repair to start at once, for a starter of this type to be useful. (And provision is usually made for quick removal of the starter and use of a rope, in case of necessity.)

The starter motor drives a hollow cone, leather or cork lined, which engages a mating unlined

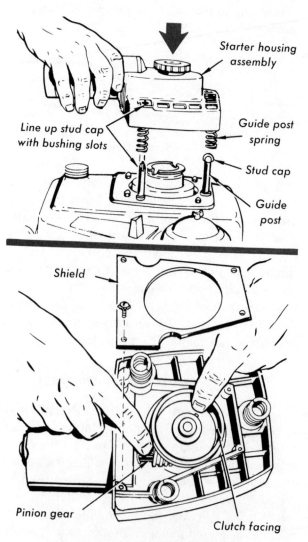

An electric friction-clutch starter.

cone on the crankshaft end. Springs hold the entire motor unit up (cones disengaged) when not in use. To start, the operator depresses this unit until the cones are engaged. He simultaneously closes a conveniently located button switch to start the motor. It is a characteristic of this type of starter that the tiny motor develops sufficient torque by operating at exceptional speed to drive the hollow cone through a geared-down worm drive.

Occasional service is required to keep the leather-cone facing and the motor worm gears properly lubricated. The motor is usually a sealed type replaceable only as an assembly, but other parts are separately replaceable. Full instructions for operation, maintenance and repair are always packaged with such units.

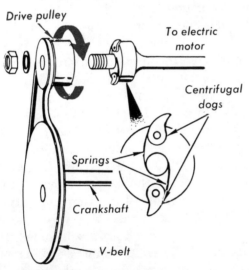

An electrical centrifugal starter.

ELECTRIC-CENTRIFUGAL AND BENDIX-TYPE STARTERS

The *centrifugal-type starter* employs spring-loaded centrifugal dogs (illustrated) on the starter shaft, and a hollow drive pulley with internal ratchet teeth for the dogs to engage. Initial starter torque swings the dogs out to engage and rotate the drive pulley, which, in turn, drives the large crankshaft pulley through a *slack* V-belt. As long as the drive pulley has more momentum than the crankshaft pulley, it will lift the slack out of the belt to rotate the crankshaft pulley. As soon as the engine starts to give superior momentum to the crankshaft pulley, the belt slack will be equalized at the two sides, allowing the belt to slip on the small drive pulley until the cessation of starter operation permits the dog springs to withdraw the dogs and complete the starter disengagement. Thereafter, the freed drive pulley is rotated through the belt by the crankshaft pulley.

Long the standard starter for automotive use, the *Bendix-drive type* utilizes a small (traveling) pinion gear on a screw-thread shaft, and a heavy coil spring. The threaded shaft and pinion float on the starter motor shaft; they are not directly connected to it. Connection is made through the spring, one end of which is bolted to the end of the threaded shaft while the other end is bolted to a drive screw that also floats on the end of the motor shaft. There is a sleeve keyed to the motor shaft (inside of the spring) that has a slot which is normally engaged with the wedge-shaped drive screw end. Therefore, the two will rotate as one and transmit shaft torque through the spring to the threaded shaft.

The initial starter-motor torque thus transmitted to the threaded shaft screws the pinion gear inward on the threaded shaft to engage it with a large ring gear on the flywheel housing (like the gear pictured in "A Geared Rewind Starter"). As long as the starter motor continues to rotate the flywheel, this pinion will remain engaged; but as soon as the engine starts and the flywheel velocity increases, the ring gear will impart a reverse motion to the pinion which

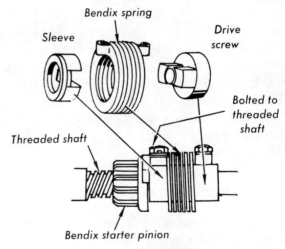

An electric Bendix-drive starter.

screws it back out of engagement. The usual violence of this reversal will throw the pinion to the end of its threaded shaft, where its lugs will engage the spring end and unwind the spring. In unwinding, the spring will spread out sufficiently to move the drive screw out of engagement with the slotted sleeve, thus momentarily disengaging the whole assembly from the starter motor shaft and preventing any harmful kickback through the starter motor. Or, if the pinion should become temporarily hung up in the ring gear, this same thing will happen, so that the engine cannot run the starter motor.

NOTE: Larger-horsepower engines start with such force and velocity that some such provision as just described, or as described for the centrifugal-type starter, must be made to protect the electric motor. Otherwise, the resulting speeded-up rotation of the motor would damage its brushes and commutator.

Bendix starters are termed *inboard* (the type illustrated) or *outboard*, depending on whether the pinion travels in or out to engage the ring gear. In automotive use the starter motor current may be controlled by a switch attached to the accelerator, plus a *vacuum switch* which breaks the circuit through manifold vacuum when the engine starts—thus permitting normal use of the accelerator thereafter. Automobiles also use *coincidental* starters, which are variations of the Bendix principle, embodying use of an electric solenoid to move the small pinion, plus use of a vacuum switch to ensure de-energizing of the solenoid to disengage the pinion. Small engines, however, employ only straight Bendix drives, as described.

While any type of electric motor may be used to operate either of the two preceding starters, the usual motor is a 6- or 12-volt dc type operated from a generator-battery ignition system. This is replaceable as a unit, as are all other starter parts. Lubrication is required per manufacturer's specifications, and parts must, of course, be in good repair.

ELECTRIC STARTER MOTORS

As with generators (Chapter 8) the major overhaul of an electric motor is a specialized field. The electrical characteristics of motor construction are beyond the scope of this text. We are, however, interested in the mechanical characteristics and those limited services which can be performed by the average mechanic.

NOTE: We do not include small, sealed motors of the type used with friction-clutch starters. Though these (both ac and dc) have brushes, etc., it requires special equipment to reassemble new brushes into such a motor. These have to be returned to the factory for exchange or repair.

The standard starter motor is a series-wound dc type. In principle it is like the simplified motor shown schematically in the accompanying illustration. Current flows through the wires of the stationary field to set up a permanent magnetic field with a *north* and a *south pole*. From the field, the current flows into the armature through one brush and out the other, then on back through the circuit to the battery. In flowing through the armature, it sets up a magnetic field having north and south poles, with the north pole adjacent to the field north pole and the south pole adjacent to the field south pole.

But like poles repel each other while unlike poles attract each other. Therefore, the armature rotates 180° to match up the poles properly. As it turns, the brushes move off the commutator segments they were contacting onto the opposite segments, to reverse the current through the armature. This is timed (by the spacing of the commutator segments) so that the poles are reversed just as the armature com-

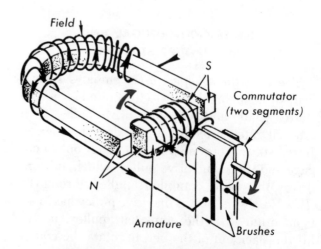

Principle of a series-wound dc motor.

pletes its 180° revolution. Its momentum carries it past dead center and it consequently goes on to make a second 180° rotation, etc.

An actual starter motor (accompanying illustration) has two field poles, two brushes in brush holders, and the armature is wound so as to produce a number of separate magnetic fields, each with its own north and south poles. The commutator is divided into twice this same num-

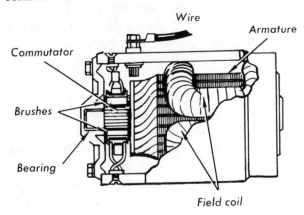

A typical dc starter motor.

ber of segments (commutator bars). Each brush contacts two bars simultaneously, and the wiring is arranged so that the two brushes together will direct current through two of the armature fields having poles 90° apart. These poles are spaced so that while one set is being attracted toward the field poles the other is being repelled at the opposite side. Thus the turning force is double; moreover, the multitude of field choices in the armature permits shifting of poles to keep

the magnetic activity centered close around the field poles, where the attraction and repulsion are naturally strongest. A motor of this type will develop up to 20 ft-lb of torque.

Motor bearings, positioned at each end of the housing to hold the armature shaft may require lubrication (usually motor oil applied through oil holes or cups). Only a few drops at a time in each hole, at occasional intervals, should be applied. Overoiling is almost as harmful as underoiling. Excess oil will spray the field and armature wires to soak them and short them and will short out the brushes and commutator to cause arcing and burning. If a motor has been seriously overoiled, it should be disassembled, cleaned, dried and reassembled.

Brushes, especially the carbon brushes, and commutator bars will wear. Replacement brushes are available and are generally easy to replace. Simply detach the pigtail lead from each, lift up the spring wire that holds the brush in contact with the commutator, and remove the brush. After installing new brushes the motor should be run, if possible, without load, until all arcing ceases and the new brushes are properly seated. If commutator bars are blackened, they can be brightened—with the motor running—by touching them lightly with fine sandpaper (*never* emery paper, which contains metal particles that could short them out). A badly worn commutator has to be rerounded and undercut (to correctly reduce the heights of the mica separators between the bars).

POWER TRANSMISSIONS

In Chapter 1 it was mentioned that mechanical energy is force, the momentum of an object put to mechanical use to move another object. We also distinguished between this and mechanical power by saying that the latter is the output of a machine capable of producing continuous mechanical energy. Then we went on to explain that mechanical power is rated in terms of horsepower (hp) and, in following chapters, how an engine produces mechanical power. We have also explained that a portion of an engine's mechanical power is consumed within the engine

(to run itself, so to speak) leaving, however, an overage of output power, which can be made to do other work.

In the following we will discuss how manufacturers rate the engine power and how the output power may be put to work.

TERMS USED TO DEFINE ENGINE OUTPUT

Torque and RPM

As previously explained, *torque* is twisting force. In an engine it is the rotating force delivered to the crankshaft, measured in foot-pounds (*ft-lb*) or inch-pounds (*in-lb.*). Motion is not required. A bolt, for instance, can be tightened to 50 ft-lb torque and, although it remains stationary, no less than 50 ft-lb of torque applied in the opposite direction will loosen it. Engine torque, then, is equivalent to the total force required to bring the crankshaft to a dead stop.

When motion is added, the velocity of this motion is measured in terms of revolutions (of the crankshaft) per minute (*rpm*). By developing torque continuously at some rpm, an engine produces power that can be put to work.

Momentum increases as the speed of motion of a body increases. Therefore, it would seem that increasing the rpm of a revolving shaft should always increase the amount of work that the shaft will do. Remember, however, that momentum equals mass times velocity. When a small shaft is revolved by an external force of considerable magnitude, the mass of the shaft itself is inconsequential. It is the magnitude of the external force that counts. For such a case, then, we can say that the output momentum of the shaft equals the force revolving it times the velocity. If this force decreases sufficiently, even though the velocity increases, the output momentum will decrease. In considering the output momentum (that is, the output power) of an engine's crankshaft, we have to consider both its rpm and the engine torque.

The torque in an engine will increase with increasing rpm as the throttle is opened, as long as the additional power obtained from the burning of larger amounts of fuel is converted into torque. The conversion of fuel power into torque does, however, reach a maximum beyond which the burning of still larger amounts of fuel will add nothing to the torque—which will, in fact, reduce the torque. This point is reached when the additional force obtained from additional fuel is totally expended within the engine to overcome increased friction losses (which mount rapidly with the increased combustion heat after a certain temperature is reached) and the increased compression load and other factors. In short, in any engine there is a point of diminishing torque, up to which increased rpm

will increase the power output; beyond this point, increasing the rpm will decrease the power output. Even before this maximum power rpm is reached, the engine's efficiency (its power output per unit of fuel burned) will start to decrease.

Horsepower and Kilowatts

As we know (Chapter 1), in this country a horsepower (hp) is the unit for measuring power. It is the unit that indicates how much work a machine that produces mechanical power (like an engine) can accomplish. For instance, an engine developing 100 hp can lift a 4000-lb automobile a vertical distance of 825 feet in 1 minute. The formula is: *hp = ft-lb of work done ÷ (550 × time in seconds)*.

But the output of an engine depends on quite a few factors—its torque, rpm, internal power loss, etc.—so that there are, unfortunately, quite a few different ways an engine may be rated as to its horsepower. A few of these are discussed in the following.

In the metric system, the *watt* is used as the basic unit for measuring power. In this country we customarily use this metric unit when designating electric power. The watt is equal to about 10,200 gram-centimeters per second. Or, to translate this into our system, it takes *746 watts to make one horsepower. A kilowatt (kW) is 1000 watts.* Hence our 1 hp equals approximately ¾ kilowatt. A 20-hp engine, then, is approximately equal in power to a 15-kW engine or electric motor.

Brake Horsepower

This is the *actual* usable output of an engine at a given rpm, as measured by an acceptable measuring device. As already explained, it will vary with the rpm; but it will not vary according to any fixed formula inasmuch as the other factors (heat, friction, compression load, etc.) will vary for different engines and even at different times for the same engine. As an engine is accelerated to a certain rpm, torque will increase more rapidly than brake horsepower, since the lack of velocity will keep the output momentum of the shaft at relatively lower values. Beyond this certain rpm, the relatively high shaft velocity will cause the output momentum (that is, the brake horsepower) to increase faster than the torque—up to the point of diminishing torque, at which both the torque and the brake horsepower start to decrease.

Either a *prony brake* or a *dynamometer* may be used to measure brake horsepower. The brake machine applies a braking horsepower (friction) force to a wheel mounted on the crankshaft, to measure the work being done to overcome this resistance and record the result on a calibrated dial. A dynamometer is a dynamo (like a generator) set up to convert *all* the mechanical energy of the crankshaft into electrical energy, and to measure this in terms of kilowatts or horsepower.

When torque and rpm are known, the brake hp can be calculated from the formula: (*torque* $\times$ *rpm*) $\div$ *5252* $=$ *brake horsepower* (the 5252 being a constant factor).

Rated and Estimated Horsepowers

The S.A.E. formula for rating gasoline engines is $hp = d^2 n \div 2.5$, where $d =$ cylinder diameter in inches and $n =$ number of cylinders. This arbitrary rating is used in most states for auto-licensing purposes.

An experimental formula for rating the maximum horsepower of a four-stroke-cycle gasoline engine is: $(d^2 \times n \times s \times rpm) \div 11,000 =$ *estimated hp.* (*d* and *n* are the same as in the preceding formula; *s* = stroke; and 11,000 is a constant which must be changed as fuels, etc., are improved.) For a two-stroke-cycle engine, substitute the constant, *9000* in place of 11,000 and use the same formula.

Indicated and Friction Horsepowers

Indicated horsepower is the amount of power an engine should produce if all the power (pressure) delivered by each combustion could be realized at the crankshaft. *Friction horsepower* (friction loss) is the horsepower consumed by the engine in running itself. In short, it is the indicated horsepower minus the brake horsepower.

Indicated horsepower can be calculated or can be obtained by measuring the pressure in the cylinder(s). Friction horsepower can be obtained by using a dynamometer to measure the power needed to motor (rotate) the engine with the ignition off, at full throttle and at the rated rpm.

Observed and Corrected Horsepowers

Observed horsepower has become an advertising gimmick. It is stated to be the amount of power delivered at a certain time. Advertisers usually select a time and conditions when the sample engine is at its best, then take the best brake-horsepower reading for advertising purposes. This is not a true comparative rating, since an engine will run much better when the air temperature is low and the barometric pressure is very high—on a cold, dry, crisp day.

Realizing the need for standardization the S.A.E. has said that, for comparative ratings, the testing must be done at 60°F with the barometer at 29.92 (inches of mercury). This is then called the *corrected horsepower,* and it can be calculated from any observed horsepower by using a correction formula.

Continuous Horsepower

This is the horsepower which an engine can deliver constantly and continuously consistent with maximum life expectancy of the engine (as rated by its manufacturer). In general, *continuous horsepower* is around 80% of maximum corrected horsepower.

Volumetric Efficiency

This is the ratio between the quantity of mixture actually entering the cylinder(s) and the amount that would enter if the cylinder(s) were completely filled. For instance, if only 60 cubic inches per second should enter an engine that could hold 120 cubic inches per second at rated rpm, the volumetric efficiency would be 50%. It is affected by the manifold, valve and carburetor restrictions to air flow, and by the inertia of the air.

Brake Mean Effective Pressure

This is the average combustion pressure (in pounds per square inch) that would have to be exerted on each piston in order to produce the stated brake horsepower.

THE FACTORS INVOLVED IN POWER TRANSMISSION

Friction

Up until now we have discussed friction but briefly, and all in terms of its disadvantages. Friction does also have its useful applications; it is a mechanical factor involving certain commonly used terms, rules, and formulas.

Friction is defined as *the resistance to any force trying to produce motion.* It is partly the result of adhesion between molecules, but it is principally due to mechanical interference between the molecular bumps on one surface and those on the other surface. Hence, friction between two rough surfaces is much greater than between two smooth surfaces.

Friction is classified as either fluid or solid. *Fluid friction* (the friction within a fluid or between a fluid and solid) is much less than *solid friction* (between two solids). This is why we use lubricants. However, fluid friction is nearly proportional to the square of the velocity. For instance, air (in this sense) is a fluid, and air resistance increase greatly with velocity. An auto traveling at 20 mph encounters only one-fourth the air resistance it will encounter at 40 mph; at 80 mph the resistance is 16 times as much as at 20 mph.

Solid friction is divided into *rolling friction* and *sliding friction.* Rolling friction is considerably less than sliding friction—which is why we use roller and similar bearings. Even with sliding friction, however, there is a great variation depending on the materials of the two surfaces. For instance, two steel surfaces have much friction, while steel will slide over an alloy of lead and antimony (called *Babbitt* in honor of its inventor) with greatly reduced friction. Hence plain bearings (Chapter 5) are usually babbitted with the alloy of this name or some other *antifriction metal.*

We have learned from experience that sliding friction is nearly always greater at starting than afterwards; it takes more push to start an object than to keep it sliding. Also, sliding friction is practically independent of velocity. Speed has nothing to do with the amount of friction, unless heat is generated to swell one substance and in-

Fluid friction.   Solid friction.

Rolling friction.   Sliding friction.

Starting friction is greater than moving friction.

Weight increases friction.

General rules regarding friction.

crease friction, as is the case with an auto's brakes. Then again, sliding friction is directly proportional to the weight of the top object or the pressure forcing two objects together. A pile of stones is much harder to slide on pavement than is a single stone.

In order to relate the friction of one situation to another we use the term *coefficient of friction.* We define this as the force required to overcome friction divided by the weight (or pressure) holding the surfaces together (*cf = friction force ÷ weight*). For instance, a block weighs 100 lb and it requires 40-lb force to keep it in uniform motion. The *cf* of this situation is $40 \div 100 = 0.4$.

Mechanical Advantage (MA) of Basic Machines

Earlier (Chapter 5) we discussed levers and gears, and talked of leverage. In the science of mechanics leverage is called *mechanical advantage (MA)*. It is defined as the number of times by which a power transmission machine (one that simply transfers but does not produce mechanical power) will multiply force. Now an *effort (E)* applied at one (the *input*) end of a transmission machine is a force. Also, the load attached to the machine at its other (*output*) end resists movement to the extent that every object has inertia. This, then, is an opposing force, the inertia of the *load (L)*. According to its definition, the MA of a machine then is the load (L) divided by the effort (E). That is: $MA = L \div E$.

Force at E is doubled at L

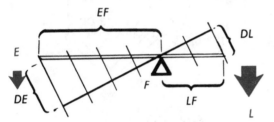

Example of an MA of 2.

Another useful formula, known as the *Law of Machines*, states that output must equal input—since no energy (Chapter 5) is ever lost—providing, of course, that there is no loss from friction. If an effort (E) has to move a *distance (DE)* in order for the machine to move the load (L) a *distance (DL)*, then this law is written: $E \times DE = L \times DL$. This is the same as saying: input (force) × distance = output (force) × distance; or, since force × distance = work, we can say: input work = output work. Since $MA = L \div E$, the last formula can be rewritten: $MA = DE \div DL$.

Now the preceding three formulas can be applied to any of the six basic transmission machines (Chapter 1)—or to any combination of them—to make it easy for you to know just how to determine what the machine's output will be when the input is so much (or vice-versa).

NOTE: In all the following discussions we are going to disregard friction and effects due to the weight of machine parts. How these two factors affect a machine will be discussed later.

The Lever

A simple lever is illustrated. The downward applied effort (E) rotates the lever on its fulcrum (F) to lift the resisting weight (force) of the load (L). Consider the formula, $MA = DE \div DL$. Here DE (whatever it actually is) is the base of an isoceles triangle of which EF is one

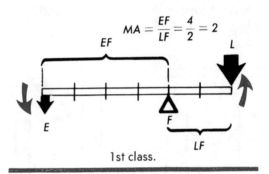

$$MA = \frac{EF}{LF} = \frac{4}{2} = 2$$

1st class.

$$MA = \frac{EF}{LF} = \frac{6}{3} = 2$$

2nd class.

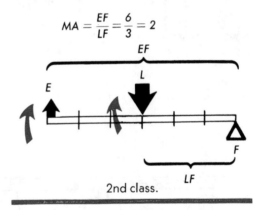

$$MA = \frac{EF}{LF} = \frac{3}{6} = \frac{1}{2}$$

3rd class.

Types of levers.

side, while *DL* is the base of another isoceles triangle having *LF* as one side. The angles at *F* of these two triangles are equal. Hence, by geometry, the ratio of *DE* to *EF* is the same as the ratio of *DL* to *LF*. We can substitute *EF* for *DE* if we also substitute *LF* for *DL*. We then have: $MA = EF \div LF$. In short, to find the *MA* of any lever, divide the length of the effort arm by the length of the load arm.

The accompanying three views show the three classes of levers (ways in which levers can be used), and illustrate how the *MA* is calculated in each case. Note that in the first two classes the *MA* is greater than 1, but in the third class it is less than 1.

The Pulley

A stationary pulley (view 1) is actually a first-class lever free to continue rotating. Any effort (*E*) applied to the input rope end will act as if it were rocking an imaginary lever (*E'L'*) to lift the load (*L*). Since the fulcrum (*F*) is the axis of the pulley, the two lever arms *E'F* and *L'F* are equal, each being a radius of the circle. Using the lever formula we have $MA = E'F \div L'F$, or $MA = 1 \div 1 = 1$. There is *no mechanical advantage in a single stationary pulley* (actually a small loss due to friction). All that is accomplished is to reverse the direction (*E* travels down while *L* travels up).

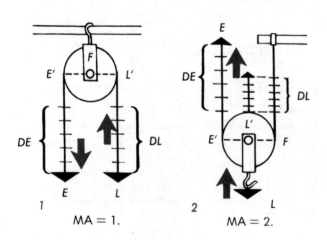

1 MA = 1.

2 MA = 2.

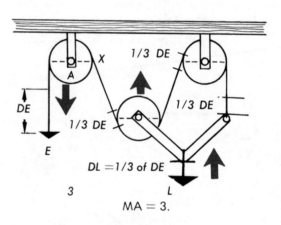

3 MA = 3.

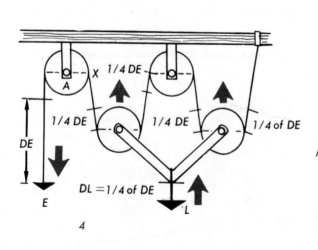

4 MA = 4.

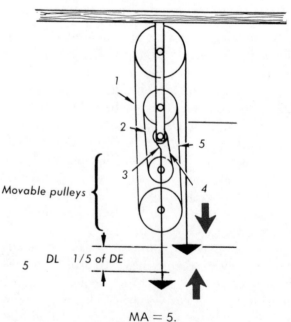

5 MA = 5.

Pulley MA's.

However, *with a single movable pulley* (view 2), we do gain a mechanical advantage. Here the fulcrum (F) is at the point where the anchored rope end starts around the pulley. An effort (E) has to lift the entire effort arm $E'F$. But the load (L) is below the pulley axis, so that the load arm is only the length of $L'F$. Now, $E'F$, a diameter, is twice $L'F$, a radius, so the $MA = 2$. The action is that of a class 2 lever.

All other pulley arrangements are combinations of stationary and movable pulleys. In view 3 there is one movable pulley and there are three strands of the rope which support this movable pulley and the load (L). The fourth rope strand is where the effort (E) is applied. Whatever distance (DE) the fourth strand moves, the other three strands must between them move this same distance (to feed the rope continuously over pulley A). Being three, they divide this distance into thirds. This means that the strand holding the load moves only its one-third share of the distance. Distance DL is one-third of DE, so $MA = 3$.

In view 4 there are four strands supporting the two movable pulleys (and the load attached to them). There are four strands to divide the distance DE, and DL is one-fourth of DE so that $MA = 4$. View 5 illustrates the same rule except that the two movable pulleys are joined together and the three stationary pulleys are joined together, and five strands of the rope (numbered) support the movable pulleys and the load. Distance DL is one-fifth of DE; the $MA = 5$.

We could go on indefinitely, but it should be obvious by now that the MA will always equal the number of rope strands used to support the movable pulleys and the load. Since this will always be the same as the number of pulleys, we can find the MA by counting the pulleys. (Any two or more pulleys on the same axis—sometimes used for two or more ropes pulling as one—count only as one.) With P representing the number of pulleys, $MA = P$.

NOTE: This formula ($MA = P$) obviously applies to a single stationary pulley. The only exception is the single movable pulley. Here we have to count the rope strands (two) which support the pulley and its load.

The Wheel and Axle

Again (with the wheel and axle) we have, effectually, a simple class-1 lever. From the accompanying illustration it should be obvious that the $MA = EF \div LF$. To prove this, if the wheel makes one revolution so does the axle. An effort that revolves the wheel once travels one circumference of the wheel, while the load travels one circumference of the axle. Consequently, $MA = cir.$ *of wheel* $\div cir.$ *of axle;* or (from the geometry of a circle) $MA = dia.$ *of wheel* $\div dia.$ *of axle;* or, $MA = radius$ (EF) $\div$ *radius* (LF). You can find the MA by measuring the circumferences, the diameters, or the radii—whichever is most convenient.

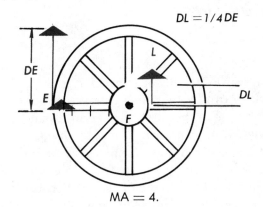

$DL = 1/4\,DE$

$MA = 4.$

Wheel and axle MA.

If the input is at the axle instead of at the wheel, then the fraction in our MA formula has to be reversed. Hence, $MA = cir.$ (etc.) of axle $\div cir.$ (etc.) of wheel. To make a general rule, we should say for all cases that $MA = $ *input cir.* (etc.) $\div$ *output cir.* (etc.)

The Inclined Plane

While the inclined plane varies somewhat in application from the lever, the same formulas apply. Consider view 1. If the load (L) is to be moved to the top of the incline by effort (E), the effort must be maintained throughout a distance (DE) equal to the distance the load slides (as if it were "walking" the load up). At the same time the load is raised only the distance (DL) straight up from the bottom to the top. $MA = DE \div DL$, or, in this case, $MA = $ *length of incline* $\div$ *height* (straight up) to top.

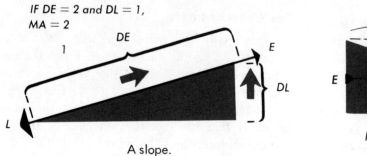

IF DE = 2 and DL = 1,
MA = 2

1
DE
E
DL
L

A slope.

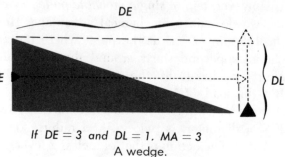

DE

E
DL

If DE = 3 and DL = 1. MA = 3
A wedge.

Inclined planes.

View 2 shows the direction and distance (DE) in which an effort (E) must be applied to drive the wedge and elevate the load (L) a distance equal to the wedge height (DL). With this type of inclined plane, $MA = height\ of\ wedge \div length\ of\ wedge$.

The Screw

Whether a screw is used to raise a jack (view 1), as the threads of a machine screw (view 2) which can be tightened to move or hold something, or in the form of a worm driving a gear (view 3)—or however it may be used, one turn of a screw will move it up or down (as the case may be) exactly the distance from the top of one thread to the top of the next. This distance is called the *pitch* of the screw. Input is always applied to the shaft end.

With a jack, effort is applied at the end of a bar handle. To raise the load (L) one pitch (DL) the effort (E) must rotate the bar one full turn—and point E will describe a circle having a radius equal to the handle length. Using our MA formula, MA equals the circumference

of the circle divided by the pitch. By applying the formula $C = 2\pi R$, we find that E travels a distance equal to $2\pi DE$. If we let P stand for the screw's pitch, then $MA = 2\pi DE \div P$. To learn the MA we have to know or measure the screw-thread pitch and know or measure the circumference, diameter, or radius of the circle the effort will describe.

In the case of a machine screw driven by a screwdriver, the effort is applied to twist the screwdriver handle, which must make one revolution for each pitch movement of the screw. It might be easier to measure the diameter of the

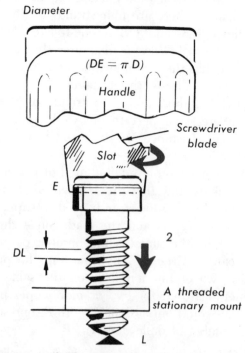

Diameter

($DE = \pi D$)
Handle

Screwdriver blade

Slot

E

DL

2

A threaded stationary mount

L

If πD = 3 in. and pitch = $\frac{1}{16}$ in.
MA = 48 (from screwdriver handle through screw).

MA of a machine screw.

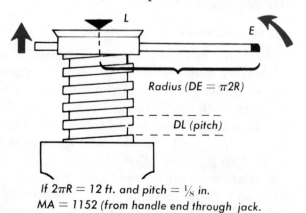

L   E

Radius (DE = $\pi 2R$)

DL (pitch)

If $2\pi R$ = 12 ft. and pitch = $\frac{1}{8}$ in.
MA = 1152 (from handle end through jack.

MA of a jack screw.

handle—in which case our formula is: $MA = \pi D \div P$. Or, if you want to disregard the screwdriver and find the MA of the screw alone (as if you were going to turn it with your fingers), measure the length of the screw-head slot, instead, and substitute this for D in the preceding formula.

> NOTE: A rotating machine thread screw that travels a nut from end to end of the screw transfers the same MA to the nut that it (the screw) develops. That is, the MA output at the nut is the MA of the screw.

With a worm and worm-gear arrangement we have a special application of the screw, due to the addition of the worm gear and its shaft, which is, in effect, a wheel and axle. We could first figure the MA from the shaft through the worm gear (which is the same as the MA from a threaded machine screw to a traveling nut, discussed previously) and then add on the MA of the wheel-to-axle portion. But there is an easier method. Whatever the pitch of the worm, in rotating it has to rotate the worm gear and the worm-gear shaft a distance of one worm-gear tooth for each full revolution of the worm shaft.

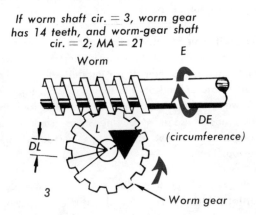

If worm shaft cir. = 3, worm gear has 14 teeth, and worm-gear shaft cir. = 2; MA = 21

MA of a worm and worm gear.

If an effort (E) moves a distance (DE) equal to the worm-shaft circumference, then the worm-gear shaft will move a load (L) a distance (DL) equal to the worm-gear shaft circumference divided by the number of teeth in the worm gear. That is, $MA = DE \div DL$ = worm shaft cir. $\div$ (worm-gear shaft cir. $\div$ no. of worm-gear teeth). That is, $MA = $ (*worm shaft cir.* $\times$ *no. of worm-gear teeth*) $\div$ *worm-gear shaft cir.* Alternately,

you can use the diameters or the radii of the two shafts instead of the circumferences.

Gears, Sprockets, and Sheaves (or Pulleys)

In Chapter 5 we discussed gears and sprockets, showed how a gear is essentially a lever, and described gear ratio as being the number of teeth in the larger gear divided by the number of teeth in the smaller gear. The gear ratio *is* the MA of a simple machine composed of two gears, if we do not have to take into consideration the diameters of the shafts. This will be clearer if you note the accompanying illustration.

> NOTE: If the input shaft is larger in diameter than the output shaft (or vice versa), the difference in shaft sizes introduces a new factor that will make the simple gear machine into a compound machine. The relationship between the larger shaft and the smaller one is the same as the relationship between a wheel and axle—and the MA of this wheel-and-axle relationship would have to be added to the gear MA to obtain the MA of the compound machine. Therefore, in speaking of gear MA we will *assume both shafts to be the same diameter.*

The teeth of a gear are spaced evenly around its circumference. Therefore, any fractional part of one full revolution that it makes will be equal to the number of teeth that pass by a given point (like "X" in accompanying illustration) divided by the total number of its teeth. Moreover, since the shaft circumference is concentric with the gear circumference, a point on the shaft must move this same fractional amount of the shaft's total circumference. In our illustration, the distance (DE) that an effort (E) moves is the circumference of the small gear shaft. But the two shafts are equal in circumference. Consequently, any distance (DL) that the load (L) moves is a fraction of the distance (DE) equal to the number of large gear teeth that pass by point X divided by the total number of large gear teeth. If the total number of large gear teeth is NL and the number of teeth that pass point X is Y, we can say: $DL = Y \div NL$, if $DE = 1$ (one full revolution).

For two gears to mesh properly all their teeth must be equal in size and spacing. Consequently, if $DE = 1$ (and our small gear therefore makes

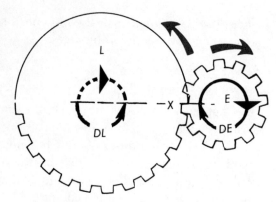

*Large gear is twice dia. of small one
(has twice as many teeth) — so*
$$MA = 2$$
(Gear ratio is 2:1)

MA of two gears.

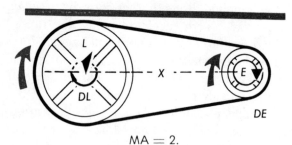

$$MA = 2.$$

MA of two sprockets.

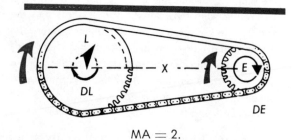

$$MA = 2.$$

MA of two flat-belt pulleys.

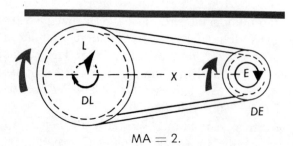

$$MA = 2.$$

MA of two V-belt sheaves.

one full revolution), all its teeth will pass by point X—and the number of large gear teeth simultaneously passing by point X must be the same. In short the Y in the preceding formula is the number of small gear teeth (represented by NS). We can say that if $DE = 1$, $DL = NS \div NL$. Using the formula $MA = DE \div DL$ we have: $MA = 1 \div (NS \div NL)$. Therefore, $MA = NL \div NS$, whenever the effort (E) is applied to the smaller gear shaft.

If effort (E) is applied to the larger gear shaft, the reverse of the preceding is true; $MA = NS \div NL$. That is, the MA always equals the number of teeth in the gear at the output side divided by the number of teeth in the gear at the input side. By having NO represent the former and NI the latter, we can write an all-purpose formula thus: $MA = NO$ (*output gear teeth*) $\div NI$ (*input gear teeth*).

NOTE: As previously shown (Chapter 5) two externally meshed gears always reverse the direction of rotation, while two internally meshed gears do not. If three or more gears are meshed, each meshed pair must be considered as constituting one basic gear machine. There will be as many basic machines (as many separate MA's to figure) as there are meshed pairs.

The shaft sizes of intermediate gears in a train of three or more gears does not matter if these are simply idler gears. Nor does it matter whether idler gears rotate around or with their shafts. In figuring MA we are only concerned with shafts used for input, and these must be the same size, or the element of a wheel and axle is introduced, per previous note.

Two sprockets are the same as two gears, except that the direction of rotation always remains unchanged. The fact that they rotate with a chain instead of meshing together makes no difference, since the chain simply transmits the motion of one gear to the other, without adding or subtracting anything from the motion it transmits.

This same is also true of two V-belt sheaves, or of two flat-belt pulleys. In either case, the belt neither adds to nor subtracts from the motion transmitted, providing, of course, there is no slippage. With sheaves or pulley, however, there are no teeth to count. Instead of the number of teeth, we have to compare either the two circumferences, the two diameters, or the two radii. The diameters are usually easiest to measure, so the best formula is: *MA = dia. of output sheave ÷ dia. of input sheave.*

> NOTE: In dealing with flat-belt pulleys it is only necessary to measure either the diameters, radii, or circumferences of the two, since the belt actually grips the two pulleys at their outer circumferences. A V-belt, however, does not grip its sheave at the outer circumference. It grips it down inside the sheave V (at the sides of the belt). To measure properly, install the belt on the two sheaves, and then mark on each sheave where the top of the belt lies in the sheave V. A circle parallel to the circumference at this point will be the effective circumference of the sheave for use in the MA formula.

Velocity Advantage (VA) of Basic Machines

In all basic machines, whenever a mechanical advantage (*MA*) is gained, a *velocity advantage* (*VA*) of equal proportion has to be lost. That is, *if force is increased, speed is reduced proportionally (and vice versa).*

Let's refer to our basic input-output formula, $E \times DE = L \times DL$. If L is greater than E, then obviously DE must be greater than DL by exactly the same amount—to keep the two sides equal. If L is twice E, then DE must be twice DL, etc.

We express this fact algebraically by saying that *VA* is inversely proportional to *MA*: that is $VA = 1 \div MA$. Since $MA = DE \div DL$, we can also say: $VA = DL \div DE$. Therefore, if we do not know the MA but do know the values of DL and DE, we can find the *VA* by using any of our *MA* formulas, by simply inverting the fraction. For instance, the *MA* of a screw is: cir. of shaft ÷ pitch; therefore the *VA* of a screw is: pitch ÷ cir. of shaft. The *MA* of a lever is $EF \div LF$; so the *VA* is $LF \div EF$. The *MA* of a gear is $NO \div NI$; so the *VA* is $NI \div NO$, etc.

Typical Basic Machine Applications That Are Not Easily Recognized

The transmission of power from one gear shaft to another gear shaft, from a wheel to an axle or vice versa, or from a rope to the hoist created by a pulley arrangement is easily seen and understood. But, in modern machinery, it is not always easy to see just what basic machine principle is involved in the transmission of power from one part to another. We can't identify all the possible arrangements here, but we shall list a few of those which are most likely to be encountered.

(1) A *crank* is a wheel and axle. The crank arm is the radius of the wheel, the output shaft

---

USEFUL MA AND VA FORMULAS

1   *LEVER*

$$MA = \frac{\text{Length Effort Arm (EF)}}{\text{Length Load Arm (LF)}}$$

2   *PULLEY*

MA = No. of Pulleys (P)
(all except single movable pulley)

3   *WHEEL AND AXLE*

$$MA = \frac{\text{Input (Wheel or Axle) Cir., Dia., or Rad.}}{\text{Output (Axle or Wheel) Cir., Dia., or Rad.}}$$

4   *INCLINED PLANE*

$$MA = \frac{\text{Length of Incline (or of Base)}}{\text{Height}}$$

5   *SCREW*

$$MA = \frac{\text{Cir. of Shaft}}{\text{Pitch (P)}}$$

6   *WORM AND WORM GEAR*

$$MA = \frac{\text{Cir. of Worm Shaft} \times \text{No. of Gear Teeth}}{\text{Cir. of Gear Shaft}}$$

7   *GEARS AND SPROCKETS*

$$MA = \frac{\text{No. of Output Gear Teeth (NO)}}{\text{No. of Input Gear Teeth (NI)}}$$

8   *FLAT BELT PULLEYS*

$$MA = \frac{\text{Output Pulley Cir., Dia., or Rad.}}{\text{Input Pulley Cir., Dia., or Rad.}}$$

9   *V-BELT SHEAVES*

$$MA = \frac{\text{Effective Output Sheave Cir., Dia., or Rad.}}{\text{Effective Input Sheave Cir., Dia., or Rad.}}$$

10   *TO FIND VA*

Either: $VA = \dfrac{1}{MA}$

Or: $VA = \dfrac{DL}{DE}$

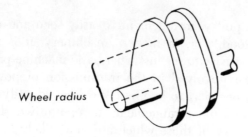

Crank and journal.

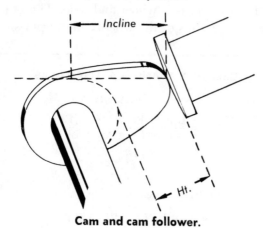

Cam and cam follower.

In a Bendix starter drive, the pinion acts as a traveling nut until it reaches the thread end and engages the ring gear. When the nut stops traveling, the relation between the starter shaft and the ring-gear shaft (the crankshaft) becomes the relation between two gears.

(5) An ordinary *balance scale*, a *valve rocker arm*, a *front-wheel steering arm*, or any other similar, pivoted device employs the principle of a lever.

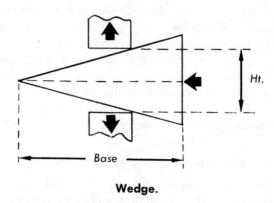

Wedge.

end is the axle, and the crank handle merely serves as a convenient place to apply the effort. This nomenclature applies to any similar arrangement such as the throw of a crankshaft and the crankshaft, whereby linear motion is converted into rotary motion, or vice versa. In the latter case the throw is the wheel, while the straight part of the shaft is the axle; the connecting rod transmits the piston effort to the throw through a second wheel and axle formed by the piston pin and rod.

(2) A *hoist drum* or *windlass* is a wheel and axle. The shaft is the axle, and the drum (or the rope spindle) is the wheel. The rope simply transmits power, either directly to the load or to the load through a (second machine) pulley arrangement.

(3) A *cam* and *its cam follower* constitute an inclined plane. As the cam revolves, its follower is forced up the incline at one side to the high point, then down an identical incline on the opposite side. The *MA* of power transmission to the follower is calculated the same as for a wedge.

(4) As previously noted, a *screw with a traveling nut* is simply a screw. Input is at the screw shaft, and output is the movement of the nut.

(6) A *wedge* used to separate two parts utilizes the principle of an inclined plane.

(7) Some basic machines are reversible; effort may be put in at either end and output is taken from the opposite end. Others, however, are either not reversible or are seldom used in reverse. The latter are: the pulley, inclined plane, screw and worm, and worm gear.

(8) In common practice the reversible basic machines are usually referred to as having a small end and a large end. In all cases "small" indicates the end with the smaller dimension. With a lever or a wheel and axle, the small end is the one from which output is taken if the *MA* is greater than 1; but with gears, sprockets, and belt pulleys or sheaves, it is the large gear from which output is taken if the *MA* is greater than 1.

(9) The nonreversible machines (No. 7, preceding) are practically always used for the sole purpose of gaining *MA*. All the reversible machines, however (being reversible), are often used to gain *VA* rather than *MA*. With machines it is also true that the direction either must, or can, be reversed. In a lever the direction must be reversed, but with gears, sprockets, and belt pulleys or sheaves the direction can

remain the same or be reversed, depending on the arrangement. Two sprockets or two belt pulleys or sheaves normally run in the same direction; but crossing the chain or belt will change direction, though this is usually impracticable unless the chain or belt is sufficiently long.

(10) Any spring that is compressed, bent, or stretched by an input effort so that it will later expand, contract, or bend back to transmit an output force acts as a lever having an $MA = 1$. It isn't, of course, a lever, but it is similar in the respect that the effort moves it about a fulcrum point (the limit to which it is compressed, etc.) and then moves about this same fulcrum point to re-create a force equal to the original effort. Transmission of the effort can be delayed by holding the spring compressed, for example.

Compound Machines

Very few of our modern power transmission machines consist of just one basic machine. Most are combinations of two or more (sometimes quite a few) basic machines. Here are a few common examples:

A gear train uses two or more basic machines (with gears and/or sprockets, etc.)

A steering mechanism uses a wheel and axle, a screw, another wheel and axle, and levers.

A crane uses a gear train, a wheel and axle, and pulleys.

A reel-type self-propelled lawn mower uses a gear train and two wheels and axles (for the cutting reel and for the drive wheels).

An outboard motor uses gears and a wheel and axle (the prop on its shaft).

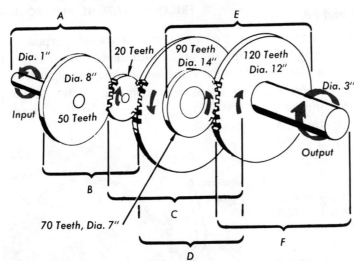

A) $MA = \dfrac{Dia. (1'')}{Dia. (8'')} = 1/8$

B) $MA = \dfrac{Teeth (50)}{Teeth (20)} = 2\,1/2$

C) $MA = \dfrac{Teeth (20)}{Teeth (90)} = 2/9$

D) $MA = \dfrac{Dia. (14'')}{Dia. (7'')} = 2$

E) $MA = \dfrac{Teeth (70)}{Teeth (120)} = 7/12$

F) $MA = \dfrac{Dia. (12'')}{Dia. (3'')} = 4$

TOT. $MA = 1/8 \times 2\,1/2 \times 2/9 \times 2 \times 7/12 \times 4 = 70/216$ (Approx. 1/3)

The MA of a gear train.

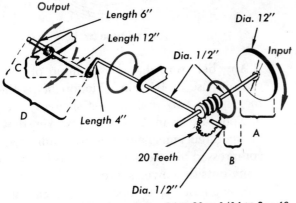

A) $MA = \dfrac{Dia. (12'')}{Dia. (1/2'')} = 24$

B) $MA = \dfrac{Dia. (1/2'') \times Teeth (20)}{Dia. (1/2'')} = 20$

C) $MA = \dfrac{Dia. (1/2'')}{Dia. (2 \times 4'')} = 1/16$

D) $MA = \dfrac{Length (12'')}{Length (6'')} = 2$

Total $MA = 24 \times 20 \times 1/16 \times 2 = 60$

The MA of a steering apparatus.

230

A motorbike uses a gear train, sprockets (or sheaves), and a wheel and axle.

To find the *total MA* of any compound machine, trace the transmission of force through it, from the input to the output end. Separate it into its basic machines. Keep in mind which is the input and which is the output end of each basic machine. Finally, multiply each *MA* by the following *MA*, from start to finish, and the result will be the total *MA*.

Total *VA* may be found in similar manner. To find the direction the output will take, start with the direction of input, and trace the direction of movement through each basic machine.

Two typical examples of calculating total *MA* are illustrated. In each, the lettered brackets each enclose one basic machine, for which the formula (identically lettered) is also given.

How Friction and Weight Affect MA and VA

Friction is created between two moving parts. We can often reduce friction to a negligible amount by lubrication and/or use of antifriction metals, roller bearings, etc., but we cannot entirely eliminate it. Since friction opposes movement, it necessarily detracts from the *MA* or *VA* of a machine by an amount equal to the force of the friction. There are no general-use formulas for calculating friction loss. The only way to learn the friction loss for a particular machine is to measure how much less the *MA* (or *VA*) is than the amount it should be (according to preceding formulas).

Weight loss is defined as the loss resulting from any necessity within the machine of lifting a part against the pull of gravity. If a very heavy lever has one arm much longer than the other, whenever an effort moves the long arm up some effort will be used simply in lifting the excess weight of this long arm. On the other hand, when this arm moves down, its excess weight will add some force to the effort.

A different situation is created by a balanced wheel rotating on a horizontal axis. It will create no weight loss as one side is as heavy as the other. Most machines either have balanced parts, or they operate continuously so that every part is returned periodically to the position from which it started. We can conclude, then, that

while weight loss is a possibility to be considered, we will probably not have to be much concerned with weight loss in any power transmission machine.

NOTE: There is the possibility that some transmission machine might be affected by weight loss, if turned on its side or upside down in an unintended position. And there is a greater possibility that overturning a machine in this manner might increase friction by causing a weight thrust between parts where no provision for lubrication has been made.

Disregarding possible weight loss we can say that the efficiency of any power transmission machine is equal to its calculated *MA* minus its friction loss. *Efficiency = MA (by formula) − friction.*

FRICTION CAN BE PUT TO USE

Friction serves a useful purpose in keeping the tires of an automobile from spinning on the roadway surface, by preventing a tightened bolt or screw from loosening, and in many similar ways. We make use of friction in power transmission.

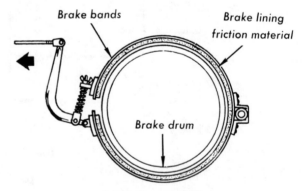

A band-type brake.

Any belt furnishes an example of the most obvious use of friction. Friction helps the belt to *grip* its two pulleys (or sheaves). Another obvious use is in *braking*. However they may be operated (mechanically, hydraulically, or by compressed air pressure; by hand or foot, or by some outside power source), all brakes—except the magnetic types—operate by creating a friction force sufficient to overcome the momentum of the moving part to which the friction force is applied. Either brake bands or shoes (depending on how they are shaped) are forced by the braking action against a drum (suitable for the purpose) that is integral with the moving part. The bands or shoes are lined with a friction-creating material which will also withstand any heat generated by the friction.

NOTE: A magnetic brake is actually an electric motor having its armature integral with the moving shaft—so that, when turned on, the motor will tend to rotate the shaft in an opposite direction, until it stops. A reversible direction electric motor can be used both for driving and for braking. And a generator—which uses the force applied to revolve its armature in order to generate current—can also be arranged so that the operator can increase its output (and the amount of force being used) at will, to make the generator serve as a brake for limited usage.

The third use of friction in power transmission is in *clutching*. A clutch is an apparatus that is set in between two parts of a power transmission machine and arranged so that the two parts may be engaged or disengaged, as desired. Two gears or pulleys that could be moved together or apart would constitute a clutch. However, such an arrangement is impractical if more than a very small amount of

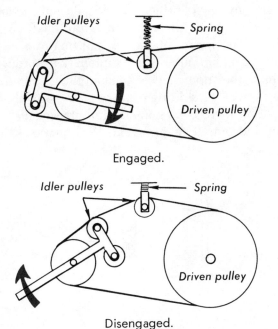

Engaged.

Disengaged.

One type of belt transfer clutch.

power is being transmitted. A more practical arrangement often used is a belt that can be made to slip by moving one of its pulleys, or be made to transfer its drive from the normal driven pulley to an idler pulley.

The hollow, leather-lined cone and mating metal cone described under "Electric Friction-Clutch Starters" earlier in this chapter is an example of a manually operated clutch that is intended to be disengaged most of the time. In power transmission machines the clutch is usually intended to be engaged most of the time (while power is flowing through it). Clutches of this type are spring-loaded to keep the two or more clutch parts in firm contact. A throwout collar (or bearing or sleeve) is provided so that the operator can move one clutch part away from the other(s) when desired.

Cone-type clutches are often used. In this case, either the inner or the outer cone is lined with leather or some similar friction-type material.

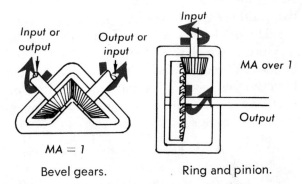

Bevel gears.   Ring and pinion.
Two types of right-angle drives.

Disc clutches are more often used, however. There may be one disc—lined on both sides—between two clutch plates, or there may be two discs (twin-disc type). The disc(s) serves simply to friction-grip the two plates together, when squeezed between them. It is not tied to either plate. One plate is fixed in position; the other (called the pressure plate) is mounted on a fixed-position cover and spring-loaded to stand out from this cover in order to squeeze the disc. A throwout collar, which can be slid along the shaft by the operation of linkages, can be positioned to disengage the pressure plate by moving it to compress the springs.

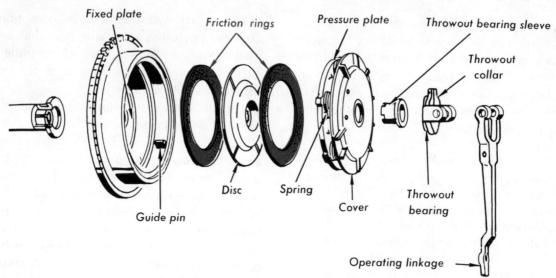

Fixed plate — Friction rings — Pressure plate — Throwout bearing sleeve — Throwout collar — Guide pin — Disc — Spring — Cover — Throwout bearing — Operating linkage

A typical automotive single-disc clutch.

A FEW OFTEN-EMPLOYED TRANSMISSION MACHINES

A right-angle drive—This consists of two shafts joined by gears (within a housing) so that the output rotates at a 90° angle to the input. If the two gears are equal the MA is 1; if one is larger, the MA or the VA will be greater than 1, depending on which is the input end. Other angles besides 90° are also used. The usual arrangement is to use two bevel gears, but a ring and pinion gear may be used (in which case input is always at the pinion end).

NOTE: In the accompanying schematic illustrations only the gears, etc., are shown. Required bearings, seals, retainers, lubrication fittings, etc., are not shown.

A gear box (or gear train)—A gear box is usually any grouping of gears (located for conven-

ience of manufacture and lubrication within a housing, called a box), arranged so that an input enters at one side while one or several outputs come out at other sides. The one or more outputs each has its own fixed relationship to the input. This may involve an MA (increased force), a VA (increased velocity) or simply a change in direction (either the direction of shaft rotation or the direction of drive, as in a right-angle drive).

A shifting-gear transmission—A gear box having one input with three or more gear combinations, any one of which can be selected to vary the MA and/or direction of output shaft rotation. Selection is obtained by shifting of a gear (or gears) to alter the combination through which the input drives the output. A *neutral* position of the shifting gear (or gears)—in which position there is no connection between the input and output—is provided. Each gear set de-

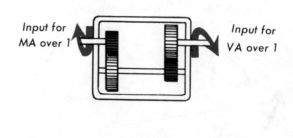

Output-VA over 1 — Input — Output-MA over 1 — Output-Change direction — Complex. — Input for MA over 1 — Input for VA over 1 — Simple.

Typical gear boxes.

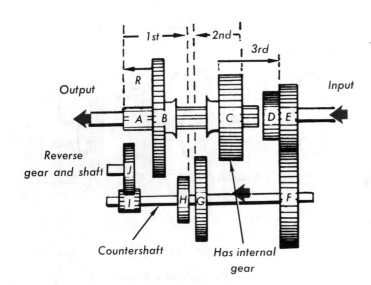

<u>Neutral.</u> *As shown. E rotates F and the countershaft; but there is no connection between input and output.*

<u>1st.</u> *Gear B slides on spline A to mesh with gear H. Drive is from input through E, F, countershaft, H, B, and A to output.*

<u>2nd.</u> *Gear C slides on spline A to mesh with gear G. Drive is from input through E, F, countershaft, G, C, and A to output.*

<u>3rd.</u> *Gear C slides on spline A to mesh with dog gear D. Drive is from input through D, C, and A to output.*

<u>Reverse.</u> *Gear B slides on spline A to mesh with gear J. Drive is from input through E, F, countershaft, I,J,B, and A to output.*

A shifting-gear transmission.

velops a certain *MA* and direction of rotation. By changing the *MA* the *VA* is, of course, also changed. The latter is, often as not, of equal or greater importance than the changing of the *MA*.

All gear combinations which operate to drive the attached vehicle or machinery in its normal direction are termed *forward speeds,* and there are usually several of these. Any one that reverses this direction is termed a *reverse speed,* and there may be one or more. The shifting is usually arranged to occur in an orderly sequence of steps, beginning with the smallest forward speed on through to the greatest forward speed —with similar provision for any reverse outputs. The various forward outputs are termed *low* speed and *high* (if only two), or *low, 2nd, high* or *1st, 2nd, 3rd,* etc. (if more than two). If there are more than one, the reverse outputs are similarly labeled. Between shifts, each shifting gear has to be returned to its neutral position. Between forward and reverse the input and output shafts have to be brought to a stop. Shifting is accomplished by a *yoke* (fork) that slides the shifting gear on its spline shaft. If there are two shifting gears (and forks), a separate handle which can engage one or the other fork is provided.

NOTE: While shifting, the teeth of the shifting gear must be at a standstill with respect to the teeth of the gear with which it is to mesh. Between forward and reverse this can be accomplished only by bringing both gears to a complete stop; but for any other shift it can be accomplished with the two gears running, if the gears are synchronized (to run at the relative rpms they will have when meshed).

In a transmission like the one illustrated, the operator must feel when the gears are synchronized. With a *synchromesh transmission,* two or more of the shifts are automatically synchronized. Common practice is to use a cone-type clutch (synchronizer), the two halves of which become engaged during the shifting operation and just prior to the meshing of the two gears. A separate synchronizer is required for each pair of gears that are to be synchronized in this manner. However, two units are often contained in one double-acting assembly.

Many modern automobiles still use synchromesh (standard) transmissions; but there is also a variety of newer transmissions, most of which incorporate fluid (hydraulic-pressure) coupling of various designs. None of these types are used with small engines.

Free-Wheeling Drive—In some cases incorporated in a transmission, this simple clutch-

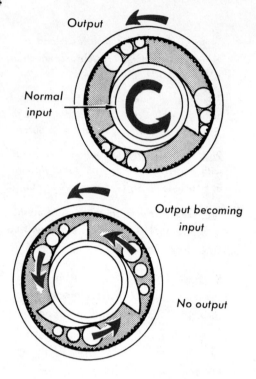

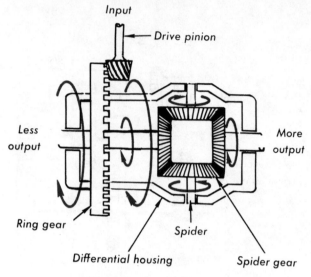

Principle of a free-wheeling drive.

Principle of a differential.

type device also has other power transmission uses. It is a 1-to-1 (straight through) drive which permits the input shaft to rotate the output shaft, but prevents the output from rotating the input. Intended only for one direction of shaft rotation, it cannot be operated in reverse (if shaft rotation is reversed, input won't rotate output, though output could rotate input). It is also called an overriding clutch or "pineapple."

Universal Joint—This is another 1-to-1 driving device used to join two shafts which are to rotate as one. The two shafts terminate in yokes which are disposed at 90° to each other and joined by a central piece called a *cross*. Each yoke has a hinge action on its arms of the cross. Since the two hinge actions are at 90°, if the two shafts are not concentric (are angled to one another), one or the other of the hinge actions

will compensate for the angle as the two shafts rotate. The amount of angulation that can be compensated without undue strain on the parts will depend on the smoothness built into the hinge actions (with bearings); but safe angulation (the angle between the two shafts) is seldom less than 135°.

Differential—A differential is a gear box having one input and two outputs so arranged that either output (alone) can produce the maximum *VA* obtainable, if the other output produces none, and will lose *VA* in proportion to the increase of *VA* at the other output. That is, the two outputs can run at equal rpm's, but, if one is slowed down by a load increase, then the other must increase its rpm to compensate.

Note the accompanying illustration. Input is from the drive pinion to the ring gear, which is attached to and rotates the housing, which is not connected to either of the output shafts. This housing rotates the spider with its spider gears—there are four, though only two are shown. If the two output shafts are equally loaded, the spider gears will serve merely as wedges between the spider and the gears of the output shafts (the gears will not revolve on their axles), and the spider will simultaneously rotate the two output shafts. If, however, one output shaft is held stationary, the spider, in order to revolve, will have to rotate around it, thereby causing the four spider gears to rotate on their axles. The rotation of these gears is

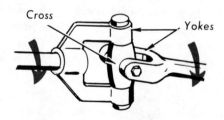

A simplified universal joint.

transmitted to the opposite output shaft gear and, at the same time, the rotation of the spider is also transmitted to this one gear. Consequently, the opposite shaft gear is rotated not only with the *VA* at which the housing and spider are revolving, but also with the additional *VA* obtained through the rotation of the spider gears.

The additional *VA* transmitted through the spider gears will always be proportional to the slowdown of the overloaded shaft. If the overloaded shaft is stopped dead (as above), the additional *VA* of the spider gears will double the rpm of the other output shaft; but if the first shaft is slowed only to half its normal rpm, the additional *VA* will increase the other shaft rpm by 50% (more rpm). The normal *VA* (equal rpm's of the two output shafts) of a differential is the *VA* of the drive pinion to ring gear; maximum *VA*, with one output stopped, is just double the normal *VA* output.

A sun gear—The sun gear is very much like a differential in principle, but different in application. The four planetary gears, like the spider gears, have their axles retained in common by a bracket to which a shaft is attached. Revolu-

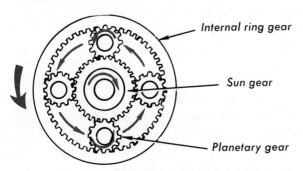

Principle of a sun gear.

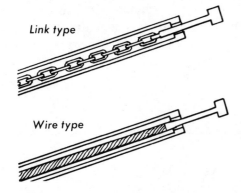

Typical flexible-shaft drives.

tion of this shaft will cause the four gears to describe a circle around the sun gear at center. Each planetary gear is meshed with both the sun gear and the ring gear. If the ring gear is rotated while the sun gear is held stationary, it will travel the planetary gears around the sun gear to rotate their bracket shaft in the same direction. If the ring gear is rotated while the bracket shaft is held stationary, it will rotate each of the planetary gears to rotate the sun gear (in reverse direction). Power can also be put in at either the bracket shaft or the sun gear shaft.

A flexible shaft—Used in motorbikes and other vehicles to drive the speedometer, it is simply a 1-to-1 drive through a shaft that is flexible enough to curve around corners, etc. Two kinds of shafts are used—the link shaft, which is a series of chainlike links employing the same principle used in a universal joint, and the wire shaft, which is simply a heavy woven wire of extreme flexibility. Either type is stretched taut from end to end, to avoid kinking, and enclosed in a lubricant-retaining housing. Small-diameter (light-duty) shafts can be curled into fairly tight circles without interruption of operation; but no shaft can ever be bent at any point.

An Adjustable-Speed Sheave Drive—Since the *VA* of a V-belt drive depends on the relative effective circumferences of the two sheaves, varying the circumference of one (or both) sheaves will vary the *VA*. A typical, variable effective circumference sheave is shown. One flange is fixed on the sheave shaft, while the other is threaded onto the shaft so that it can be screwed closer to or farther from the stationary flange. It is held at any position selected by tightening of the set screw.

As changing the circumference of one sheave will loosen or tighten the V belt, some provision

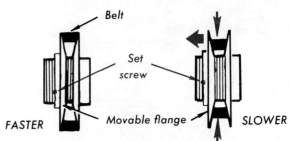

Principle of an adjustable-speed sheave.

Small Gasoline Engines Training Manual

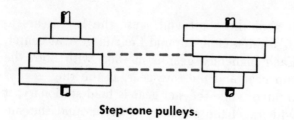

Step-cone pulleys.

is usually made with this type of drive for keeping the belt under proper tension. Having the shaft of one sheave mounted on a spring-loaded platform will (for instance) accomplish this purpose.

Step-Cone (Adjustable-Speed) Pulleys—An arrangement such as illustrated is often used with flat-belt drives for varying the *VA*. The two step-cone pulleys must be identical, and each is in a fixed position on its shaft. *VA* is varied by shifting the belt from one set of pulley cones to another set.

A Friction-Clutch Variable-Speed Drive—This variable *VA* drive also employs the principle of varying the circumference of one of the two wheels involved. Generally, the plate is a flat metal disk, while the clutch wheel is a friction-type wheel encircled by a rubber tire or similar friction-producing material. Some provision is made (by lever-operated yoke, or otherwise) for moving the clutch wheel toward or away from the center of the plate. *VA* is decreased as the clutch wheel moves out to increase the effective circumference of the plate when the input is through the clutch wheel. If the input is through the plate, this is reversed.

By moving the clutch wheel past the center of the flat plate, the direction of drive is reversed, as the clutch wheel will then rotate the flat plate in the opposite direction. At dead center of the flat plate the clutch wheel is in a

neutral position. Provision may also be made for lifting the clutch wheel off the flat plate to completely disengage the drive.

Friction Declutching Drives—There is a great variety of drives which depend on the friction between a belt and pulley or sheave, or between two plates or cones to transmit the power from one part to another.

One of the most common in use with small engines is an idler pulley (or sheave) declutching device. The belt is stretched taut enough (to transmit power) by an idler pulley, usually held in position by a stout spring. This pulley is mounted on a bracket which can be rotated, by a convenient lever, to move the idler pulley out of the way and allow the belt to become slack. When slack, the belt will not transmit power, since the drive pulley (or sheave) will slip and revolve freely within it.

Another device used with small engines is a centrifugal (automatic) clutch, which will disengage whenever the input rpm drops below a given amount. There are several methods of accomplishing this. In one type, weighted friction shoes are mounted on the input shaft inside of a drum attached to the output shaft. These weighted shoes act as the centrifugal weights of a governor. They expand outward (against retaining springs) to engage the drum, when a specified rpm is attained. In other types the centrifugal weights are wedges which engage internal spline teeth of the drum, or are in control of a simple cone clutch, which they slide into engagement when expanded.

TRANSMISSION MACHINE MAINTENANCE

Aside from the obvious need to keep all parts (usually replaceable as separate items) in good

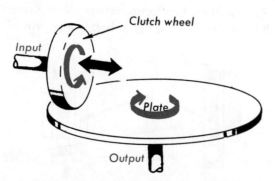

Typical friction-clutch variable speed drive.

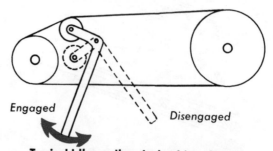

Typical idler-pulley declutching device.

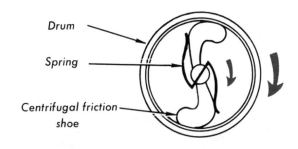

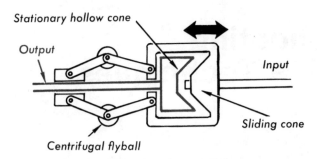

Automatic friction clutches.

repair and well-lubricated per manufacturer's instruction, there is little need for maintenance with this type of equipment. A few important notes follow.

Gears must mesh properly. If pressed too tightly together the teeth will bind and wear rapidly. The undue friction will cause power loss. If meshed too loosely, the resultant play (freedom) between teeth and the vibration (slapping) will eventually crystallize the metal and result in broken teeth.

Any metal part that is subjected to excessive vibration and slapping can crystallize and break. The play in all parts should be exactly as specified.

Springs that are flexed frequently (or overstretched) will in time fatigue and require replacement.

Belts that are allowed to accumulate dirt (particularly oil or grease) will become glazed and slip, or brittle and break. Pulleys and sheaves, too, can become glazed.

Wearable parts like clutch facings and liners obviously have a more or less limited service life. Oil or grease will serve as a preservative for most of the materials used for liners—however, check specifications on this.

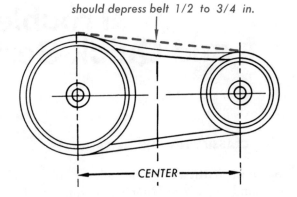

Measuring belt tension.

All chains and belts must be properly tensioned. If too loose, a chain may jump a sprocket tooth and snap. A belt will slip and wear. If too tight, either a chain or belt will pull inward on its sprockets (pulleys or sheaves) and wear out their shaft bearings. When properly tensioned, a chain should sag at the center of the top strand when the two sprockets are held to draw the bottom strand taught. The sag should be no less than one-quarter nor more than three-quarters of a link length. A belt is properly tensioned when easy thumb pressure at the center of the top strand will depress it one-half to three-quarters of an inch.

CHAPTER 11

Troubleshooting
and Quick-Service Techniques

CLASSIFYING ENGINE TROUBLES

Science, particularly the science of metallurgy, and our ever-advancing production skills are responsible for building greater and greater trouble-free service life into each new year's quota of gasoline engines. Today's engine, large or small, measured against its counterpart of yesteryear is a truly remarkable machine. It possesses built-in durability and performance characteristics unheard of a few years back. And tomorrow's engines will be even better.

But even the best of machines will, at times, act up; they will, in time, give out. And they will develop the symptoms of old age much sooner if they have been put to abusive use.

So it is with an engine. Every engine has a *normal life span,* a time during which, given reasonable care and attention, it will function well. Afterwards, when it has run out its normal life span or has become decrepit sooner from abuse and neglect, there are, from a purely economical standpoint no practicable remedies to revive it.

It isn't our concern here to define the normal life span of any engine, or to state what degree of neglect and abuse can make an engine *not worth repairing.* We wish, simply, to state the fact that there is such a time for every engine. The fewer dollars it takes to replace an engine with a new one, the sooner this time will arrive.

While it is true that any machine with replaceable parts could be made to last forever, the cost of parts and repair labor should be weighed against the results to be obtained. For the owner this is a matter of dollar savings; for the serviceman it is a matter of providing the customer satisfaction that will bring future business.

Old age in an engine is *not* a matter of years. An engine run once a year for 10 years is younger than one run 24 hours a day for six months, or one left out to rust. Then, again, an engine that has been run without proper lubrication or cooling for even a short time may be so badly damaged that it isn't worth repairing. Engine age, therefore is a matter of *service record.*

An owner knows his engine's service record; but the uninformed serviceman cannot know until he has made a possibly costly diagnosis by completely disassembling the engine. It is no more unrealistic for an owner to kid the serviceman than it would be for him to withhold information from his doctor. Good service begins with as much real information as there is available, and with an honest appraisal of probable costs offset against probable benefits to be derived. There are certainly times when an engine's true service record, accompanied by the evidence of a few quick checks, will indicate many advantages to be gained all around from replacing the engine with a new one.

These Are the Serviceable Problems

Disregarding those engines that are ready for the junk pile, all engines (even brand new ones) develop troubles. All such troubles can be traced to three causes:

1. Owner's failure to follow operating and maintenance instructions.
2. Normal deterioration due to gradual wearing of parts and gradual accumulation of the by-products of combustion and the impurities in gasoline.
3. Abnormal deterioration due to parts breakage.

Surprisingly enough a very high percentage of all small-engine failures occur when the engine is first put into service, and *are due to owner's failure*. We cannot overstress the importance of explicitly obeying the manufacturer's instructions. Every serviceman who is himself in contact with owners should reemphasize the importance of so doing. The owner who is not encouraged to follow instructions properly may be a good source of income for a short time, but he certainly won't become a "spreader of good will" regarding the product or the service. And the fact that his attitude may be entirely unfair won't help.

Troubles due to parts breakage—whether resulting from a defect or an accident—are usually obvious. Also, these troubles usually require a minor or major overhaul for correction. Those resulting from owner failure or normal deterioration are less obvious, but can often be corrected in a matter of minutes.

We Are Concerned With the Quick-Service Problems

In covering a diagnosis of failure causes we necessarily will have to cover many of the more series troubles which can be remedied in full only by minor or major overhaul. Due to the differences in engine designs, what may prove to be a simple remedy in one engine may turn out, in another-make engine, to require much disassembly of interfering parts and become as lengthy a procedure as any minor overhaul.

There is no hard-and-fast dividing line for all engines between *quick* service and longer service, if time alone is to be our yardstick. So, we are not going to define quick service in terms of time. Rather, we are going to say that quick service embraces everything which can be done with simple tools, without need of shop equipment, special tools, or jigs. It also includes any troubleshooting which can be accomplished in a matter of a few minutes, if the proper testing equipment is available. Whether or not the diagnosis and remedy which we describe come within any other classification of quick service for a particular-make engine will depend on the design features of that engine and the equipment at hand.

SHORTCUTS TO GET A BALKY ENGINE GOING

NOTE: The following shortcuts are intended merely for the benefit of owners or others who may not have the proper tools and instruments to make a thorough check. Most of these shortcuts are superficial remedies which will serve only as temporary "stopgap" measures.

Whenever any engine part is giving trouble, the part should be permanently repaired or replaced at the first opportunity. In addition, it is generally advisable to properly check engine performance, and to repair or replace those associated parts which also may be on the verge of malfunctioning.

1. *CHECK THE OBVIOUS*
   a. *Ignition ON?*
   b. *Out of Gas?*
   c. *Have You Followed Starting Instructions?*
   d. *Engine Overheated?*
      If overly hot, let it cool until you can hold your hand on it and then try again. But before trying, correct any of the following appropriate immediate causes for overheating:
      (1) Fill radiator with water.
      (2) Clean away accumulated dirt and grease that has been blocking free air circulation.
      (3) Tighten or replace fan belt as necessary.

(4) Clean and strengthen or replace any clogged, bent, or broken baffles designed to direct airflow over engine.

(5) Add oil as required—in crankcase (four-stroke-cycle engine) or in gasoline (two-stroke-cycle engine).

(6) Lighten the load, if overloaded. If engine starts and none of the preceding was to blame for the overheating, watch carefully for further overheating. Refer to the section, "Diagnosing Troubles," later in this chapter.

NOTE: We assume that if you have an electric starter, it is functioning satisfactorily. If it isn't (if starter is either not turning over or is draggng), you must first of all find the cause of this. (Refer to "Diagnosing Trouble.") Or, in an emergency and if ammeter shows a discharge when ignition is on, try starting by hand cranking.

2. IGNITION FAILURE

a. *No Spark.*

Pull off the wire from the spark plug. If possible, bare the wire clip (at end of wire); if not, insert a key or coin inside the clip cover to make contact with the clip. Position the clip (or the metal inserted) about 1/8 to 3/16 inch from the engine and, with the ignition on, crank the engine. A fat spark should jump the gap. If it does, the trouble is in your plug; if it doesn't, you'll have to check further.

CAUTION: Don't hold the clip (or metal insert) barehanded. You could receive quite a jolt.

Don't hold the clip (or metal insert) farther away than $\frac{3}{16}$ inch and don't run an engine very long with the wire held in this manner. Too much gap puts a strain on the magneto coil or induction coil and, if continued too long, can result in breakdown of the coil.

b. *Bad Plug.*

Remove the plug and examine it. If it is obviously fouled or with gap out of adjustment, cleaning and readjusting the gap should suffice. However, if the plug looks clean and properly adjusted, it may be broken down (electrically); the same is true if it is corroded and/or the insulator tip is discolored (dead gray) or cracked. Install a new plug. In any event, however, read "Diagnosing Troubles" before cleaning the old plug or installing a new one.

NOTE: Gap should be carefully calibrated to manufacturer's specifications. In an emerengency and lacking the specifications, however, you can be guided temporarily by this: Most magneto systems and battery systems used on stationary engines having constant loads use a spark plug gap in the range of .020–.030 inch. All others use a gap in the range of .030–.040 inch. Now, .030 inch is approximately the thickest of a dime's edge. Too little gap is better than too much. Widening the gap may temporarily fatten the spark, but will result in other system failures as noted above.

c. *System Grounded.*

Check for dampness or shorts everywhere in the electrical system starting with the plug(s). Look for moisture affecting wire insulations, especially in the plug-to-magneto or plug-to-distributor wire—and especially if the insulation looks cracked or frayed. Look for moisture inside the spark-plug clip covering, in distributor tower recesses around wire ends (if there is a distributor and it is accessible), and in the insulation of any other system wires. Look for any place where a wire might be pressed again a metal part and shorting through bad insulation to this part. If necessary, remove the distributor cap and look for internal shorts evidenced by moisture (condensation) inside, or by carbon streaks between any two of the metal conductors. Dry off, tape up, or clean off the moisture as a temporary cure, but replace defective parts at the first opportunity.

d. *Loose or Bad Connection.*

Check the connection to the spark plug and all other connections throughout the system. All must be clean and tight. Pay particular attention to high-tension wire connectors, to see that these are properly soldered onto wire ends, are clean, and are seated firmly in their terminals. Check distributor, if there is one and it is accessible, for dirty electrodes and/or poor contact between the rotor and its central contact. All such contacts should be bright. If rotor has a spring-type conductor at top, check to see that it is bent up high enough to ensure firm contact with the cap terminal.

e. *Defective Switch.*

If the system has an ignition switch, disconnect one wire from the switch and connect it to the other switch terminal to jump the switch. Should this end the trouble, replace the switch at the first opportunity.

f. *Defective Points.*

If the engine has a distributor it is possible to check and correct this by simply removing the distributor cap. Rotate the engine slowly until the breaker arm rests on a low part of its cam (points closed). With ignition ON, open and close the points by hand. A tiny spark should accompany each opening. If it does not, clean the points by filing lightly while pressing them closed over a thin nail file or piece of sandpaper (*never* use emery), or use a knife. The spark will now appear if dirt was the trouble.

While rotating the engine slowly, watch the action of the breaker arm. It should ride firmly against the cam. If it doesn't, bending its spring will temporarily correct this; tension must be set to specifications for permanent use, or undue wear will result. Also, points should open to a maximum gap of approximately half the thickness of a dime. If they don't, readjust the stationary point temporarily to this gap width, but remember that exact correct adjustment must be made later.

NOTE: In magneto ignition systems the points are quite often too inaccessibly located to warrant the preceding. If not, however, these can be checked and corrected as outlined.

g. *Defective Condenser.*

If practicable, remove condenser mounting screw and lift the condenser clear of its mount (but leave its pigtail connected as it was). Slip a scrap of paper between the points so that there is no electrical contact between them. With ignition (but nothing else) ON, scratch the condenser body against a metal part of the engine. If any sparks appear, the condenser is defective. Another condenser of approximately the same capacity (procurable at any engine supply or repair shop, at many service stations, or at any radio shop) will suffice temporarily; but one of the exact same capacity must be installed at the first opportunity (or points will quickly go bad).

h. *Defective Impulse Coupling.*

In a magneto system so equipped you can hear the click of the impulse coupling as it releases each time a plug is to be fired during the cranking process. Absence of this click will indicate a defective coupling, though chances are that by now you will have had at least one kick back while cranking to tell you that the spark is too far advanced for starting. If the coupling is jammed or the stop pin is bent or broken, you may be able to correct the trouble. Otherwise, a new, properly installed coupling is required. The only possible way to effect a start with a defective coupling is to turn the engine over at considerable speed —remembering that a kickback is most likely.

i. *Other Causes.*

If none of the foregoing produces a proper spark, the trouble must be in your magneto (or ignition coil), or in the timing of the spark. There is no stopgap service that can be performed to correct such troubles.

3. *FUEL FAILURE*

   a. *Too Much or Too Little Gas or Water?*

   (1) If there is a distinct odor of gasoline, if the carburetor is damp (especially around the gasket edges), if the engine has coughed out a cloud of gray smoke, or if the starter has been spinning the engine with intermittent hesitation—the probability is that the carburetor is overchoked. Wait 5 minutes, turn the choke off, and try starting again without choking. If choked

condition persists, refer to the remarks under "Overchoked," which follow.

NOTE: With a two-stroke-cycle engine overchoking will sometimes result in fouling of the spark plug, especially if there is too much oil in the gas. You may have to remove and wipe off the plug, or wait a considerable time for it to dry.

It is occasionally difficult to determine which way to turn the choke off (if marks are illegible). With most carburetors it is off when handle is down toward the throttle. If in doubt, remove the air cleaner and look in to see.

(2) If the fuel strainer bowl is dry, if the engine fires intermittently and/or sputters out, if the starter seems to turn the engine too freely then to slow down when the choke is held on, or if persistent choking cannot produce an overchoked condition, then gasoline is not getting to or through the carburetor. Try these suggestions:

(a) Remove gas tank cap. If it is tight and/or a hissing sound of air rushing into tank accompanies its removal, the vent is clogged. Opening the vent will cure the trouble.

(b) Remove the air cleaner. Cup your hand tightly over the carburetor air-horn opening and crank the engine a few times. This may dislodge any tiny speck of dirt blocking gas flow, and remedy the immediate trouble. The probability is, however, that more trouble will follow later unless gas system is thoroughly cleaned throughout. Before reinstalling air cleaner to try a start, see (c), following.

(c) Inject a teaspoonful of raw gasoline into the air-horn opening (or, if the spark plug is removed, inject the gasoline into the cylinder) to prime the engine directly. Try to start the engine. If the engine runs, the immedi-

ate trouble is remedied as previously stated. If it runs a few seconds and quits, you will have to check fuel system further. If it refuses even to fire once, the trouble is probably more serious (lost compression or incorrect valve or spark timing).

(3) If there is no engine response at all to persistent cranking, if the liquid in the fuel strainer bowl is discolored or has globules of other color liquid floating in it, or if persistent choking does not produce a distinct odor of gasoline it is most likely that water has got into the gasoline. You may be able to remedy the trouble simply by removing and cleaning the fuel strainer bowl (and, also, for safety, by draining and refilling the tank if there is any chance that the gasoline in it is full of water). Or, if there is no fuel strainer, disconnect the hose from the carburetor and drain the carburetor while also letting some gasoline run out the hose to flush it. If necessary, proceed as stated in precedings steps (b) and (c) to help get the engine started. But if these suggestions do not suffice, the entire system will have to be drained, cleaned, and refilled.

a. *Overchoked.*

(1) Overchoking can result from a completely stopped-up air cleaner. An examination will show this quickly. Clean it per applicable instructions,

Overchoked.

or replace it if it is too dirty. Do not operate the engine without it or you will be increasing your troubles.

(2) If the choke is automatic, the butterfly could be jammed, or any of the linkages could be bent or damaged. Such a condition will be obvious, and it can easily be corrected. If the thermostat control is broken, however, it will have to be replaced. By disconnecting it and tying the choke open you can operate the engine to some extent, but it will lack full power.

(3) Another cause of overchoking is a stuck (open) carburetor float valve. If it is simply jammed, easy tapping on the side of the bowl will dislodge it; otherwise, the carburetor must be serviced.

b. *Dirt in Fuel System.*

(1) The first place to look is in the fuel strainer bowl (if so equipped). Discoloration of fluid in bowl indicates dirt. Remove the bowl, clean, and place it back. To do a more thorough job, proceed also as follows.

(2) Disconnect the hose from the carburetor. Drain the carburetor, and drain up to about 1/4 cupful of gas through the hose (if this much will run out). Reconnect the hose, and then try to start the engine.

(3) Open idle and high-speed adjustment valves (or the one valve, if so equipped) on carburetor one full turn each. Remove the air cleaner, cup your hand over the air-horn opening, and crank the engine a few times. Replace the air cleaner, screw the valves back in to original settings, and try to start engine. Any dirt in the valves should now be dislodged.

(4) If all of preceding fails, you will have to drain, clean, and refill the entire system.

c. *Carburetor Maladjustment.*
It is possible that the idle-mix adjusting (or the only one adjusting) screw has be-

come overly tightened or loosened. Tighten the screw finger tight. Back out one full turn (or, preferably, make adjustment in accordance with manufacturer's instructions), then try to start. One full turn out will place most carburetor idle-mix adjusting screws (or single screws) at a suitable position for starting. Finer adjustment will have to be made later.

d. *Leaky or Kinked Fuel Lines.*
Check the fuel lines and the connections from the carburetor back to the tank. Tighten any loose connections; remove any obvious blocks to gasoline flow (such as kinks or flattened areas).

e. *Faulty Fuel Pump.*
Disconnect the gas hose from the carburetor. Crank the engine briefly. About a teaspoonful of gasoline should squirt out of the hose end with each revolution. If not, the pump is malfunctioning. Sometimes, removing the pump and reinstalling it temporarily without its gasket will provide enough extra stroke to make it function for a while. In any event, a new pump or service on the old one is required.

f. *Vapor Lock.*
This may occur in hot weather, at high altitudes, or following a period of engine operation under heavy load. The symptoms will be much the same as those for water in the fuel. Let the engine cool. Wet a cloth, cool it in the air, and apply it to the fuel pump. If the fuel line is close to the exhaust manifold at any point, slip a sheet of tin, cardboard, or anything suitable, in between to deflect the heat from the line. When it is thoroughly cooled, try again to start the engine.

g. *Other Causes.*
If none of the preceding proves to be adequate, either the entire system is clogged up or the carburetor or fuel pump requires servicing. If there is a vacuum tank, it also may be malfunctioning. Should the engine have been standing idle for some time, the cause can very well be stale (gummy) gasoline.

4. ENGINE FAILURE

If an engine will not start or, having fired sporadically, will not continue to run for more than a few strokes, and if the trouble cannot be traced to any of the foregoing causes, then the engine itself is most probably in a very poor condition. These can be its troubles:

(a) *Poor Compression.* This is indicated by almost complete lack of resistance to cranking, even during that part of the cycle when piston is on its compression stroke. An extremely bad ring condition, blown head gasket, a cracked cylinder, cylinder head, or piston is required to reduce compression to such a poor state that an engine won't start or run.

(b) *Bad Valve(s).* This condition is also indicated by lack of resistance if the valve(s) are stuck open, or by constant resistance to cranking should valve(s) be broken and stuck shut. In a two-stroke-cycle engine so equipped, all that is required is to clean, repair, or replace the reed or rotary valve. In a four-stroke-cycle engine the valve spring may be broken, the adjustment may have loosened, or the valves or any part of their operating mechanism may require servicing.

(c) *Water in Engine.* This is indicated by water in the crankcase and, probably, by water dripping from exhaust when cranking, or (in an extreme case) by inability to turn engine over at all. This, of course, can only occur in a water-cooled engine. It results from an internal cylinder block or cylinder head crack, or a partly blown gasket.

(d) *Clogged Exhaust.* This is indicated by difficulty in cranking and probably by hissing exhaust sounds in an extreme case. This is especially likely in a two-stroke-cycle engine since it is possible for carbon deposits, which result from too much oil in the gas, to accumulate inside exhaust pipe and/or muffler. In a four-stroke-cycle engine only a bent pipe

or muffler, or mud, etc., lodged in the end could be the cause.

(e) *Other Causes.* These are indicated by a jammed engine or unusal noises when cranking. A complete overhaul will doubtless be required.

DIAGNOSING TROUBLES

What You Can Learn From Spark Plugs

NOTE: Examination must be made immediately after removing plug from engine.

Normal Plug. A healthy plug is one that is not built up with too many carbon deposits. The insulator tip is brown or tan in color. There is little or no pitting of the electrodes. The indication is that this is the correct heat plug for the engine and its operating conditions and that engine is functioning normally.

Oily Plug. One that is oil wetted indicates:

(a) In a two-stroke-cycle engine that too much oil is being used in the gasoline, or that the fuel mixture is too rich at normal running speed. If an extra-rich mixture is required, use a hotter plug.

(b) In a four-stroke-cycle engine that the engine is using more than a normal amount of oil, but that leakage is not serious. A decarbonizing job may correct this. Otherwise, substitute a hotter plug.

Wet, Sludgy Plug. This is not likely to occur in a two-stroke-cycle engine. In a four-stroke-cycle engine, flaky, smudgy carbon deposits indicate very poor oil control. Either the rings or the valve stem guides are badly worn, indicating further service.

Dry Carboned Plug. Hard, dry carbon deposits excessively built up on plug indicate either that the fuel mixture is too rich at normal running speed, or that too cold a plug is being used. In a two-stroke-cycle engine, this condition will have been preceded by an oily plug condition. If condition require the mixture to remain as rich as it is, use a hotter plug. Otherwise, correct the adjustment, if possible. If impossible, look for worn carburetor jects, excessive fuel-pump pressure, or too high a fuel level in the carburetor bowl.

Normal.   Oily.   Wet, sludgy.   Dry, carboned.   Burned, pitted.

Spark-plug conditions.

Burned, Pitted Plug. If the electrodes are burned and/or pitted out of shape and the insulator tip is dead gray in color, the plug has been operating too hot. This may simply be due to improper plug selection, but it can also be the result of too high a compression due to internal carbon deposits or improper head gasket, to faulty cooling, or to consistent engine overloading. In some cases, too lean a fuel mixture at normal running speed is responsible, but the chances are that any such fault will have developed other symptoms more noticeable than the plug condition. In any event, preferably check and correct the fault, if there is one. Otherwise, substitute a cooler plug.

NOTE: Each plug is indicative of condition of its cylinder only. With multicylinder engines there are conditions which warrant using different heat-range plugs in the different cylinders.

What You Can Learn From Cranking

With the ignition OFF, turn the engine over by hand as slowly as possible. You should be able to feel each compression stroke.

No Compression. If you cannot feel any compression (or none in a certain cylinder) the chances are that the exhaust valve is sticking open or completely out of adjustment. However, a rumbling or hissing in the carburetor would indicate that it is the intake valve, instead. In a two-stroke-cycle engine, of course, the trouble has to be with the single crankcase valve.

Light Compression. If the feel indicates very poor compression, either one of the valves is poorly seated, the rings are bad, the cylinder or cylinder head is cracked, or the head gasket is blown. In case of cracks or a blown gasket you may be able to hear the hissing of the lost compression.

Good Compression. When the compression is good, if you release the crank just at the end of the compression stroke the crankshaft should be revolved backward to some extent. Or, if you stop and hold the crank right there, the strain should remain constant for at least a few minutes. You can judge the quality of the compression by the amount of crankshaft reversal, or by the time required for the compression to be dissipated so that the strain is gone.

Irregular Compression. The compression strain should last approximately half a crankshaft revolution, building up during this interval from nothing to maximum and back to nothing. If the periods are shorter and too closely spaced together for the engine type and/or number of cylinders, bad valve timing is indicated, due to actual out-of-time condition or to maladjustment of valve clearances. In a two-stroke-cycle engine with a reed valve this can't happen; but it can if the engine has a rotary or poppet valve. In a four-cylinder engine, it is difficult to detect, but it can be felt in a two-cylinder, four-stroke-cycle engine, or in any single-cylinder engine.

Constant Compression. If the feeling is one of continuous strain, as though the piston(s) is (are) always under some degree of compression, the cause could be a clogged exhaust or clogged intake—look for dirt, carbon, etc., in the exhaust or for a damaged pipe or muffler, or for a stuck throttle or choke valve, or very clogged air cleaner. In a two-stroke-cycle engine another cause could be a stuck (shut) reed valve. In a four-stroke-cycle engine (or a two-stroke-cycle engine so equipped) there is also the chance that the camshaft has ceased to operate due to a broken camshaft gear or similar cause. In a water-cooled engine, water leaking into the cylinder(s) may also produce the same effect. On the whole, the degree of strain and accompanying noises may help you to diagnose the trouble.

NOTE: If an engine cannot be turned over at all, or if it drags abnormally, it must be disassembled for inspection.

What You Can Learn From a Running Engine

1. ENGINE HARD TO START

If the engine can be started, but with difficulty, two of the likeliest causes are improper idle speed or improper idle-mix screw adjustment. Should there be an impulse coupling, this may not be functioning correctly. Should there be an automatic choke, check its operation. In addition, stale gasoline in the tank, or any of the following No. 2 causes may be to blame.

2. MOTOR DIES OR JERKS AT IDLE

In any engine the most probable cause is improper idling-screw adjustment, or a dirty, loosened, or worn carburetor (adjustments can't be stabilized). With a magneto system the spark, which is weaker at idle speed, is also suspect. Look at the spark plug electrodes (fouled or too wide a gap) and at the points (dirty) or at all connections throughout the system (loose or dirty). If there is a distributor, check the rotor and electrodes for proper contact. With a battery-ignition system, check the contacts (dirty) and all connections including distributor contacts. With any system, check the spark setting (too advanced). The last trouble will, however, make itself known by backfiring of the engine.

NOTE: In a battery-ignition system the secondary circuit current delivered to spark plug may be of positive or negative polarity. The voltage required to jump the plug gap can be as much as 35 to 45% less when the polarity is negative. Therefore it is desirable to connect the coil primary to the distributor and to ground in such a manner as to produce negative polarity in the secondary. A coil tester and voltmeter are required for proper checking, but sometimes a considerable difference in spark strength can be visually noted when connections are switched.

3. ENGINE IDLES BUT WON'T ACCELERATE

If engine conks out with a spitting and sputtering during or soon after acceleration from idle speed, something is blocking the normal gas flow. The likeliest cause is a malfunctioning fuel pump (leaking diaphragm or loose linkage); but also check for dirt in carburetor, for leaky lines, and for air-cleaner obstruction.

4. ENGINE MISSES DURING ACCELERATION

This condition indicates likelihood of leakage in the ignition system secondary circuit. Check the spark plug (electrically unsound). Check the condenser, all wires, and the distributor for leakage. Check the coil for a malfunction. Other causes could be the same as in No. 8 following.

5. ENGINE DIES WHEN THROTTLED DOWN TO IDLE

If engine idles okay except when throttled back, a flooding condition is indicated. Check the carburetor adjustment. If there is an automatic choke, check this—it may fail to open (release) fast enough. Check the carburetor float (set too high or sticking). Otherwise, trouble may be in the automatic spark control with the advancement curve started at wrong setting or cam or breaker plate sticking and failing to retard.

6. ENGINE BACKFIRES REGULARLY

Either a sticking (open) intake valve or improper valve or spark timing is indicated.

7. ENGINE SPITS OUT OF THE EXHAUST

Either a sticking (open) exhaust valve or improper valve or spark timing is indicated. Sometimes, too lean a mixture will also produce this result.

8. ENGINE MISSES AT RUNNING SPEED

If engine has accelerated properly, the likeliest causes are carburetor maladjustment (too lean) or a slight obstruction in the fuel system. A failing fuel pump (some leakage through a diaphragm or too loose a linkage), could also be the cause. Then, too, a slightly malfunctioning valve (intermittently sticking) or a weak spark from a slightly fouled plug, a loose connection, or leakage as in No. 4 preceding could be possible causes.

9. ENGINE WON'T REV UP PROPERLY

Should operation appear to be satisfactory in all respects except that the engine will not attain sufficient speed when throttle is opened, check these: (1) Governor malfunctioning—spring too slack or too tight, linkages bent or incorrectly connected, governor weights (or valve) jammed, or improper adjustment setting. (2) Throttle butterfly loose or damaged. (3) Automatic choke malfunctioning—thermostat spring broken, linkages bent or not properly connected, obstruction preventing operation, etc. (4) Carburetor adjustment too lean. Also see No. 10, which follows.

10. ENGINE LACKS POWER UNDER LOAD

If you could not detect any noticeable failure of engine to rev up satisfactorily when not loaded, but it seems to lose speed and lack power when loaded, check these: (1) Carburetor adjustment too rich. (2) Spark setting too retarded or, if automatically controlled, the automatic control is not functioning properly—cam or breaker plate stuck, centrifugal weights jammed, vacuum control spring(s) too tight, etc. (3) Too cold a plug is being used for required richness of mixture. (4) Exhaust partially blocked. (5) Valves not seating properly—slightly out of adjustment, spring or springs weak, or valve(s) carboned or worn. (6) Rings slightly bad.

Quite often it is impossible to make any distinction between this symptom and symptom No. 9, in which case the causes listed under No. 9 must also be checked if the engine lacks power.

NOTE: Plug selection can be very important to a two-stroke-cycle engine. Too cold a plug can result in loss of power (due to the fact that such an engine is often run on an overly rich mixture). On the other hand, too hot a plug can actually burn out the engine by causing piston disintegration, etc.

11. ENGINE SPEED IS UNSTABLE

This is probably due to a malfunctioning governor (bad spring), dirt in the fuel line, a sticking carburetor float, a worn carburetor (orifices and needle seats out of calibration), or an erratically operating fuel pump. Look also for leaks in the gas line (air seeping in as gas leaks out) and for an overly dirty air cleaner. Another cause can be use of old (stale) gasoline which has been left standing too long in the tank.

12. ENGINE OVERHEATS

In a two-stroke-cycle engine, most probable cause is too little oil in the gas. Lack of oil in the crankcase in a four-stroke-cycle engine can also cause this. In an air-cooled engine, look also for an excess of dirt on engine or for obvious damage to air baffles, etc. In a water-cooled engine, check the condition of the cooling system and operation of the water pump. In any engine, excess of carbon accumulation inside or in the exhaust system will also cause overheating.

13. ENGINE NOISY

If the noise is a knocking (pinging) related to the firing of the cylinder(s), it is a preignition knock and can be due to: (1) too hot a spark plug; (2) too high a compression ratio as the result of carbon accumulation inside; or (3) use of too volatile a fuel—winter-blended gasoline used during summer's heat.

If the noise is a clacking or jarring sound, check the flywheel to see if it is bolted on tightly.

If the noise is a heavy pounding sound, there is a good chance that the crankshaft main bearing(s) or one or more of the connecting-rod bearings are worn.

14. ENGINE GASPS TO A STOP

This is not likely to happen just when you are checking performance, but if the engine's service record indicates that it is a frequent occurrence during operation, look for causes of vapor lock (such as a gas line too close to exhaust, or if there is an automatic heat control, an improperly functioning control). Other causes could be use of too volatile a fuel or installation of engine in a position of poor air circulation.

A FEW QUICK-SERVICE HINTS

Carburetor Adjustments and Service

The simplest carburetors have just one speed (mixture) adjusting screw. With this the mixture is always made leaner when the screw is rotated clockwise.

Most carburetors have two speed (mixture) adjusting screws—an idle-speed screw for speeds up to approximately 1200 rpm, and a high-speed power adjusting screw for all other speeds. The two, however, do not work entirely independently of one another. It is generally necessary to adjust one, then the other, then readjust both in order to smooth out the acceleration from low to high speed.

When there are two screws, the high-speed screw always leans the mixture when rotated clockwise. Unfortunately, however, there is no uniformity with idle-speed screws; some rotate clockwise to lean the mixture, others rotate counterclockwise. The only way to tell is to turn the screw all the way in. If the reset kills the engine, this direction has leaned the mixture, but if the engine simply lopes and runs erratically, this direction has richened the mixture.

If in doubt as to which screw is which, rev the engine up to high speed. Turn the screw you believe to be the high-speed screw clockwise one turn. If the engine slows, or if continued turning of this screw stops the engine, it is the high-speed screw. If turning the screw one or even more turns has little effect, this is the idle-speed screw.

Lacking other (proper) instructions for a particular engine, you can generally achieve satisfactory adjustment as follows: Open (richen) both screws approximately two full turns from closed (lean) positions. Start the engine with the throttle half open. Hold this speed and adjust the high-speed screw for the smoothest running. Close the throttle to idling position, and adjust the idle screw for smoothest running. Now accelerate the engine. If it accelerates smoothly, the adjustments are correct. If it rolls and lopes—an indication of too rich a mixture—slightly lean the idle screw; but if it coughs and gags—an indication of too lean a mixture—slightly enrich the idle-screw setting. If a quarter-turn of the idle screw will not correct the gagging, readjust the high-speed screw to a slightly richer setting. But, if more than a quarter-turn of this screw is required, something is wrong with the carburetor.

The bulk of all carburetor service consists of cleaning, inspection, and adjustment. Unless a carburetor has been abused by jamming shut the adjusting screws, or has been in service a long time or under very dusty conditions, its calibration generally will be good after cleaning. Manufacturers provide several repair kit selections for minor and major overhauls—and other parts

Quick service.

not in kits are also available. Whenever a carburetor is disassembled for service, at least the minor overhaul kit should be used, and all gaskets should be replaced with new ones from this kit.

Magneto Adjustments and Service

On some engines the points can be reached for service quite easily, and timing adjustments can be made without disassembling the engine. On others, engine disassembly is required even to adjust the points.

Preferably, check out all other parts of an ignition system first before resorting to magneto inspection and servicing, especially if the magneto is hard to get at. Install new spark plug(s), if at all in doubt about the originals. In checking any magneto, make certain that the points are functioning properly, that the timing is correct, and that the condenser is good before labeling the magneto as requiring major overhaul. Install a new condenser and points, if at all in doubt about them. Servicing of the coil, permanent magnets, or structural parts generally does require a major overhaul of the engine.

Battery Ignition Adjustments and Service

All parts of any battery ignition system are readily accessible for service. Removal of a distributor, however, will require retiming procedure for its reinstallation.

If a general inspection fails to reveal the cause of trouble (no spark or too weak a one), start a more thorough inspection by installing new spark plug(s) if at all in doubt about these. Do the same with the condenser. Next, thoroughly check out the distributor without removing it. Then check out the coil if test equipment is available, or replace it. Timing, unless symptoms have clearly pointed to a malfunction, is not likely to be off. This can be the final thorough inspection.

A complete test for a distributor must include: checking for electrical leaks in cap and body; careful inspection of rotor and electrode contacts including tension of rotor spring contact; checking of contacts and of breaker-arm action including tension of its spring and wear of the arm cam-following surface and cam; and checking of all automatic mechanisms. If the contacts are badly burned, the causes may be: improper condenser capacity, gap too small, voltage too high (voltage regulator malfunctioning), or oily vapors condensing on the points.

If the system includes an ammeter, the latter can be of help in isolating trouble. With the ignition on, crank the engine slowly. No indication of current through the ammeter means that the primary circuit is grounded; a fluttering ammeter-hand movement means that primary circuit is good, and the trouble must be in the secondary circuit.

TUNE-UP TOOLS AND EQUIPMENT

All the preceding has been written with the assumption that checks, adjustments, and repairs (as suggested) will necessarily be made without benefit of proper shop equipment, by a quick-service serviceman or an operator. In the modern shop, however, proper testing and adjusting equipment, properly used, can eliminate *all* of the guesswork—can pinpoint most troubles in short time and make adjustments, etc., rapid and certain. Following are the principal tools and their uses:

NOTE: Any piece of equipment will operate only as well as the operator's knowledge and skill can make it perform. Every piece of equipment is issued with accompanying instructions for proper operation. These instructions must be followed.

Vacuum Gage

Connected properly to the intake manifold, such a gage, properly read and interpreted for the specific engine, will tell much about engine condition. If the engine is operating perfectly, a steady, high reading will result. A leaking cylinder (compression loss) will be shown by a quick jump of the gage needle each time the cylinder fires. Misfiring of a cylinder will have the same effect; but if misfiring is general, the jumps will not follow in regular sequence. When the mixture is too rich, the gage reading will be abnormally low and will pulsate slowly and rhythmically.

The gage can be used to perfect idle mixture adjustment. Used correctly with an idling engine it can even detect sticking valves (through irregularities in valve overlap time). If connected to the vacuum side of a fuel pump, it will quickly indicate fuel pump condition.

Compression Gage

Used to measure actual compression in a cylinder, this gage or its hose end is screwed (or squeeze fitted) into the spark plug opening. Its correct use requires a set procedure that ensures proper readings. If compression varies more than 10 psi from engine manufacturer's specification, the engine requires overhaul rather than a simple tune-up.

Pressure Gage

Any gage calibrated to show pressures from 0 to 15 psi may be used to advantage to check fuel pump operation.

Exhaust Analyzer

Properly used, this tool will quickly demonstrate even to an inexperienced person the need for tune-up of an engine that is operating at less than maximum efficiency. By the same token, it will also show that a tune-up job has been well and satisfactorily done.

Magneto and Distributor Test Stands

A variety of test stands are available for checking and adjusting magnetos and distributors. Depending on quality, a test stand will afford some or all of the many detailed procedures required to fully check out and adjust point settings, spring tensions, timing settings, etc., and to test condensers and coils. Separate coil testers are also available.

Spark Plug Testers and Cleaners

Setups are also available for quick testing of spark plugs and for sandblast cleaning of dirty plugs.

Battery Testers and Chargers

Combination testers and chargers, quick charging, as a rule, are in general use. The testing equipment will quickly indicate battery condition and reveal impending failure due to cell breakdown.

Miscellaneous Tools

Voltmeters, ammeters, and ohmmeters, and their usefulness were discussed in Chapter 8. While it is possible to track down shorts, loose connections, etc., without these instruments, their use certainly speeds up the job.

Other useful special-purpose tools are a spring scale for measuring spring tension, feeler gages for gap settings, and a torque wrench for correct tightening of bolts.

Index

A

Accelerating
   -pump system, 136-138
   systems, 136-137
   -well system, 136-137
Accessory systems, 39
Active forms of energy, 14
Adhesion, 21
Adjustable
   -air idling system, 130-131
   -mixture idling system, 131
   -speed sheave drive, 235
Advance curve, 164
Advancing the spark, 49
Air
   cleaners, 112
   -cooled engines, 207
   gap, 175
Airfoil, 121, 141
Airplane wing airfoil, 121
Alnico magnets, 167
Alternator, 147
   -battery system, 153-154
Ammeter, 158, 250
Angle of obliquity, 84
Apex seals, 85-86
Armature, 150
Atmospheric pressure and density, 32
Automatic
   friction clutches, 237
   mix lubrication system, 190

B

Back-suction principle, 133
Baffle
   plates, 183
   -type muffler, 183
Band-type brake, 230
Basic
   engine refinements, other, 52
   substances, 21

Battery systems, 148
Bearings
   and bushings, 58
   roller type, 57-58
   sleeve type (plain), 58
Belt transfer clutch, 231
Bendix-type starter, 215-216
Bernoulli's principle, 104-105
Bevel gears, 62
Bleeder air, 131
Boiling point, 22
Brake
   horsepower, 218-219
   mean effective pressure, 219-220
Breaker contacts, 164
Breakerless ignition, 177
Brush pigtail leads, 151
BTU, 201
Bushing, typical, 59
Bypass passage, 136

C

Calorie, the, 200
Cam dwell, 163
Capacitor discharge magneto system, 147
Carbon
   dioxide, 100
   monoxide, 100
Carburetion, 107
   how engine speed affects, 122-123
Carburetor
   misadjustment, 243
   requirements of engines, 120
Carburetors, types of, 120
Casing—head gasoline, 98
Centrifugal
   advance control, 164
   force, 115
   water pump, 209
Chains and sprockets, 63
Chatter marks, 87

Check valves, 138
Chemical
   analysis, 20
   charge, 24
   energy, 14
Choke
   system, 120
   systems, 123-128
   valve, 123-124
Choking
   effect of, 124
   to adjust mixture, 125
   to start, 124-125
Chromium, 66
Clark, Sir D., 8
Coefficient of
   linear expansion, 201
   superficial expansion, 201
Cohesion, 21
Coil, 149
   saturation, 163
Color coding, 149
Combination generators, 151
Combustion, 75
   degree of, 100
   engines, 7
   internal, 26-27
   lag, 48-49
     provision for, 50-51
   spontaneous, 26
Commutator, 150
   ripple, 151
Compound machines, 229-230
Compounds, 21, 24
Compression, 26, 30
   explanation of, 33-34
   gage, 250
   ratio, 54
Condensation point, 23
Condenser, 168
   typical, 160
Conduction heating, 205-206
Cone-type clutches, 231

Connecting rod, 30
  bearing, 30
  typical, 56
Constant-current battery charging, 157
Contact gap adjustment, 162-163
Convection heating, 205-206
Cooling (Wankel), 92-93
Corner seal, 86
Corrected horsepower, 219
Counterbalance function, 57-58
Counterweight automatic choke, 127
Conventional magneto system, 147
Cracking process, 98
Crank starter, 213
Crankpin, 30
Crankshaft, 30
  and counterbalance, 57-58
  design, 57
  double throw, 57
  single throw, 57
Cutout, 154-155
Cylinder
  and crankcase, 63
  position
    opposed, 37
    straight, 37
    v type, 37

D

Daimler, Gottlieb, 8
  generators, 148, 151
  starter motor, typical, 217
Density, meaning of, 31
Diaphragm-type carburetor, 120, 138-141
  comparison with bowl types, 138
  operation, 139-140
  two types of a, 138
Differential, principle of a, 234
Diode, 177, 154
Direct rewind starter, 212
Disc clutches, 231
Displacement
  piston, 53
  Wankel, 80
Distillation, methods of, 98
Distributor with condensor, 159-165
  automatic controls, 164
  construction controls, 164
  function and operation of, 159-160
  maintenance, 164
DKM design, 67
Double throw crankshaft, 57
Downstroke, 39
Driving fit, 65
Dry-sump system, 193
Dust shield, 168
Dynamic pressure, 102-103
Dynamometer, 219

E

Eccentric shaft, 71

Economizer
  jet, 133
  orifice, 133
  systems, 133-135
Edge gap, 175
Ejection-pump system, 193
Electrical
  energy, 14
  friction-clutch starters, 214-215
  ignition, 34
Electric-centrifugal starters, 215-216
Electromagnet, 149
Electronic-ignition battery system, 180
Elements, 21, 24
Energy, 13
  active forms of, 14
  chemical, 14
  electrical, 14
  heat, 14
  kinetic, 14
  latent, 15
  light, 14
  limitless, 15
  mechanical, 17
  potential, 151
  transfer magneto system, 147
Engine
  cooling, problem of, 206-207
  definition of, 9
  displacement, Wankel, 80
  external combustion, 19
  internal combustion, 19
  rotary, 12
  Wankel/NSU, 12
Engines
  combustion, 7
  stroke-type, 12
Epitrochoid, two-lobe, 73
Estimated horsepower, 219
ET magneto system, 174
Events, the five, 30
Exhaust
  analyzer, 250
  event, 30
  manifold, 182-183
  principles involved in, 34-35
  systems maintenance, 184
External combustion engine, 19

F

Feed systems, type of, 108
Feeler gauge, 163
Ferrous
  metal core, 149
  metals, 66
Field coil, 150
Filler tube and oil level arrangement, 191
Firing order, 51
Fits, types of, 65
Five events, 35
  Wankel, 77
Flash point, 25

Float
  and low-speed system, 131
  systems, 128-129
    nonadjustable, 128
  -type carburetor, 120-121, 128-138
Flooding, 119
Flow of oil, 192
Flushing plunger, 139
Flyballs, 115
Flywheel, 169-170
  alternator, 174
  design, 38
  function of, 38
  magneto, 167-172
    construction features of a, 167
    operation, 169-170
    typical, 168
Follow piston, 48
Force, 16
  centrifugal, 115
Forced
  -draft system, 208
  -feed system, 109
  fit, 65
Foreign-make carburetors, 143-146
  slide-throttle direct-proportioning
    carburetor, 143-144
    air flow, 143
    gasoline flow, 143, 144
Four-stroke-cycle engine, 35-36
Fractional distillation, 97
Free-wheeling drive, 233-234
Freezing point, 23
Friction, 220-221
  drag, 107
  horsepower, 219
Froede, Walter, 10
Fuel
  failure, 241-242
  pump, 110
    operation, 109
    typical, 110
  strainer, use of a, 110-111
  system, 97-146
    complete job and parts of, 100-101
  tank accessories, types of, 108
  tanks and feed systems, 108
Full-wave rectification bridge circuit, 177

G

Gas
  engines, history of, 7-11
  expansion, 202
  transfer velocity, 88
  venturi, 106
Gaskets, 59-60
Gasoline, 97-99
  controlling the, 120-121
Gate-controlled silicon rectifier, 177
Gear
  boxes, 232

Gear—cont
parts and types, 61
ratio, 61
Geared rewind starter, 212-213
Gears
and chains, 60-61
operating principle, 61
Generator, 147
-battery ignition system, 148-149
Governing spring, 116
Governor
adjustments, 118
maintenance, 118
principle, 114
Governors, 114-118
Gravity
flow, 104
system, 108-109

H

Half-wave rectification, 177
Heat
absorption, 203
and kinetic energy, 23
energy, 14
relationship to an engine, 200
sink, 177
transfer, 205
travel, 205
Helical gears, 62
Heptane, 99
Herringbone gears, 62
High-speed
jet, 132
pressure variations, 122
system, 120
definition of, 132
needle valve, 133
History of gas engines, 7-11
Horsepower, 29, 218
Hydraulic principle, 30-31
Hydrometer, 157

I

"I" cylinder head, 42, 52
Idle
-adjusting screw, 113, 129, 143
jet, 129
Idling
and high-speed adjustments, 137-138
systems, 120, 129-131
operation and principle parts, 129-130
Ignition, 30
coil
construction features, 159
maintenance, 159
explanation of, 34
failure, 240-241
oil, 158-159
switch, 158

Ignition—cont
system, 147-181
Wankel, 90-91
Impulse
channel, 138
coupling, 167, 175
spark, 176
Inclined plane, 223-224
Indicated horsepower, 219
Induced voltage, 159
Induction-coil principle, 159
Inertia, law of, 34
Intake
event, 30
principles involved in, 30-31
manifold, 112-113
port, 36
vacuum principle of, 33
valve, 36
Intermediate speed pressure variations, 122
Internal
combustion engine, 19, 26-27, 28-29
gears, 62
Inverted-vertical engines, 38
Iso-octane, 99

J

Jet engine, 28

K

Kick starter, 213
Kilowatts, 218
Kinetic energy, 14
KKM design, 67
Kreiskolbenmotor, 12

L

"L" head, 52-53
valve, typical, 42
variation, typical, 43
Lag angle, 176
Laminations, 172
Lapped, 43
Latent energy, 15
Law of Inertia, 34
Lead
-acid battery, 158
construction features, 156-157
maintenance, 157-158
piston, 48
Leaning angle, 84
Levers, types of, 221
Light energy, 14
Lighting coils, 174
Link blocks, 86
Liquid
expansion, 202
flow, 102
venturi tube, 105

Lubricating pad, 168
Lubrication
systems, 184-199
Wankel, 91

M

Magnematic magneto, 172
Magnetic fields, 149, 169-170
Magneto
adjustments and maintenance, 175
-dynamos (ac magnetos), 173-174
operation, three stages of, 169
system, 147, 167-177
Magnets, permanent, 169
Main bearings, 30
Makeup of matter, 20-21
Manganese, 66
Manifold
intake, 112
vacuum, 107
Manual
choke, 125
spark-advance control, 163
Mass, 16
Matter, 13
makeup of, 20-21
three states of, 19-20
Measuring
belt tension, 237
heat, 200
Mechanical
energy, 17
governor, 114
power, 15-18
process, 21
windup starters, 214
Melting point, 22
Metals, 66
Metering
jet, 135
rod, 135
system, 135
Mixture, 35-36
Modified 2-stroke-cycle engine spark plug, 166
Molecules, 21
Molybdenum, 66
Momentum, 16
Motor
oil, function of a, 186-187
oils and greases, 185
Motors, electric starter, 216-217
Movable brush holder, 153
Muffler, 182, 183-184
Multicylinder
four-stroke-cycle engine, 170
manifold, 182
Multirotor engines, 95

N

Nature conserves matter and energy, 14-15

Needle
    float, 128
    valve, 132, 141
Nickel, 66
Nonferrous metals, 66
Nonseparable conventional-type spark
        plug, 165
Nozzle airfoil, 121

O

Observed horsepower, 219
Octane rating, 99
Oil
    breathers, 198-199
    filter, 197-198
    pumps and linkages, 195
    seals, 86, 199
    strainers, 197
    types of, 185
One-cylinder
    camshaft, typical, 42
    four-stroke-cycle engine, 170
Open-draft system, 207-208
Overchoked, 242
Oxidation, 25

P

Peak of combustion, 49
Performance factors, Wankel, 78
Permanent magnets, 169
Petroleum, 97-98
Phased-piston, 48
Phasing gears, 72
Pinion gear, 62
Piston, 55
    and connecting-rod assembly, 54
    engine, 28
    -type automatic choke, 127
Plain bearings, 58
Plate, reed, 96
Polymerizing, 98
Pneumatic governor, 114-115
    principle of a, 116
Poppet valves, 42, 46
    maintenance of, 44-45
    operation of, 43
    placement of, 41
    principle, 41
Port
    designs, 89
    exhaust, 36
    intake, 36
Postignition, 88
Potential energy, 15
Power
    (economizer) system, 135-136
    mechanical, 15-18
    stroke, 28-29
        event, 30
    transmissions, 217-237
    valve, 136
Preignition, 88

Pressure, 23
    atmospheric, 32
    gage, 250
    sensing, 140
    variations, typical, 122
Primary
    circuit, 158
    coil, 159
    current, 172
    field, 169
    sealing area, 87
Priming, 104
    pump, 111
Principal engine structural assemblies,
        64
Principles involved in intake, 30
Process, mechanical, 21
Pulley, 222-223
Pulse
    current, 177
    transformer, 177
Push fit, 65

Q

Q angle, 84
Quick charging, 157

R

Radiation heating, 205-206
Rated horsepower, 219
R/e ratio and
    compression ratio, 83-84
    sliding angle, 84-85
Rectifier, 183
Reed valves, 46
Requirements of engines, carburetor,
        120
Retarded spark, 49
Rich and lean mixtures, 118-120
Right-angle drives, 231
Rings, 55
Ripple filter, 151
River flow, 105
Roller
    chain, 63
    -type bearings, 58-59
        parts, 59
Rotary
    combustion, 12
        engine, 12, 67-96
            advantages of, 67-70
    type valve principle, 40
    valves, 46
Rotating-coil magnetos, 172-173
Rotor
    adjustments, 175
    -face
        depression, effect of, 83
        design, thickness, and radius, 81
        (one), the 5 events for, 79
    magneto
        operation, stages of, 172

Rotor—cont
    magneto—cont
        system, components of a, 171
        shapes, three possible, 82
        Wankel, 94-95
Running fit, 65

S

SCR, 177-178, 179, 180
Screw, 224
Secondary
    coil, 159
    current, 170, 172
    sealing area, 87
Semiautomatic choke, 126
Series-wound dc motor, principle of a,
        216
Shifting-gear transmission, 233
Shorting terminal, 169
Shrink fit, 65
Shunt generator, 151
Side seals, 85-86
Silicon, 66
Single
    -cylinder
        engine manifold, 183
        vs multicylinder engines, 37-38
    throw crankshaft, 58
Sleeve-type
    bearings, 58
    valve principle, 40
Slide-type throttles, 114
Sliding angle, 84
Slow-speed pressure variations, 122
Solid
    boss, 56
    expansion, 202
    -state ignition circuits, 177-181
Spark
    advance adjustment, 162
    lever, 163
    plug
        conditions, 245
        construction features, 165
    selection, 166
Specific
    gravity, 157
    heat, 200-201
Spiral bevel gears, 62
Spitback, 86
Splash system, 192-193
Spline and socket, 62
Split
    boss, 56
    single-cylinder, 48
Spontaneous combustion, 26
Spun fit, 65
Spur
    bevel gears, 62
    gear, 61-62
Starter
    need for, 210

Starter—cont
    requirements of, 211-212
    types, 212
Static
    pressure, 102
    spark-advance, 175
Stator, 194
Step-cone pulleys, 236
Story of carburetion, 121
Stroke
    definition of, 35
    power, 28-29
    -type engines, 12
Structure of elements and compounds,
    23-24
Substances, basic, 21
Suction
    feed system, 109
    -type carburetors, 120, 141-143
        adjustments, 142-143
        operation, 142
        principles of a, 142
Sulfation, 157
Sulfuric acid, 157
Sun gear, 235
Swept volume, 80
Switch, ignition, 158
Synthetic gasoline, 98
Syphoning, 104

T

"T" head, 42, 52
Tachometer, 118
Taper-charge, 157
Tappet, 41
Thermistor, 178
Thermometers, 200
Thermostatic spring, 126
Thermosyphon systems, 209
Third brush generators, 151-152
Three states of matter, 19-20
Throttle
    and linkages, 113
    valve
        airfoil, 121
        and control lever, 113
    valves and governors, 113-118

Tickler, float, 128
Timing, 43-44
    gears, 43, 63
    mark, 163
    the five events, 40-51
    Wankel, 87-91
Tolerances
    and fits, 64
    meaning of, 64
Torque, 217-218
    tightening, 65
    wrench, 65
Transfer
    port, 91-92
    wave, 88
Transmission machine maintenance,
    236-237
Trickle charging, 157
Trigger module, 179
Troubleshooting, 238
Tune-up tools and equipment, 249
Tungsten, 66
Turbine engine, 28
Twin-piston
    engine design and operation, 48
    operation, 49
Two
    -cylinder
        two-stroke-cycle engine, 170
        2-stroke magneto, 170
    -lobe epitrochoid, 73
    -stroke cycle engine, 36-37
Types of
    combustion engines, 19-27
    internal-combustion engines, 28-29

U

Uniflow, 48
Universal joint, 234
Upstroke, 39
Useful kinetic energy, 19-23

V

Vacuum
    brake control, 164
    gage, 249

Vacuum—cont
    passage, 136
    system, 109
Valve
    exhaust, 35
    -in-head, typical, 42
    intake, 35-36
    principles, 40
    spring, 41
Vanadium, 66
Vane-type automatic choke, 127
Vapor lock, 99, 243
Variable
    cam, principle of, 171
    stroke oil feed pump, 190
Velocity advantage, 227
Voltage
    maintenance, 156
    regulator, 148, 154-155
Volumetric efficiency, 219
Velocity, 16
Venturi
    -principle applied to a carburetor,
        106
    -principle applied to gases, 106

W

Wankel
    engine, 12
        housing assembly, 70
        mechanical design of, 70
    Felix, 9
    KKM, 12
Water
    channels, 209
    -cooled engines, 208-209
    venturi tube, 104
Wheel and axle, the, 223
Windup rope starter, 212
Wiring harness and wires, 148, 167
Work, 17
Worm gear arrangement, 62
Wristpin, 30, 56

Z

Zener diode, 178-179